Nature, Value, Duty

The International Library of Environmental, Agricultural and Food Ethics

VOLUME 8

NATURE, VALUE, DUTY
LIFE ON EARTH WITH HOLMES ROLSTON, III

edited by

Christopher J. Preston
University of Montana, Missoula, MT, U.S.A.

and

Wayne Ouderkirk
Empire State College, Saratoga Springs, NY, U.S.A.

Springer

A C.I.P. Catalogue record for this book is available from the Library of Congress.

ISBN-10 1-4020-4877-7 (HB)
ISBN-13 978-1-4020-4877-7 (HB)
ISBN-10 1-4020-4878-5 (e-book)
ISBN-13 1-4020-4878-4 (e-book)

Published by Springer,
P.O. Box 17, 3300 AA Dordrecht, The Netherlands.

www.springer.com

Printed on acid-free paper

CONTENTS

BIOGRAPHIES OF AUTHORS

Allen Carlson is Professor of Philosophy at the University of Alberta, Edmonton, Canada. His teaching and research interests include aesthetics, environmental philosophy, and the aesthetics of nature, landscape, and architecture. He has published a number of articles in various journals, as well as co-editing two collections of essays. He is the author of *Aesthetics and the Environment: The Appreciation of Nature, Art and Architecture* (Routledge, 2000).

Victoria Davion is Head and Professor of Philosophy at the University of Georgia. She is the founding and current editor of the international journal *Ethics and the Environment*. Davion is co-editor (with Clark Wolf) of *The Idea of a Political Liberalism: Essays on Rawls* (Rowman & Littlefield, 2000). Davion specializes in ethical theory, applied ethics, womens' studies, and social and political philosphy and publishes in a variety of journals such as *Social Theory and Practice, Hypatia, Public Affairs Quarterly*, and the *Journal of Social Philosophy*.

Eugene Hargrove is the Founding Editor of the journal *Environmental Ethics* (1979), the director of the Center for Environmental Philosophy, the author of *Foundations of Environmental Ethics* (1989), and the editor of *Beyond Spaceship Earth: Environmental Ethics and the Solar System* (1986), *Religion and Environmental Crisis* (1986), and *The Environmental Ethics/Animal Rights Debate* (1992). He has established an M.A. and Ph.D. graduate program at the University of North Texas focused primarily on environmental ethics.

Brenda Hausauer is an Environmental Consultant and mom, specializing in research, writing, planning (and 4-year-olds). She has written the state of Vermont's Comprehensive Energy Plan and Greenhouse Gas Action Plan, and numerous reports for environmental organizations on green taxes, land use planning, and environmental issues. She studied with Holmes Rolston, III, at Colorado State University.

Ned Hettinger is Professor of Philosophy at the College of Charleston, SC. He has published several dozen papers including articles in *Philosophy and Public Affairs, The Boston College Environmental Affairs Law Review, Environmental Ethics, Environmental Values*, and *Ethics and the Environment*. His paper "Justifying Intellectual Property" was chosen for inclusion in the *Philosopher's Annual* as among the ten best philosophy articles published in 1989.

Keekok Lee's research interests include environmental philosophy, philosophy of technology, as well as the philosophy of genetics, conservation biology and medicine. Her more recent publications are: *The Natural and the Artefactual* (1999), *Philosophy and Revolutions in Genetics* (2003, 2005), *Zoos: A Philosophical Tour* (November 2006). She is affiliated to the Institute for Philosophy and Public Policy at the University of Lancaster as Honorary Chair in Philosophy.

John Lemons is Professor of Biology and Environmental Science at the University of New England, Biddeford, ME, USA. Dr. Lemons has authored over 100 articles and book chapters on topics ranging from nuclear waste, climate change, national parks, the use of science in environmental policy, sustainability, and environmental ethics. He is the editor of eight books, including *Scientific Uncertainty and Environmental Problem Solving* by Blackwell Science, Inc. and *Sustainable Development: Science, Policy, and Ethics* with Donald Brown. Dr. Lemons is a past editor-in-chief of the journal *The Environmental Professional* and is an elected fellow in the American Association for the Advancement of Science.

Andrew Light is Associate Professor of Philosophy and Public Affairs at the University of Washington in Seattle. He is also a Research Fellow at the Institute for Philosophy and Public Policy at Lancaster University (UK), and a Faculty Fellow at the Center for Sustainable Development in the School of Architecture at the University of Texas at Austin. In addition to authoring over seventy articles and book chapters he has edited or co-edited sixteen books on environmetnal ethics, philosophy of technology, and aesthetics, including *Environmental Pragmatism* (Routledge 1996), *Moral and Political Reasoning in Environmental Practice* (MIT 2003), *Animal Pragmatism* (Indiana 2004) and *The Aesthetics of Everyday Life* (Columbia 2005). He is also co-editor of the journal *Ethics, Place, and Environment* (Routledge).

Katie McShane is an Assistant Professor of Philosophy at North Carolina State University. She specializes in environmental ethics and ethical theory, with a particular interest in the moral significance of our emotional engagements with the natural world. She works in areas related to ecosystem health, the place of environmental concerns in theories of value, and the structural differences among valuing attitudes.

Clare Palmer is Associate Professor in Philosophy and Environmental Studies at Washington University in St Louis. She is author of *Environmental Ethics* (ABC-Clio 1997) and *Environmental Ethics and Process Thinking* (Oxford University Press 1998). She is the editor of the journal W*orldviews: Environment, Culture, Religion* published by Brill Academic Press. Most recently she co-edited, with J. Baird Callicott, *Environmental Philosophy*, a 5 volume collection of papers published by Routledge in 2005. She is currently completing a book on animal ethics.

Wayne Ouderkirk teaches environmental philosophy at Empire State College (SUNY). He is the editor (with Jim Hill) of *Land, Value, Community: Callicott and Environmental Philosophy* (SUNY Press, 2002) and the author of essays and book reviews in a variety of publications in the field. Currently, he is researching and writing about the nature/culture relationship. An avid hiker, he plans to complete his climbs of all 115 four-thousand foot peaks in the Northeastern U.S. by the time this volume is published.

Christopher J. Preston is a part-time Assistant Professor in the Department of Philosophy at the University of Montana, Missoula, MT. He is the author of *Grounding Knowledge: Environmental Philosophy, Epistemology, and Place*. (Athens: University of Georgia Press, 2003). He is also the guest editor of a journal special issue on "The Epistemic Significance of Place" in *Ethics and the Environment* 10 (2)(2005). He has over a dozen publications in the areas of environmental philosophy, ethics, and the philosophy of mind. He is currently working on an intellectual biography of Holmes Rolston, III with support from the John Templeton Foundation. He teaches classes in ethics, environmental ethics, ecofeminism, and contemporary moral issues.

James W. Sheppard is an Assistant Professor in the Department of Philosophy and a co-founder of CAPE (The Center for Applied and Professional Ethics) at the University of Missouri-Kansas City. Sheppard also serves as a city commissioner on the Environmental Management Commission and on the Race Relations Commission for Kansas City, Missouri. Sheppard's main research focuses on philosophy and public policy with an emphasis on environmental ethics, environmental politics, and environmental policy as well as urban politics, urban affairs, and urban policy.

Lisa Sideris is Assistant Professor of Religious Studies at Indiana University where she teaches courses on ethics, environmental ethics, and science and religion. Her book, *Environmental Ethics, Ecological Theology, and Natural Selection* (Columbia, 2003) examines the way in which much of Christian environmental ethics misconstrues, or simply ignores, Darwinian theory, and the problems this creates for developing a realistic ethic for nature and animals. Sideris is currently co-editing a collection of interdisciplinary essays on the impact and legacy of Rachel Carson.

Mark Wynn is a Senior Lecturer in Philosophy of Religion and Ethics at the University of Exeter. His publications include *Emotional Experience and Religious Understanding: Integrating Perception, Conception and Feeling* (Cambridge UP, 2005) and *God and Goodness: A Natural Theological Perspective* (Routledge, 1999).

WAYNE OUDERKIRK AND
CHRISTOPHER J. PRESTON

INTRODUCTION

A PHILOSOPHER GONE WILD: HOLMES ROLSTON, III, AND ENVIRONMENTAL PHILOSOPHY

The field of environmental ethics would not be what it is today without the contribution and influence of Holmes Rolston, III. More than any other single figure Rolston has been central to its genesis and development over the last thirty years. Given the environmental problems the world faces and the reorientation within philosophical ethics towards applied ethics that was happening in the 1970s, it seems likely that environmental ethics would have eventually emerged without Rolston's contributions. But it would not have emerged with the same force or with the same emphases. Quite simply, he has put his stamp on the whole field as it has grown from its modest beginnings.

It is especially significant to emphasize that the groundbreaking work done by Rolston in this arena is not just about applying an ethical theory. It is about doing philosophy in the broadest possible sense. From his earliest writings, Rolston realized that offering ethical guidance about how we humans ought to interact with our environment would require something far beyond the mere application of ethical theories developed for inter-human ethics, the approach that remains paradigmatic in the fields of business and medical ethics. Thinking about human relationships with the nonhuman world quickly took him beyond the categories and concepts of traditional moral theory. The problems that environmental philosophy addresses are indeed often ethical, but complete answers to those problems also require revisiting some central metaphysical and epistemological categories, in order to situate the moral categories in relation to them. Although Rolston was not the only thinker to realize this, and not the only one to start breaking apart those traditional categories, he has always recognized, and forced others to recognize, that environmental ethics demands a significant re-orientation of philosophy.

After more than thirty years of significant contribution and influence, we think it is time to devote a whole volume to a contemporary consideration of Rolston's impact and his thought. Because he has published so prolifically in so many venues and with so many varying emphases, we knew it would be impossible to achieve a complete analysis of his work. In light of the fact that many of the discussions and evaluations of his work have been staples in the environmental philosophy literature for many years and are widely republished elsewhere, we decided to approach this review by collecting a set of contemporary reflections on Rolston's work. Some of the essays that follow are by relatively new contributors to the field; others are from established figures. But all of them are original to this collection, and all of them assay Rolston's work to some degree with an eye towards advancing the field, whether through friendly developments of his ideas or through vigorous intellectual disagreement with him.

ROLSTON'S BIOGRAPHY

Holmes Rolston, III, was born in 1932 in Staunton, Virginia. Both his father and his grandfather were Shenandoah Valley preachers. Rolston and his two sisters grew up in an environment typical of depression era rural Virginia. Water was drawn from a cistern, chickens scratched in the yard, and a large vegetable garden supplied the family with a sizeable portion of their dietary needs. The young preacher's son rambled barefoot across the fields and forests of the lightly settled Shenandoah, fishing and bathing in the creeks, and watching with fascination as the water wheels on the nearby Maury River milled flour from local grain.

When Rolston enrolled at Davidson College in 1950, he was interested in understanding the physical world in its most fundamental dimensions. This meant studying physics, astronomy, and mathematics, all of which he did with talent and enthusiasm. In the immediate aftermath of World War II, the secrets of the atom appeared to hold the key to the world's future. Though he majored in physics and mathematics, Rolston could not help also finding himself drawn to the mysteries of biology. Biology spoke strongly to the immediate experiences of the natural world that he had grown up with in the Shenandoah Valley. A key moment in Rolston's college education came at the end of a spring break field trip to the Florida Everglades when his biology instructor brought back an insect that later turned out to be a species new to science. As Rolston gazed down the microscope at the new creature, he gained a vivid and lasting sense of the drama that biology could reveal.

Despite this burgeoning interest in biology, Rolston's theological leanings now stepped in. Within a few years of leaving Davidson he had graduated from Union Theological Seminary in Richmond with a divinity degree. Now married to Jane Wilson, he swiftly completed a doctorate in theology at the University of Edinburgh in Scotland. On their return to the United States, Rolston and his wife moved back to their geographical roots when he became a Valley of Virginia pastor at the Walnut Grove and High Point Churches outside of Bristol, Virginia. It was during this time that Rolston's fascination with natural history started to become a vocation. Rolston's insatiable wonder at the complex ecology within which he lived drew him first to biology classes at nearby East Tennessee State University and eventually to leave his job as a preacher and enroll in a masters program in the philosophy of science at the University of Pittsburgh. Rolston graduated in only a year and then, with some sadness at leaving his native landscape, left Appalachia and headed west to teach philosophy at Colorado State University in Fort Collins, thinking that he might stay out west for just a couple of years. He has been there ever since, earning the accolade of being named University Distinguished Professor, a honor reserved for only twelve in a faculty of twelve hundred.

Even before he left Virginia to start his career as a philosopher, Rolston had begun to write philosophical essays about our place in the natural world for magazines such as *Virginia Wildlife* and *Main Currents in Modern Thought*.[1] Profound questions about earthly roots, belonging, and moral obligation were already surfacing in essays such as "Meditation at the Precambian Contact." Here Rolston states "[I]f I can recollect my prenatal past, my gestation in the geological womb, my genealogy, then I shall know who I am and where I am."[2] Mystified by the idea of ethical "oughts" that bore no connection to geological and evolutionary roots, Rolston complained about a sheriff he encountered in the woods looking for moonshiners:

> [i]t was as though he pursued his moral question independently of his origins, as though there were now superimposed on the bedrock of Earth a novel, ethical traffic. He came and went preaching and enforcing his "Thou shalt not," as though his sermon and authority were derived elsewhere than from the dust of which he is composed.[3] (1986, 237).

He remained curious about the connection between the geological sciences and ethics, but by the time he moved to Colorado he had begun to sense that evolutionary theory and the biological sciences contained more promise for instructing us about our place in the world. In *The Pasqueflower*, an early essay that Rolston still ranks among the closest to his creed, his turn to biology for inspiration becomes clear. "We love the landscape, the sunset, the night sky," he affirms:

> [y]et greatly exceeding the geophysical, mineralogical, and celestial ranges of beauty are those of the emergent structures of life, particularly as these come to their botanical apogee in the flowers of the higher plants, which so marvelously combine function and beauty, as though to mark life's reproduction with a special sign.[4]

With the publication of "Is There an Ecological Ethic?" in the philosophy journal *Ethics*, in 1975, Rolston became known as one of a handful of philosophers shaping environmental ethics into a recognizable academic discipline. In 1979, together with Eugene Hargrove, Rolston became a founding editor of *Environmental Ethics*, which provided a forum for discussion of the philosophy of nature. In 1986, Rolston published a collection of his early thoughts on environmental philosophy in *Philosophy Gone Wild*. He quickly followed this collection with one of the first—and still one of the most influential—monographs in the field titled *Environmental Ethics: Duties to and Values in the Natural World* (Temple University Press, 1988). Rolston was fast becoming one of the most recognizable names in the rapidly growing field of environmental philosophy. In 1990 he was the founding president of the International Society for Environmental Ethics. After attending the United Nations Conference on Environment and Development in Rio de Janeiro in 1992, Rolston published his most policy oriented work in environmental ethics, *Conserving Natural Value* (Columbia University Press, 1994). In 1997, Rolston was invited to give the prestigious Gifford Lectures in natural theology at Edinburgh University, immediately placing him alongside such historically influential thinkers as William James, Werner Heisenberg, Alfred North Whitehead, Karl Barth, Albert Schweitzer, and Iris Murdoch. The Templeton Prize for discoveries in science and religion followed in 2003, cementing Rolston's place on the international stage.

ROLSTON'S ENVIRONMENTAL PHILOSOPHY

As this brief biography indicates, much of Rolston's work in environmental philosophy aims at a careful negotiation of a pathway that connects theology, science, and ethics. In some senses, Rolston's environmental philosophy *is* this pathway. The negotiation has been carried out largely through an articulation of the concept of objective intrinsic value in nature. Ethicists in the western tradition had previously assumed almost universally that certain aspects of our humanity make humans alone worthy of moral consideration. In contrast, nature is only judged important for its instrumental value to humans. For example, productive fisheries, clean water and air, wide open recreational spaces, and aesthetic vistas might all be considered to make a valuable contribution to the values present in human life and culture. But before the early nineteen seventies it was uncommon to think that nature itself might have intrinsic value, value that existed independently of human needs and interests. With the help of a cohort of other early environmental philosophers such as Arne Naess, Richard Routley, and Paul Taylor, Rolston began to articulate a coherent account of such value.

This central notion of his theory, the objective intrinsic value of nature, is perhaps the biggest challenge Rolston offers to traditional western philosophy. It demands a significant change in the value status of nonhuman entities—entities that previously were thought to have no value of their own at all—and it transforms human relations with those entities into matters for moral deliberation. Rolston's justification of such values brings these entities themselves—rather than only human concerns or interests—into the moral spotlight.

How does Rolston establish or justify this radical concept? Rolston leads us through the argument with some detailed observations in ecological and biological science. There we discover such features as respiration, photosynthesis, digestion, reproduction, symbiosis, predation, with individual organisms playing crucial roles in larger contexts, and with the larger contexts in turn significantly influencing their constituent members. At the level of the organism, the biological processes appear to center on the survival and reproduction of the individual. In that basic biological setting, Rolston finds a life being defended and thus valued by the individual that defends it. The biological facts reveal that the organism values its own survival even though it may not be conscious of its own valuing.

When Rolston considers further how a non-conscious organism can be said to value its own survival, he looks towards the informational features of the organism's genes. Each such genetic set, he claims, by carrying the information for the creation of individuals of that type, is at the same time a "normative set," containing the genetic directions for the achievement of that which the organism "ought to be." Rolston quickly moves from the level of the individual on to the level of the species by arguing that this genetic set is conserved primarily at the species level. He identifies a species as "a coherent ongoing form of life expressed in organisms, encoded in gene flow, and shaped by the environment."[5] While admitting that the category of species is dynamic and evolutionary Rolston insists that species have enough identity to be objects of direct moral consideration in their own right.

At the ecosystem level, where geography and ecology interact to create beneficial conditions and resources for individuals and species, Rolston observes that such benefits are both instrumentally valuable to the individual organisms, and valuable in themselves as parts of a system that nurtures and furthers life. It is the system's generativity, its ability to create billions of diverse and complex life forms over evolutionary time that Rolston finds both impressive and morally considerable. He argues that this generativity demonstrates the value of the overall biosystem, but feels that the term "intrinsic value" is inadequate at this systems level. He therefore characterizes this biological generativity as "systemic value." With lyrical writing, word plays, and seductive argumentation, Rolston makes a comprehensive case that Earth as a whole is morally considerable and worthy of a certain type of treatment by humans. "[S]ystemic nature is valuable intrinsically, as a projective system" he insists, "for its capacity to throw forward (pro-ject) all the storied natural history . . ."[6] By the time he is finished with his argument, Rolston has found value throughout biological nature, from the smallest individual organism to the largest biotic whole.

The final major piece of Rolston's ethical view is the direct connection he sees between natural intrinsic value and moral obligations. He subtitles his systematic work *Duties to and Values in the Natural World*. Clearly he sees the value discovered in nonhuman nature as entailing human moral duties: "Whatever has such resident value lays a claim on those who have standing as moral agents when they encounter such autonomous value."[7]

Many readers find in Rolston's defense of natural intrinsic value a powerful and satisfying philosophical expression of their own evaluative responses to nature. But for many thinkers, including those sympathethic to his project, his theory raises numerous complex philosophical questions. In the essays collected in this volume, the authors discuss some of those questions in depth.

CRITICALLY ENGAGING ROLSTON'S PHILOSOPHY

The anthology begins with several papers that address Rolston's signature ethical theory. Katie McShane raises two central questions that challenge the heart of Rolston's account of intrinsic natural value. Focusing on an organism's achievement of its genetically directed functions, McShane asks why, in recognizing those processes, Rolston thinks it appropriate to use evaluative language in our descriptions of them.

McShane challenges Rolston's inference from the fact that organisms exhibit goal-directed behavior to the conclusion that the achievement of their goals generates value. She argues that this inference is in need of more argumentative support than Rolston has provided for it. Her second question is whether, even if Rolston is right to use evaluative language in such contexts, those specific kinds of values are capable of generating moral obligations towards the natural entities involved. She claims that Rolston has not adequately defended the claim that they are.

Keekok Lee's essay offers a plea for Rolston to go further. Although she believes Rolston's theory about intrinsic natural values is essentially correct, Lee argues that Rolston is not radical enough. She makes a case for expanding human moral concern from the biotic realm to the abiotic, including to planets where there is no life. To make the argument, Lee elaborates several additional categories for environmental philosophers, including "trajectory," "immanent teleology," and "independent value."

Another essential part of Rolston's theory of the intrinsic value of nonhuman nature is his insistence that such value is objectively present in nature. In fact, it is this insistence that makes his view so controversial. His position has in the past provoked complaints from philosophers such as J. Baird Callicott and Bryan Norton about whether such value could really be objective.[8] Recent developments in epistemology challenge humans' ability to know objective features of the world as they are in themselves, without a strong element of human "construction." This challenge to the objectivity of knowledge applies especially strongly to the concept of objectively existing value because values have traditionally been viewed as unquestionably dependent upon human conscious experiences. In his essay, Christopher Preston presents and evaluates Rolston's responses to this kind of "postmodern" critique. Arguing that, ultimately, Rolston's responses are unsatisfactory, Preston offers an alternative way of viewing natural value, based on some recent proposals about the metaphysical status of unobservable entities found in philosophy of science.

Mark Wynn offers a generally sympathetic elaboration of Rolston's ideas in his contribution. Rolston is sometimes criticized for the strongly rationalistic approach he takes to his project; Wynn's essay counters that criticism. Wynn points out that Rolston's arguments also include discussion of the emotions. Moving from the few but favorable explicit uses of emotions in Rolston's works, Wynn uses some recent theories of the emotions to claim first, that Rolston's writings are compatible with those theories and second, that additional emphasis on the emotions, understood in light of those theories, strengthens Rolston's arguments for intrinsic natural value. In the final section of his essay, Wynn connects Rolston's ideas about the character of natural values with some theological reflections, deliberately imitating Rolston's own concerns with situating his theories within a theological context.

While it is possible to read a good deal of Rolston's work in environmental ethics without encountering any mention of a Christian God, a wider search reveals that Rolston carefully integrates his philosophical ethic with a well developed theology of nature. An important question immediately arises about the degree to which Rolston's environmental ethic is ultimately a religious ethic. Many philosophers, suspicious of religious motivations for belief, would lose enthusiasm for Rolston's position if it did not work without the addition of a divine being. For Ned Hettinger, the whole merit of the ethic hinges on exactly how Rolston articulates the connection between the deity and the intrinsic values found in nature. Hettinger is not too concerned about there being *some* religious dimension to our relationship with nature. The environmental movement often cites approvingly individuals and even whole cultures in which the appropriate attitude to nature is one of reverence. But Hettinger worries that as God's role in grounding intrinsic value increases, the ethic becomes less of a truly environmental ethic and more of an opportunity to reverence the deity.

While it is one thing to find positive moral value in nature, both philosophers and theologians have also recognized how important it is to account for so-called "disvalues" in nature. Even a casual look at the biological world reveals that it is full of suffering, waste, and death. The problem of disvalues in nature mirrors the traditional problem for theists of how to account for the presence of evil in the world. Theodicy, the task of reconciling apparent evils such as predation, suffering, death, parasitism, indifference, waste, and struggle with a benevolent deity, has long been crucial to natural theology. In her contribution, Lisa Sideris discusses Rolston's theodicy in depth. She traces the path Charles Darwin traveled as he found himself reluctantly converting to atheism, and she illustrates how Rolston manages to steer a different course. Sideris suggests that Rolston's gift to both Darwinians and Christians is his refusal to censure either nature or God in the light of apparent evil in nature. She explores Rolston's explanation that death and disease are essential to an evolutionary process that is lured in the direction of ever greater value. She notes how Rolston insists that suffering plays a critical role in both Christianity and in Darwinism.

Another aspect of Rolston's position that attracts considerable attention is his articulation of the aesthetic value of nature. In 1949 Aldo Leopold coined what has become one of the most recognizable phrases in environmental philosophy: "A thing is right when it tends to promote the integrity, stability, and beauty of the biotic community. It is wrong when it tends otherwise."[9] In stating things this way, Leopold unwittingly illustrated just how closely connected are aesthetic values and moral values. The question of exactly how close is this link between aesthetic and moral values is a question Rolston's own account raises and one that has been a considerable bone of contention in environmental philosophy. Generally speaking, Rolston follows a fairly traditional approach in that he finds aesthetic value to be a different kind of value from morally considerable intrinsic value. The former, he thinks, depends upon an observer—probably human—to ignite it while the latter is objectively present in nature independent of any observer.

Both Eugene Hargrove and Allen Carlson explore Rolston's position on the relationship between ethics and aesthetics. Carlson points out that even when aesthetics is not the explicit focus of a particular passage of Rolston's writing, he often refers to and relies on aesthetic notions. For example, when Rolston uses ecological science to find objective intrinsic value in nature it is often the "order," "harmony," "stability," or "unity" of a system that forms the basis of the value. Carlson points out that these common terms taken from ecological science are at least in part aesthetic terms.

As he reads Rolston between the lines, Carlson pinpoints in Rolston's articulation the idea that ecological science is being used both to re-describe and to re-illuminate a scene so that values that did not appear to be present before now become visible. Rolston says that an apparently ugly scene such as maggots feeding on a rotting elk carcass or blackened stumps after a forest fire need to be reconsidered as parts of ongoing historical and ecological processes. Not only are the long range ecological processes valuable but the rotting carcass and the burned forest, seen now as essential parts of those processes, are also deemed to have aesthetic value. Carlson shows how this view about science directly illuminating ecological features of a particular scene makes possible a positive aesthetics, the view held by John Muir that everything in nature—from raging rivers to rotting elk carcasses—has at least some degree of positive aesthetic value.

Eugene Hargrove questions the asymmetry that Rolston wants to maintain between aesthetic and moral values. In his own work Hargrove seeks to ground the environmental movement firmly within the aesthetic tradition. Hargrove is no enthusiast for the idea that values exist objectively in nature; but he does think that aesthetic sensibilities can provide a secure and reliable, even objective, foundation for environmentalism. Hargrove notes that once Rolston has denied the objectivity of aesthetic values, his signature commitment to the objectivity of moral values depends entirely upon establishing an asymmetry between aesthetics and ethics. If Rolston fails to maintain the asymmetry then his signature claim is seriously undermined. Hargrove explains carefully with numerous textual examples why he thinks Rolston fails in these efforts.

Brenda Hausauer's contribution to the volume discusses for the first time Rolston's distinctive literary style. Rolston is rare in the philosophical world for blending analytic argument with literary flourishes and word plays. Hausauer compares some of the rhetorical devices Rolston uses to similar devices used by Annie Dillard in *Pilgrim at Tinker Creek*. Despite the surface similarities, Rolston and Dillard are writing for radically different audiences. Hausauer finds herself compelled to ask whether Rolston's style detracts from the philosophical content of his work or whether it is an effective vehicle to convey the same sentiments that nature writers such as Dillard are more free to employ.

Although he has worked extensively on developing and defending his theoretical account of natural intrinsic value, Rolston has always had in mind its implications for specific, concrete human interactions with nature. Indeed, we test moral theories not only through the analysis of their constituent arguments and their coherence but also through their general workability in real-life situations. No such theory can give an exhaustive account of our duties because the questions of what we ought to do arise in the context of an infinite number of actual situations of choice and action. But each theory does have implications for how we should choose, and Rolston's is no exception. Clearly, his theory admonishes us to respect the intrinsic value of nature, and in books like *Conserving Natural Value*, Rolston provides numerous concrete examples of how we are obliged to act.

Several contributors to this volume examine the practical implications of Rolston's ideas. Victoria Davion evaluates some of the implications of his theory from an ecofeminist perspective. The results are mixed. She claims that Rolston's discussions of the treatment of domestic animals, especially food animals, and of hunting are problematic when evaluated through ecofeminist lenses. In the case of food animals, she uses an ecofeminist critique of dualistic thinking to argue that

Rolston's position mistakenly relies on an unjustified dualism that permits cruel treatment of the animals in such practices as factory farming. In the case of hunting, she finds Rolston's partial defense to be based too uncritically on an essentialized and gendered concept of violent behavior. Despite these objections, Davion finds Rolston's evaluation of the dilemma of when to feed the poor and when to protect the environment to be largely consistent with an ecofeminist evaluation. She also argues that his rejection of the sociobiological analysis of altruistic behavior mirrors ecofeminist sensibilities.

In the past, there have been many disagreements between philosophers concerned about ecological systems and philosophers concerned about the wellbeing of individual nonhuman animals. Clare Palmer's essay focuses directly on the application of Rolston's theory to the treatment of animals, and one of her aims is to show that these two schools of thought need not be at odds. Although she finds Rolston's own statement of his view regarding the treatment of animals "problematic in some respects," she believes that his theory "provides tools for thinking through the complicated location of domesticated animals both conceptually and ethically." A significant and useful part of her discussion is her identification of several overlapping but differing senses of "nature" and "natural" in Rolston's work.

Rolston has high regard for the value of wild areas. These areas contain little human influence and as such contain all the levels of biological value that Rolston finds in historical nature. Consistent with his overall theoretical position, the guidance he offers regarding our treatment of such areas has primarily been to allow the wild creatures and processes in them to proceed without human interference. John Lemons, an ecologist and environmental professional, questions the adequacy of that advice and, by implication, the helpfulness of the theory. Focusing on the U.S. National Parks, Lemons claims that many of Rolston's case studies omit the huge complexities facing park administrators and managers. Thus, even if those professionals want to make use of environmental ethics in their decisions, Rolston has not provided a theory that gives clear guidance for situations in which letting nature take its course is not an option. Despite this weakness, Lemons argues that Rolston's promotion of the intrinsic value of nature is a valuable perspective that can help motivate ethical reflection on decisions regarding wilderness in national parks.

Rolston's fondness for wild areas is the source of the criticism leveled by Jim Sheppard and Andrew Light. Sheppard and Light complain that Rolston takes an unnecessarily negative view of urban environments. They argue that Rolston's work has been fairly typical of New World environmental ethics in its prejudice against anthropocentric forms of value. They suggest that his anti-urban geographical bias is at best unargued and at worst misanthropic. While lamenting this aspect of his work, Sheppard and Light see possibilities within Rolston's ethic for remedy. They highlight the presence of some often ignored spontaneous natural values in urban environments, investigating urban soil, water, and geological formations for their natural values. In addition to these sources of spontaneous natural values, they find in cities evidence of numerous other values that Rolston champions, including systemic natural values and mixes of natural and cultural values. While acknowledging that Rolston will probably never come to embrace the urban environment in the way they would like him to, Shepard and Light attempt to show that Rolston's position need not be so exclusive of the urban environment as it currently appears.

ROLSTON REPLIES

At the end of the volume Rolston himself provides a commentary on each of the preceding papers. He engages each of the authors dialogically in order to advance the conversation he helped begin over three decades ago. At times he acknowledges the sting of some of their criticisms; at other times he shows that he and his critics agree; at still other times he argues that his critics have misread him or are simply mistaken. But at all times in his response, he affirms an ethic that he has maintained with few modifications over a lifetime of reflection on natural science, philosophy, and religion.

Following the reply essay, we have included an abbreviated bibliography of Rolston's major works organized by topic area for readers that wish to delve further into Rolston's writing.

NOTES

1 A number of these early essays are reprinted in the closing section of Rolston (1986).
2 Rolston (1986), 233.
3 Ibid, 237.
4 Ibid, 257.
5 Rolston (1988), 136.
6 Ibid., 198.
7 Ibid, 86.
8 Norton, "Epistemology and Environmental Values" *Monist* (1992): 208–226. J. Baird Callicott, "Intrinsic Values, Quantum Theory, and Environmental Ethics," *Environmental Ethics* 7 (1985): 357–375 and "Rolston on Intrinsic Value: A Deconstruction" *Environmental Ethics* 14 (1992): 129–143.
9 Leopold (1949), 224–225.

REFERENCES

Leopold, Aldo. 1949. *A Sand County Almanac and Sketches Here and There*. Oxford: Oxford University Press.

Rolston, Holmes, III. 1986 *Philosophy Gone Wild*. Buffalo, NY: Prometheus Books.

Rolston, Holmes, III. 1988. *Environmental Ethics: Duties To and Values In the Natural World*. Philadelphia: Temple University Press.

Rolston, Holmes, III. 1994. *Conserving Natural Value*. New York: Columbia University Press.

KATIE McSHANE

1. ROLSTON'S THEORY OF VALUE

ROLSTON'S VIEW

Holmes Rolston, III was one of the first philosophers to take on the task of constructing a theory of value adequate to environmental ethics. He proposes a theory of value according to which, he explains, "value is not anthropogenic, it is biogenic."[1] His view of value is one on which value judgments can be justified independently of appeals to human interests or preferences, and on which nature—including organisms, ecosystems, and non-living things—can and does possess value in its own right.

As Rolston describes it, his theory of value is "objective" rather than "subjective," which for him means that it locates value in mind-independent parts of the world rather than in mind-dependent appearances, sensations, attitudes, etc.[2] He begins his account of what this objective value is and where it comes from by describing living organisms in quasi-Aristotelian teleological terms:

> Something more than causes, if less than sentience, is operating within every organism. There is information superintending the causes; without it the organism would collapse into a sand heap. This information is a modern equivalent of what Aristotle called formal and final causes; it gives the organism a telos, "end," a kind of (nonfelt) purpose. Organisms have ends, although not always ends-in-view. All this cargo is carried by the DNA. . . .
>
> . . .
>
> [T]he genetic set is a normative set; it distinguishes between what is and what ought to be. This does not mean that the organism is a moral system, for there are no moral agents in nature apart from persons, but that the organism is an axiological system, an evaluative system. So it grows, reproduces, repairs its wounds, and resists death. We can say that the physical state the organism seeks, idealized in its programmatic form, is a valued state. Value is present in this achievement.[3]

The general line of reasoning seems to be this: we can understand an organism's DNA as containing a set of instructions that direct the organism to seek certain states and avoid others. That is to say, according to the instructions encoded in the organism's DNA, some states are to-be-attained states and others are to-be-avoided states. We can think of these as proto-preferences, for the to-be-attained states and against the to-be-avoided states. The satisfaction of these proto-preferences constitute the

C.J. Preston and W. Ouderkirk (eds.), Nature, Value, Duty, 1–15.

organism's (nonconscious) purpose or goal—in Aristotelian language, its telos. Thus from the point of view of the instructions encoded in the organism's DNA, the telos is a "valued state." Because the telos is a valued state, Rolston reasons, the fulfillment of the telos involves the realization of value. This value is what Rolston refers to as "natural value."[4]

On this view, value would be present whenever some aspect of an organism's telos is realized. Although this would allow for value to be present in the absence of humans, it would also limit the possession of value to states that are valued by the genetic instructions of individual organisms. Perhaps with this in mind, Rolston expands upon his original analysis. First, he argues that species have their own telos, for the genetic set is as much a property of species as of individual organisms.[5] Thus the achievement of a species' genetic goals also has natural value. Second, he argues that organisms, species, ecosystems, and even nonliving geological and astronomical bodies have another kind of value, which he terms "systemic value." Things have systemic value in virtue of being a part of what Rolston calls "projective nature," the elements of and systems in the world that produce and support the teleological processes of life.[6] For Rolston, then, objective value comes from entities either achieving some aspect of their telos or contributing to the production or support of entities that have a telos. This value is objective since it can be generated regardless of whether any minds exist in the world.

While Rolston rejects the claim that all value is subjective, he does allow that subjective mental states of appreciation are essential to humans' experiences of value (i.e., essential to their valuing). However, he insists that appreciating value must not be confused with conferring value. He explains,

> We humans cannot know the value of anything in the natural world without some feeling about it, but it does not follow that the value is just how we feel about it. The value comes mediated, communicated by our experience, but it does not follow that the value is just the experience.[7]

CRITICISMS OF ROLSTON'S VIEW

However, there are a number of concerns that one might have about such a theory. In what follows, I will focus on one of these: doubts about using the teleological structure of organisms as a basis for ethical obligations. There are two questions that we need to ask about Rolston's account of natural value. First, why should we think it is appropriate to use the language of "value" and "valuing" to describe what's going on when genes direct the development and behavior of an organism? And second, even if we were to understand the behavior of an organism's genetic set as a type of valuing, is this the sort of valuing that is relevant to ethics—i.e., the kind that ultimately generates "oughts" for moral agents?[8]

Genetic sets valuing their goals

Let us consider the first question first. Many systems, including most notably organisms, seem to exhibit goal-directed behavior. To say that behavior is goal-directed in this sense is not to say that it involves conscious or intentional purposes. Rather it is to say that the system operates in ways that systematically move toward some states and away from others, that the system exhibits what we might call "seeking and avoiding behavior." Many philosophers of science have tried to offer accounts of goal-directed behavior, explaining which features a system needs to have in order to count as goal-directed. Arturo Rosenblueth, Norbert Wiener, and Julian Bigelow, in their classic 1943 article, define goal-directed behavior as "behavior controlled by negative feed-back."[9] Others have tried to refine this model, claiming that the behavior of a system also has to be characterized by plasticity (the ability to reach the same goal in a number of different ways), persistence (the ability to compensate for obstacles that get in the way of reaching the goal), or self-regulation (the ability to reach the same goal despite a range of different changes in the external environment).[10] Still others reject the Rosenblueth et al. model, opting for a different type of analysis altogether. Larry Wright has offered what is perhaps that most popular alternative in this regard. He claims that a behavior is goal-directed if it is the case both that the behavior tends to bring about the goal, and that the behavior occurs because it tends to bring about the goal (where "because" is meant to indicate a causal relation).[11]

Rolston's description of the goal-directedness of organisms is much closer to the negative feedback model offered by Rosenblueth et al. than it is to the model offered by Wright. In fact, Rolston's description contains many of the same features mentioned by Rosenblueth et al. and their later defenders—not only negative feedback, but also suggestions of plasticity, persistence and self-regulation. Consider, for example, how Rolston describes the " 'genius' of life" that we find "encoded into genetic sets":

> There is some internal representation that is symbolically mediated in the coded "program" of the goal that is held forth. There is motion toward the execution of this goal, a checking against performance in the world, by means of some sentient, perceptive, or other responsive capacities with which to compare match and mismatch. Organisms measure success. On the basis of information received, the cybernetic system can reckon with the vicissitudes, opportunities, and adversities that the world presents.[12]

This account, like that of Rosenblueth et al., is a description of how the system operates—of what it does and how. In this way it differs from Wright's account, which offers a causal-historical explanation of why the system operates as it does.

In any case, it is worth noticing that neither Rosenblueth et al. and their defenders nor Wright use the language of "value" or "valuing" to describe the relationship between the system and its goals. But Rolston does want to do so in the case of organisms and the goals of their genetic sets. This leaves us with the following questions: Under what conditions is it appropriate to describe seeking and avoiding behavior as valuing? And have these conditions been met in the case of the genetically-directed seeking and avoiding behavior of organisms?

Clearly not everything that systematically moves toward certain states can be counted as valuing those states. If I throw marbles in the air, they will systematically move downward toward the ground—they do this in many different kinds of environment (indoors, outdoors, on the moon), despite changes we might make to their environment (introducing wind, raising or lowering the temperature), and often in spite of obstacles we might try to put in their way (blocking their downward path with a stick or a rock). But it would not be right to describe this by saying that the marbles value being on the ground. Similarly, the thermostat in my office, like the bodies of warm-blooded animals, has all sorts of mechanisms and processes that operate to keep the office temperature within a certain range. Why then would we be entitled to use talk of "values" and "valuing" in describing the temperature-regulation of warm-blooded animals but not the temperature-regulation of thermostats?

I am not sure that Rolston gives us a good answer to this question.[13] His discussion of artifacts suggests that he thinks it critical that valuing systems be the sort of systems that could be generated from "spontaneous nature," that their goals not be goals that were given to them by humans, and that they have "nonderivative, genuine autonomy (though environmentally situated) as spontaneous natural systems."[14] This set of criteria may be enough to distinguish thermostats from warm-blooded animals: thermostats aren't generated from spontaneous nature, and their goals were given to them by humans. Rolston's discussion of organisms also suggests that he thinks it critical that valuing systems be cybernetic systems. This may be enough to distinguish the operation of genes in organisms from the operation of gravity or other physical laws: the laws of physics are not cybernetic systems.

But while all of this might help us to distinguish genetic systems from other systems, it's not clear what any of these distinctions have to do with valuing. Why should we think that valuing can only be done by things that were generated by "spontaneous nature"? What is it about valuing that makes it the case that things that had their goals given to them by humans cannot do it? Why should the goals of systems that operate cybernetically get the special status of "valued states" while the goals of other systems do not?

Answering these questions requires us to get a clearer picture of what valuing is and how it is different from mere systematic goal-directed behavior. It is worth noticing that this isn't just a technical theoretical matter. The goals that we count as valued states, the seeking and avoiding behaviors that we count as instances of valuing, Rolston wants to claim, are ethically special. They make ethical claims on us that other goals do not. Their achievement generates value, and value generates ethical duties.[15] So what our ethical duties are will depend on which instances of systematic goal-directed behavior we count as instances of valuing.

Contemporary value theorists usually claim that to count as a case of valuing, a seeking or avoiding behavior must be in some way psychological—i.e., that valuing requires subjectivity of some kind.[16] Rolston, as well as a few biocentrists, question this claim.[17] After all, why privilege seeking and avoiding behavior that operates through psychological mechanisms as opposed to other kinds of mechanisms? As Kenneth Goodpaster points out, in some sense it is just evolutionary luck that we have evolved ways to "maintain, protect and advance" our lives that rely on conscious representations of our environment, while other organisms have evolved ways to do this that rely on nonconscious representations, or perhaps in some cases, don't rely on any representations at all.[18] From an evolutionary perspective, it seems arbitrary and unfair to pick out psychological mechanisms as morally important while ignoring other mechanisms that do the very same thing (i.e., help the organism gather and process information about its environment and respond in light of this information) in a different way.

But we can and should raise this same sort of question about Rolston's own position. Why privilege the seeking and avoiding behavior of organismic systems as opposed to other kinds of systems? It doesn't seem fair to pick out goal-directed systems that operate cybernetically and without human intervention as morally important while ignoring other systems that do the very same thing (systematically seek some states and avoid others) but in different ways. After all, what's so morally significant about systems that operate cybernetically? It may be an amazing feature of the world that some things are capable of gathering, storing, and processing information of various kinds. But our world has many other amazing features too—the formation of black holes, the growth of crystals, the "life" cycle of stars. We need some reason for thinking that cybernetic systems are special in some way that makes them specially deserving of our moral attention. Likewise, why should we think that things that arise from "spontaneous nature" are morally important in a way that other things aren't? What's so great about arising spontaneously as opposed to being formed by a thoughtful and creative human designer? These are all questions that we should want answers to before accepting Rolston's claims (and the ethical implications they bring with them) about which behaviors are to count as instances of valuing.

The value of achieving genetic goals

But let us suppose that Rolston can provide an account of what makes it appropriate to use the language of "value" and "valuing" to describe the relationship between organisms and their genetic goals, but not the relationship between other goal-directed systems and their goals. (Nothing I've said argues that he cannot provide such an account, only that so far, I believe, he has not.) The next worry is whether this would be enough to ground the kinds of ethical claims he wants to make—claims about how we ought to behave toward things that engage in valuing and the states that they value. To answer this question, we first need to get a clearer picture of what role the concept of value plays in Rolston's overall ethical theory.

Perhaps most clearly, Rolston takes value to be a normative concept—as he puts it, "[v]alue generates duty."[19] So whatever value is, it is the kind of thing that generates "oughts" for moral agents. Insofar as a thing is valuable (i.e., "has value"; "is a good thing") in this way, moral agents have an obligation to behave in certain ways toward it—for example, to protect it, promote it, endorse it, achieve it, respect it, etc., depending on what kind of thing it is and the particular way(s) in which it is valuable. This claim about value isn't unique to Rolston; this is how the concept of value is typically understood within ethical theory. It explains why theories about what we ought to do have an interest in the question of which things in the world have value.

We can think of this as a description of the general ethical concept of value—or "ethical value" for short. The question we are asking of Rolston's theory, then, is this: what is the relationship between what Rolston calls "natural value" and what I am calling "ethical value"? More precisely, what reasons do we have for thinking that if something involves the achievement of genetic goals (i.e., if it has natural value), it is thereby something that moral agents have a duty to protect, promote, endorse, etc. (i.e., it thereby has ethical value)? What reasons has Rolston given us for thinking that anything with natural value would have to have ethical value?

In order to answer this question, we need to understand what role natural value plays in Rolston's overall theory of value. Specifically, we need to know whether this account of natural value is supposed to be a conceptual explication of the concept of ethical value itself—i.e., a claim about *what ethical value is*, or whether it is supposed to be a (partial) story about which things fall under that concept—i.e., a claim about *which things are ethically valuable*.

If Rolston is offering us an account of what ethical value is, then we can think of the account of natural value as a proposal for re-defining "ethical value" as "the achievement of genetic goals." On this understanding of Rolston's project, he would be arguing for a naturalistic reduction of value; an analysis of what all ethical value, at its most basic level, is. If Rolston means for his account of natural value to be providing a definition of ethical value, then accepting his account would require us to accept that the claim "everything with natural value has ethical value" is a tautology, for ethical value on this view *just is* natural value.

However, I think it is fairly clear that in describing natural value Rolston does not mean to be offering us an account of what ethical value is. His account of natural value is not a definition or a conceptual explication of the concept of ethical value, but rather a description of one particular way that things can have ethical value. Rolston is a pluralist about value: he believes that value comes in different kinds and not just in different amounts. Natural value is one of the many kinds of value that he describes (others include aesthetic value, religious value, historical value, and systemic value, just to name a few).[20] Natural value is best understood as a subset within the more general category of ethical value, for it is just one of many ways that a thing can come to have ethical value. If this is right, then Rolston should be understood as offering an account of which things have ethical value rather than an account of what ethical value is. His claim, so understood, is that having natural value is a sufficient condition for having ethical value—not by definition or as a conceptual claim, but just as a matter of fact.

If this is what Rolston is claiming, then we must next ask what reasons there are for thinking that he is right—that anything with natural value will in fact have ethical value. Why should we think that every time a genetic set's goal is achieved, this is a good thing? One possible explanation might go like this: the attainment of an organism's genetic goals is good for that organism—i.e., it increases that organism's well-being. Since increasing a thing's well-being has ethical value (i.e., it is a good thing), one might argue, the attainment of genetic goals has ethical value.

The key claim in the above line of reasoning is that the attainment of an organism's genetic goals is good for that organism. Although this may be true in many of the cases we typically encounter, there are still many cases in which it is not true. It is only when an organism is in an environment to which it is well adapted that the ways that its genetic set directs it to operate will increase or at least sustain its well-being.[21] When an organism is in an environment to which it is not well adapted, the very same behaviors can instead make it worse off. Consider as an example the preference for sweet-tasting foods encoded in humans' genetic instructions. Such a preference might have been beneficial to our ancestors in the environment in which they evolved—by leading them to eat nutritious ripe fruits rather than less nutritious and/or toxic plants. But in our present environment, where refined sugars are plentiful, this trait has become detrimental—it leads to obesity, diabetes, and other diseases.[22] Cases of maladaptation such as this can happen for a number of reasons: sometimes environmental change happens more quickly than evolutionary mechanisms of adaptation can keep pace with; sometimes there are constraints (structural constraints, for instance) that make certain adaptations impossible; sometimes adaptation would require tradeoffs (e.g., smaller size) that would be even more detrimental to selective fitness.[23] It may be because so much of human history has taken place in an era of relative climatic stability that we tend to think of cases of maladaptation as the exception rather than the rule. The organisms we come across tend to be fairly well adapted to the environments in which we find them. But it is worth noticing that this needn't be the case, and in fact it may not continue to be the case for very much longer. If the current climate change models are correct, maladaptation may soon be much more widespread.[24] So we cannot say that the achievement of an organism's genetic goals has ethical value because it is good for that organism, since the achievement of its genetic goals may not *be* good for the organism.

But if we cannot say that the achievement of an organism's genetic goals is good for the individual organism, perhaps we could say that it is good for the species. After all, in cases where it is detrimental to an organism to achieve its genetic goals, isn't this often beneficial to the species as a whole? That organisms with maladapted sets of genetic instructions don't fare so well is essential to the operation of natural selection. It might be bad for the individual organisms that have these maladapted genetic instructions, but it is ultimately good for the species because it is what allows advantageous traits to become more frequent and disadvantageous traits less frequent in the population.

However, as intuitively plausible as this might sound, cases of maladaptation also show that the achievement of organisms' genetic goals needn't contribute to the well-being of the species. In cases where a species either has a beneficial trait present in the population or can produce one through mutation and natural selection so as to become better adapted to its current circumstances, it might be true that maladapted genetic instructions will kill the individual organism but leave the species better off for having done so. But not all cases are like this. If the environmental context of a species shifts too quickly for genetic mutations to arise that might do the species any good, then the member organisms' pursuit of their genetic goals may well lead the entire species to extinction. If it were the case that natural selection always resulted in better-adapted species, then perhaps the attainment of genetic goals would always be good for the species. But natural selection does not always result in better-adapted species; sometimes it results instead in species' extinction. In these cases, we cannot say that the attainment of genetic goals was good for the species. If this is right, then we cannot explain why things with natural value have ethical value by claiming that the attainment of an organism's genetic goals increases the well-being of the organism's species.

It is worth noting that the above claim about species could be made about any other holistic entity (ecosystem, biosphere, etc.) as well. For any living system that is disposed (via the genetic instructions encoded into the DNA of organisms) to do certain things, if circumstances changed quickly and dramatically enough, its doing those things could lead to its own destruction. Facts about what it takes for a living system to flourish or even just to survive are dependent upon facts about its environmental context. When the environmental context shifts, so may facts about what is required in order for a system to flourish within it. Since genetic instructions cannot always keep pace with such changes, it will always be an open question whether following those instructions will make the system better or worse off. For this reason, it will always be an open question whether the achievement of a genetic goal will have a good effect or a bad effect on the well-being of such systems. If this is right, then there does not appear to be much hope for establishing the ethical value of things with natural value by appealing to natural value's contribution to well-being—of individual organisms, species, or other holistic entities.

That said, perhaps we should be looking not for an explanation that appeals to some claim about well-being—about what is good for this or that—but rather for an explanation that appeals directly to some claim about what is good in general. After all, not all claims about what is good or bad in the world have to be claims about what is good or bad *for* something. So perhaps the claim is not that the achievement of genetic goals is always good for something or another, but rather that the achievement of genetic goals is always just a good thing in the world. If this were the claim, then what reasons might we have for thinking that it is true? Why should we think that every time an organism achieves a genetic goal, this is a good thing?

Before answering this question, it is worth noting that the claim that something is "a good thing" is ambiguous in a very important way. In claiming that the achievement of a genetic goal is a good thing, we might mean that this achievement makes a net positive contribution to the world; that every time a genetic goal is achieved, the world is overall better off for it. I will refer to this sense of "being a good thing" as "being good overall." Alternatively, we might mean that this achievement makes *some* positive contribution to the world, though this is consistent with the overall net result being negative. On this second meaning, every time a genetic goal is achieved, some goodness is thereby added to the world, though this may not be enough to compensate for badness that is also generated as a result of the goal's achievement. I will refer to this second sense of "being a good thing" as "being good in some way."

Let us first consider whether we have reason to think that the achievement of a genetic goal will be good overall. Does the achievement of a genetic goal necessarily make the world a better place? Rolston does point out that many of the behaviors of organisms that we think of as bad (predation, selfishness, etc.) play an important role in generating things that we think of as good (species survival, biodiversity, etc.). In nature, he says, "there are disvalues as surely as there are values, and the disvalues systematically drive the value achievements."[25] So perhaps every achievement of a genetic goal, even when it doesn't seem to be a good thing, is good because it ultimately plays a role in bringing about some greater good. One might think that Rolston's "systemic value" has an important role to play in explaining why this would be so: by invoking systemic value, Rolston might be able to claim that the achievement of these goals is a good thing, for it is what keeps the processes of evolution working. Since the natural world and its processes are, Rolston claims, valuable, those things which enable them to function are also valuable.

However, even if this were true in some general sense, there are surely many particular cases in which it fails. If Jeff's cat follows her genetically-given goals by catching a bird outdoors and "playing" with it in ways that kill it slowly and painfully, this doesn't seem to make the world a better place overall. While the cat might be made happy by this, the bird has probably been made more unhappy, and the cat might have been just as happy playing with a toy as with the bird. While the bird might have disadvantageous traits that get selected out of the population this way, this needn't be the case. It could be a bird with perfectly advantageous traits that just happened to be in the wrong place at the wrong time. While the bird's decomposing body might serve as nutrients for other living things, this also needn't be the case. Jeff might put the body in a plastic bag and toss it in the garbage for deposit in the local landfill or garbage incinerator. It seems that for any contribution to evolutionary processes that we can think of, it might or might not happen. Thus while it might be good overall that predator-prey relationships exist in the world in general, or that cats have predatory instincts in general, there might not be any net value added to the world by this particular cat achieving this particular genetic goal in this particular circumstance.

At points Rolston seems to agree with this. He explicitly rejects the "panglossian" view that there aren't any disvalues in nature, that somehow everything that happens in nature is for the best.[26] By rejecting the panglossian view, Rolston gives up the general principle that might have assured us of the overall goodness of the achievement of each genetic goal. Without this principle, even if we can say in general that the continued existence of life on earth is a good thing, this won't be enough to let us conclude in any particular case that the achievement of an organism's genetic goals is a good thing—i.e., has value in general. In particular cases, the achievement of a genetic goal may or may not have systemic value—whether it does will depend on whether its contribution to the production and/or support of living systems is on balance positive.

But even if Rolston were to adopt the panglossian view and claim that the achievement of genetic goals is always a good thing because it always makes a positive net contribution to the production or support of living systems, this claim wouldn't get him very far. His account of biogenic value was supposed to answer the question, "In the absence of humans, are living systems valuable?" Rolston's answer to this question was, "Yes, because the achievement of genetic goals has value." Given this argumentative strategy, accepting the panglossian view would only leave him with a circular argument. The circular argument can be represented this way: (i) living systems are valuable because they involve the achievement of genetic goals; (ii) the achievement of genetic goals is valuable because it contributes to the production and support of living systems; (iii) contributing to the production and support of living systems is valuable because living systems are valuable.

If this is right, then we have not yet seen good reasons for thinking that the achievement of a genetic goal will always make a positive net contribution to the world. But even if the achievement of a genetic goal isn't always good overall, we may still think that it is always good in some way. It is worth noticing that this is all Rolston needs to claim in order to show that value can exist in the world independently of whether any minds exist in the world. So might there be reasons for thinking that the achievement of a genetic goal is always good in some way?

It might be useful to begin by reminding ourselves why it seemed to make sense to think of the achievement of genetic goals as a kind of value in the first place. What we saw above was that from the point of view of the "normative set" of the genetic instructions, genetic goals have value in the sense that they have the status of to-be-attained. That is to say, they are states that the genetic set instructs the organism to seek; they are what I referred to earlier as the "proto-preferences" of the genetic set. In asking questions about how this is related to ethical value, then, we are essentially asking whether we have reason to think that anything that is valu*ed* (in this case, by the genetic set) is thereby valu*able* (i.e., to be protected, promoted, endorsed, etc. by moral agents) in some way.

In claiming that that which is valued is thereby valuable, Rolston's theory is structurally similar to preference-satisfaction views in anthropocentric ethics. Preference-satisfaction views claim that the satisfaction of human preferences—i.e., the achievement of the states of the world that we in fact value—is always valuable in some way. Preference-satisfaction theorists' claims about human preferences are nearly identical to Rolston's claims about the genetic set. They claim, to paraphrase what Rolston says about the genetic set, that our set of preferences is a normative set; it distinguishes between what is and what ought to be; our preferences are valued states; value is present in their achievement.[27] The line of reasoning is the same; the only difference is whose valuings each theory is willing to count as generating the value.

Given the similarity, it might be useful to think about whether we would accept this line of reasoning in the case human valuings, since they (unlike genetic goals) at least have the advantage of being widely accepted as genuine instances of valuing. If something has the property of being valued by someone, does that fact make it valuable (not valuable overall, just valuable in some way)? Does the fact that somebody values something give moral agents a reason (not necessarily an overriding reason—just *a* reason) to protect it, promote it, endorse it, etc.? I would argue that it does not. A neo-Nazi might value what she thinks of as "racial purity," but I do not think this fact makes "racial purity" valuable—not even a little bit. It doesn't give moral agents any reason—not even a small one—to promote or endorse "racial purity." A rapist might value raping people, but I do not think this fact adds to the value (or even subtracts from the disvalue) of the act of rape. Rape is not made any better by the fact that rapists have a preference for it. Thomas Nagel might value a piece of parsley being on the moon, but this fact does make the presence of parsley on the moon valuable—it does not make it something that moral agents have a reason (even a small one) to promote or endorse.[28]

The fact is that we can—and sometimes do—value all sorts of things that aren't in fact valuable. Sometimes we, like the racist, value things that are disvaluable. Other times we, like Nagel, value things that are merely valueless. As valuers we are capable of getting it wrong—in some cases quite drastically wrong. (This is one of the well-known problems with preference-satisfaction views. People's actual preferences can be irrational, cruel, uninformed, and the like.[29]) Most of us probably can look back over our lives and identify things that we erroneously valued at one point or another. And most of us critically assess our own valuings from time to time in the hopes of avoiding these sorts of errors. However, if valuing something were enough to make it valuable, then it would be impossible to value something that is not valuable, since our valuing it would *make* it valuable. If valuing something were enough to make it valuable, our worries about falling into error would be misplaced.

I would argue that moral obligations are not generated by any instance of valuing whatsoever, but rather only by instances of *appropriate* valuing. It isn't that which is valu*ed* that generates obligations for moral agents, but rather that which is valu*able*. In fact, I think that this is the attitude that most of us quite sensibly take toward the instances of valuing we encounter. Not only do we not think we have any obligation to respect the racial preferences of racists—not even a little tiny obligation—but most of us also think that our own valuings should only have force insofar as they are not mistakes. This fact explains why we bother to critically reflect on our own valuings and revise them if we think they've gone wrong.

One might object that limiting the realm of the "valuable," and thus limiting the source of moral obligations, to instances of appropriate valuing is too restrictive. In the first place, one might argue, we often do take other people's or our own valuings to generate obligations without checking to make sure they are not mistaken. When you tell me that you'd prefer to schedule our meeting for later in the day, I don't stop to ask whether your scheduling preferences are really the best ones for you to have. I take your preference as a given and think about whether we can find a way to satisfy it. Likewise, if you tell me that Minneapolis is a charming city, I will believe you unless I have some reason not to trust your judgments about such matters. But the fact that we often don't take the time to assess the appropriateness of others' valuings is, I think, simply a testament to how much we are usually inclined to trust people to get it right. When you tell me that it's raining outside, I'll believe you unless I have some reason not to. Likewise, when you tell me that afternoon meetings are better for you, or that Minneapolis is a charming city, I'll believe you unless I have some reason not to. Most of us are willing to take the fact that people value a thing to be some evidence that the thing is valuable. Of course, this is compatible with other evidence ultimately convincing us that it is not valuable.

But one might object that even if this is right, there are many cases where we take ourselves to be obligated by the valuings of others even when we have reason to think that those valuings are wrong—democratic decision-making, for example. Yet I would contend that in cases where this is true, it is only because we have some other reason to accept the valuings of others as making legitimate claims upon us regardless of whether the valuings are mistaken. There are a number of such reasons we might have: having committed to a political arrangement whereby we agree to accept the opinions of the majority whatever they may be; having reason to think that more harm than good would come from subjecting the valuings of ourselves or others to critical scrutiny; and so on. Nonetheless in the absence of such reasons, the mere fact that something is valued by someone isn't enough to generate moral obligations.[30]

The arguments I have given so far may or may not be satisfying. These are complicated matters, not easily settled in a few paragraphs. But what I think they show is that we cannot simply assume that something's being valued by someone makes it valuable. There are some good reasons for thinking that something's being valued by someone isn't enough to make it valuable. So one wants to hear from the preference-satisfaction theorist some reasons for thinking that the mere fact that something is valued by someone *is* enough to make that thing valuable.

Turning now from preference-satisfaction views to Rolston's own view, we can see the very same worries arise. Why should we think that the mere fact that some state is valued by the genetic set of some organism makes that state valuable? Especially once we notice that genetic sets can value states the attainment of which is not good for the individual organism, the species, the ecosystem, or any other living system, we might have good reason for wondering why the mere fact that it is valued by the genetic set is supposed to make it valuable. As far as I can tell, Rolston gives us no answer to this question. He moves from the claim "the physical state the organism seeks . . . is a valued state" to "value is present in this achievement" without comment or explanation. But I do think that this inference requires explanation. Being valued isn't the same thing as being valuable; some things that are valued don't seem to be valuable (not even a little bit); and so the move from claiming something is valued to claiming it is valuable calls out for explanation.

In summary, then, we have seen that Rolston's theory contains two main claims that seem to call out for further explanation and justification. First, there is the claim that genetic sets should be counted as valuing their goals rather than merely pursuing them, as in the case of other goal-directed behavior. One wants to know what valuing is such that this claim, together with its ethical implications, should be accepted. Second, there is the claim that states valued by genetic sets are thereby valuable (in the normative, ethically relevant sense of the term "valuable"). One wants to know what positive reasons there are for thinking that this is true. What is it about being valued by a genetic set that makes something ethically valuable—i.e., the sort of thing that moral agents have a duty to protect, promote, endorse, etc.?

It is worth noting that Rolston's theory of value isn't the only possible way of accounting for the existence of value in a world that doesn't contain any humans, or even any minds. Another strategy, one that has been popular among environmental ethicists, is to focus on well-being—i.e., on what has value for things rather than on what is valued by them. There are many parts of the natural world, one might argue, that can be benefited or harmed independently of the existence of humans, or minds in general. Instead of focusing on which states are sought by the genetic set of an organism, one might focus on which states would be good for the organism (or the species, ecosystem, etc.)—i.e., on which states the organism would be better off seeking. On this sort of view, if one can make the case that the well-being of other things makes a moral claim on us, and if one can come up with an account of well-being that explains what well-being is, how things without minds can have it, why moral agents should care about it, etc., then one can get a perfectly good account of how value (of an ethically relevant kind) can exist in the world independently of minds. Many environmental ethicists have worked toward accounts of this sort (Nicholas Agar, Paul Taylor and Gary Varner, to mention a few), but there isn't room to review them here.

Rolston's theory of value remains a significant and interesting alternative to these well-being-based views. Whether its main claims can be defended in the ways described above is, I think, still unclear. But at the very least, Rolston has raised some important questions about the nature of value, and shown us why such questions are well worth trying to answer.

I would like to thank Jeff Kasser, Robert Mabrito, Christopher Preston and Wayne Ouderkirk for many helpful comments on earlier versions of this chapter. I have also benefited greatly from conversations with Elizabeth Anderson, Stephen Darwall, Dale Jamieson, Paul Moriarty, Peter Railton, Larry Simon, Blake Schlukbier, and Vicki Weafer concerning some of the issues I discuss here.

NOTES

[1] Rolston (1992a), p. 138.

[2] See, e.g. Rolston (1982); Rolston (1988), p. 4. It is this claim about the mind-independence of value that leads Rolston to think of his theory of value as a type of objectivism rather than subjectivism. However, it is important to keep in mind what Rolston means by "objectivism" and "subjectivism." In contemporary metaethics, people often use the term "subjectivism" to refer to views about the meaning of ethical statements, views on which ethical statements are just reports of the speaker's subjective states. Such a theory might claim, e.g., that when I say "O is valuable," I am telling you that I approve of O. (For an example of such a view, see Elliot (1997), esp. Ch. 1. For a general description of metaethical subjectivism, see Miller (2003), pp. 36–37.) However, subjectivism as Rolston understands it is not primarily a claim about the meaning of ethical statements, but rather a metaphysical claim about what value is (a relational property—the property of being the object of certain subjective states in valuers) and how it gets into the world (via the subjective states of valuers). Notice that understood as a metaphysical claim, subjectivism has more robust normative implications. If subjectivism is a metaphysical view, the claim "a world without valuers has value" will be necessarily false. However, if subjectivism is a claim about the meaning of ethical statements, the claim can be true—it just means "I approve of a world without valuers."

[3] Rolston (1988), pp. 98–100, emphasis omitted.

[4] For similar accounts of natural value, see Foot (2001) and Hursthouse (1999).

[5] Rolston (1992a), p. 140.

[6] Rolston explains, "The life in which these astronomical and geological processes culminate is still more impressive, but it is of a piece with the whole projective system . . . . One cannot be impressed with life in isolation from its originating matrix. Nature is a fountain of life, and the whole fountain—not just the life that issues from it—is of value." Rolston (1988), p. 197. Thus the genetic fallacy, Rolston claims, is not a fallacy at all.

[7] Rolston (1988), p. 28. Rolston does believe, however, that complex values arise out of the interaction between the value in nature and our mental states (which, he claims, also have value). See Rolston (1994), pp. 134–141.

[8] Although in what follows I will talk about the genetic set of organisms, if one accepts Rolston's view that genetic sets can also belong to whole species, the same claims that I make here will also apply, *mutatis mutandis*, to the genetic set of species.

[9] Rosenblueth, Wiener, and Bigelow (1943). They define negative feedback as follows: "the behavior of an object is controlled by the margin of error at which the object stands at a given time with reference to a relatively specific goal . . . , the signals from the goal are used to restrict outputs which would otherwise go beyond the goal" (p. 19).

[10] See, e.g., Braithwaite (1953) (plasticity), Nagel (1961) (plasticity and persistence) and Scheffler (1963) (self-regulation). For overviews of both the criticisms of Rosenblueth et al. and the attempts to vindicate their basic approach, see George and Johnson (1985) and Nissen (1997).

[11] Wright (1976). For a survey of the criticisms that have been made of Wright's view, see Nissen (1997), pp. 127–158.

[12] Rolston (1988), p. 98.

[13] For a recent attempt to answer this and related questions in the context of an overall defense of biocentrism, see Agar (2001).

[14] Rolston (1988), p. 105.

[15] Ibid., p. 41.

[16] For a discussion of different versions of this claim, see Gaus (1990), pp. 80–144.

[17] See, e.g., Agar (2001); Taylor (1980); and Goodpaster (1978).

[18] Goodpaster (1978), p. 317.

[19] Rolston (1988), p. 41.

[20] For descriptions of the other kinds of value, see Rolston (1981); Rolston (1988), chapters 1 and 6; Rolston (1994).

[21] Here I set aside the matter of altruistic behavior, which might also be thought to be a case of genes directing an organism to do something that makes the organism worse off.

[22] Logue (2004), pp. 65–74; 225–230.

[23] Nesse (2005); Crespi (2000).

[24] See Thomas et al. (2004).

[25] Rolston (1992b), p. 275.

[26] Ibid.

[27] Compare to Rolston (1988), pp. 99–100.

[28] Nagel (1970), p. 45.

[29] See Griffin (1986) and Scanlon (1998) for further discussions of problems with preference-satisfaction views.

[30] Indeed, even Kant thought that we only need to respect the rationally adopted ends of others, and more sophisticated preference-satisfaction theorists talk about satisfying people's rational preferences rather than their actual preferences.

REFERENCES

Agar, Nicholas. 2001. *Life's Intrinsic Value: Science, Ethics, and Nature*. New York: Columbia University Press.

Braithwaite, Richard Bevan. 1953. *Scientific Explanation: A Study of the Function of Theory, Probability and Law in Science*. Cambridge: Cambridge University Press.

Crespi, B.J. 2000. "The Evolution of Maladaptation." *Heredity* 84:623–629.
Elliot, Robert. 1997. *Faking Nature: The Ethics of Environmental Restoration*. London: Routledge.
Foot, Philippa. 2001. *Natural Goodness*. Oxford: Oxford University Press.
Gaus, Gerald F. 1990. *Value and Justification: The Foundations of Liberal Theory*. Cambridge: Cambridge University Press.
George, Frank, and Les Johnson (eds.). 1985. *Purposive Behavior and Teleological Explanations*. New York: Gordon and Breach, Science Publishers.
Goodpaster, Kenneth E. 1978. "On Being Morally Considerable." *Journal of Philosophy* 75:308–325.
Griffin, James. 1986. *Well-Being: Its Meaning, Measurement, and Moral Importance*. Oxford: Clarendon.
Hursthouse, Rosalind. 1999. *On Virtue Ethics*. Oxford: Oxford University Press.
Logue, Alexandra W. 2004. *The Psychology of Eating and Drinking*. 3rd ed. New York: Brunner-Routledge.
Miller, Alexander. 2003. *An Introduction to Contemporary Metaethics*. Oxford: Polity Press.
Nagel, Ernest. 1961. *The Structure of Science: Problems in the Logic of Scientific Explanation*. London: Routledge & Kegan Paul.
Nagel, Thomas. 1970. *The Possibility of Altruism*. Oxford: Oxford University Press.
Nesse, Randolph M. 2005. "Maladaptation and Natural Selection." *Quarterly Review of Biology* 80:62–71.
Nissen, Lowell. 1997. *Teleological Language in the Life Sciences*. New York: Rowman & Littlefield.
Rolston, Holmes, III. 1981. "Values in Nature." *Environmental Ethics* 3:113–128.
Rolston, Holmes, III. 1982. "Are Values in Nature Subjective or Objective?" *Environmental Ethics* 4:125–151.
Rolston, Holmes, III. 1988. *Environmental Ethics: Duties to and Values in the Natural World*. Philadelphia: Temple University Press.
Rolston, Holmes, III. 1992a. "Challenges in Environmental Ethics." In *The Environment in Question: Ethics and Global Issues*, edited by David E. Cooper. New York: Routledge.
Rolston, Holmes, III. 1992b. "Disvalues in Nature." *The Monist* 75:250–278.
Rolston, Holmes, III. 1994. *Conserving Natural Value*. New York: Columbia University Press.
Rosenblueth, Arturo, Norbert Wiener, and Julian Bigelow. 1943. "Behavior, Purpose and Teleology." *Philosophy of Science* 10:18–24. Reprinted in Frank George and Les Johnson (eds.), 1985.
Scanlon, T. M. 1998. *What We Owe to Each Other*. Cambridge, MA: The Belknap Press.
Scheffler, Israel. 1963. *The Anatomy of Inquiry: Philosophical Studies in the Theory of Science*. New York: Alfred A. Knopf.
Taylor, Paul. 1980. *Respect for Nature: A Theory of Environmental Ethics*. Princeton: Princeton University Press.
Thomas, Chris D., Alison Cameron, Rhys E. Green, Michel Bakkenes, Linda J. Beaumont, Yvonne C. Collingham, Barend F.N. Erasmus, Marinez Ferreira de Siquiera, Alan Grainger, Lee Hannah, Lesley Hughes, Brian Huntley, Albert S. van Jaarsveld, Guy F. Midgley, Lera Miles, Miguel H. Ortega-Huerta, A. Townsend Peterson, Oliver L. Phillips, and Stephen E. Williams. 2004. "Extinction Risk from Climate Change." *Nature* 427:145–148.
Wright, Larry. 1976. *Teleological Explanations: An Etiological Analysis of Goals and Functions*. Berkeley, CA: University of California Press.

KEEKOK LEE

2. BIOTIC AND ABIOTIC NATURE: HOW RADICAL IS ROLSTON'S ENVIRONMENTAL PHILOSOPHY?

1

Environmental philosophies may be said to be either conservative or radical in scope and vision. Consider the following instances with their respective foci:

1. On humankind on planet Earth;
2. On humankind on Earth and possibly on other planets (in our Solar System);
3. Not only on humankind, but also on other forms of non-human life on Earth;
4. Not only on humankind and non-human life on Earth, but also on the abiotic on Earth, the Sun and its other planets (in our Solar System).

The first and the second are essentially anthropocentric in orientation, assuming that human life is alone intrinsically valuable while non-human life and the abiotic have only instrumental value for humans. Philosophically speaking, they are the least adventurous as they simply work within the dominant framework of modern Western philosophy which, in principle, recognises no ethical (only prudential) restraints to human control and manipulation of non-human nature to further human intentions and goals.

The third is more enterprising as it challenges that very framework; it is essentially biocentric in orientation, arguing that non-human (individual) organisms are intrinsically valuable, too—such recognition would entail constraints on what humans might wish to do to plants and animals in the name of promoting human well-being or freedom. It may, however, also be ecocentric, arguing that biological species as well as their habitats and the habitats of individual organisms may have intrinsic value.

The last poses a still greater radical challenge to modern philosophical orthodoxy as it contends that even the abiotic has a value beyond being merely instrumentally valuable for humankind. Such an articulation is also prepared to confront certain philosophical/ethical problems posed by the rapidly changing character of contemporary technology which envisages, however, admittedly not in the immediate future, the terraformation of planets like Mars (in our Solar System) to transform its atmosphere—very different from that of Earth—which cannot support life as we know it on Earth to one which can. For those who defend the second conception of environmental philosophy just mentioned, the prospect and eventual realisation, some day in the future, of terraformation, however, would pose no philosophical/ethical challenge whatever, as such an enterprise could simply be justified within instrumentalist terms.

Holmes Rolston, III, environmental philosophy is radical in so far as it falls under the third category identified above; his goes beyond the mere biocentric version to embrace as well an ecocentric orientation. In that sense it may be said to be both

C.J. Preston and W. Ouderkirk (eds.), Nature, Value, Duty, 17–28.

biocentric and ecocentric. There is no conflict or contradiction, for by that one means no more than the following: Rolston celebrates the value in individual organisms, the value in species, as well as the ecosystems of which individual organisms and species are necessarily members. As such it challenges the anthropocentric framework as well as transcends the limitations of a merely individualist (biocentric) approach. His account of environmental philosophy is, therefore, a rich one, and as pioneer in this field, he richly deserves the accolades his work has attracted.

However, the focus of Rolston's writings is in the domain of the biotic, and he has paid, at best, very little explicit attention—that is, in proportion to the volume of his total publications—to the domain of abiotic nature per se.[1] For him, on the whole, the abiotic plays a necessary, though nevertheless, a supplementary role to the biotic in the unique (as far as evidence up to now demonstrates) drama of life on Earth—the biotic, of course, is impossible without the abiotic, but all the same, the value of the latter lies in being instrumental to the former, if one may use the term in this context. At the same time, in celebrating life on Earth, especially from an ecosystemic orientation, Rolston is celebrating that indivisible whole of which the biotic as well as the abiotic are constituent parts. As far as most of Planet Earth is concerned, this perspective holds true and works fine.

The scope of his environmental philosophy is essentially Earth-bound. As a result, by and large, Rolston has not felt impelled to confront the more challenging question posed directly by the fourth, the most radical type of environmental philosophy, namely, whether the abiotic, on its own, in the absence of the biotic, could be said to have value per se independent of whatever value it may have in association with biotic nature in specific contexts, such as that which obtains on Earth. It follows as a consequence of the orientation of Earth-bound environmental philosophies, such as Rolston's, that in circumstances where biotic nature is absent, abiotic nature is valueless, as such philosophies recognise only instrumental value (of the abiotic for the biotic), intrinsic value which resides either in the individual organism, in the species, or in the ecosystem of which the individual organism, its species and abiotic elements such as water, air, light, and minerals etc., are a part.

However, Rolston is also interested in Antarctica, a continent with sparse biological species and individual organisms compared with the other continents and the deep oceans, and has no human habitation—large scale and permanent—in the normal understanding of the term.[2] He finds that he confronts the limits of his Earth-bound environmental philosophy when he looks at Antarctica, the seventh continent. There, he finds, that it is obvious that his most distinctive concept for theorising about the intrinsic value of the other continents, namely, that of an ecosystemic ethic, has little application; nor has the concept of biocentrism and for the same reason, as Antarctica, part land but mostly ice, at best, shelters very few species of life, and certainly almost no higher flora. Rolston has, therefore, argued that "(a)n appropriate ethics for the seventh continent differs radically from that for the other six" (Rolston 2002, 115).

Paradoxically, a "soft" anthropocentricism may be able to cope better—by "soft" anthropocentrism is meant the view which sees non-human nature as being instrumental to humans in aesthetic, rather than in straightforward utilitarian economic terms of digging for minerals, etc. On this view, if people find the icescape of Antarctica beautiful or sublime, then this has value for present or future humans. It follows then that Antarctica ought to be preserved. Rolston, however, does not feel completely at ease with this perspective. He writes: "enjoying getting in on nature's show seems to presume something going on worth watching." What is it, though, which is worth watching? It is this "something" which seems to elude Rolston's grasp, given the usual concepts which he deploys in his environmental philosophy. Yet, he is clear that Antarctica has value even if its value cannot be readily caught within the framework he has developed and which works well for the other continents on Earth.

> Antarctica is a "wonderland." The humanists will insist that people must be there to "wonder." The naturalists will wonder that such experience is generated in the presence of something remarkable, worthy enough to induce our wonder, a natural wonderland that generates duties when moral agents encounter it. (Rolston, 2002, p. 132)

The task of the present chapter, as tribute to Rolston's vision of environmental philosophy, is to go beyond it, to explore in outline some of the issues mentioned under the rubric of the fourth category, an environmental philosophy which may be able to do justice to the peculiar problems posed by Antarctica which Rolston has identified. It is also an environmental philosophy which is not merely Earth-bound, but whose scope extends, at least, to the other planets of our Solar system. It is intended as an attempt to complement Rolston's own environmental philosophy whose primary remit is biotic nature, although it is obvious, as we have seen, that Rolston has great respect, too, for abiotic nature, even though he might not have devoted the major portion of his life's work to its defence.[3] However, what this contributor cannot hope to do is to emulate his style which adds a quasi poetic dimension to philosophising.

The outline of such a defence will include the following elements[4]:

1. The notion of trajectory.
2. The distinction between "in themselves," "for themselves," "by themselves."
3. Recognized-articulated values and mutely-enacted values.[5]
4. Three theses of teleology: external teleology, intrinsic/immanent teleology, extrinsic.
5. The notion of independent value.

2

Every individual entity, whether biotic (like a human being, an oak tree or a lion) or abiotic (like a mountain or a river) may be said to have a trajectory. This is to say that each of these entities has its own particular history, given the kind or type of entity it is, which in turn implies that their trajectories are necessarily different. To make the latter point obvious, take a human individual, like Hitler. Hitler's trajectory is very different from that of the oak in the forest or the lion in the savannah. Hitler lived for fifty odd years, and met his end by committing suicide. The particular oak in mind probably has a much longer life span. The particular lion might not live quite as long as Hitler. But what Hitler did in his lifetime is very different from what the oak and the lion each respectively does in its own. Hitler plotted and planned to found the Third Reich, which was intended to last for at least a thousand years; he conquered other nations in Europe and sent millions to concentration camps and to their deaths in specially constructed gas chambers. None of these things for which he was responsible could be attributed even in principle to what an oak or a lion might have done in their own lives, as it would make no sense to do so. For a start, the action of a lion could not be informed by ideological constructs like that of ensuring the triumph of the Aryan race. True, the lion might kill a gazelle for food, but it is (conceptually) incapable of murdering a gazelle either for food, for *lebensraum*, or for preserving the supremacy of the species lion in the name of leonine purity. Of course, the lion kills for food so that it can survive and reproduce, thus perpetuating some of his own genes. However, the lion lacks the kind of awareness, consciousness and language, which humans possess and which enables them to build ideological, religious or philosophical theories and systems to inform their intentions, goals and actions.

The oak lacks consciousness of any kind. However, this does not mean that it cannot take steps to protect its own interests or to satisfy its own needs. The lion, in search of water, could roam about to find a source. True, the oak cannot uproot itself in search of water in a drought, but this does not prevent it from growing deeper roots to tap any water that might be further down in the soil, or to close up the pores on its leaves to conserve water by cutting down on its rate of evapo-transpiration.

The trajectory of the river is again different. Its lifespan as well as its size are determined by its source (how much water feeds into it via rainfall or snow-melt), the kind of terrain it passes through (granite or limestone), the action of the sun, wind, plants and animals on its banks—bio-geo-physical-chemical processes at work—in transforming the landscape.

The peculiar characteristic of human consciousness has impressed modern (western) philosophers to such an extent that they draw three main conclusions from it: first, that only humans are intrinsically valuable, second, that all non-human natural entities are only, therefore, of instrumental value to humans, third, in virtue of their intrinsic value, humans are superior to all non-human natural entities. These three inferences are contestable. The first could be said to beg the question, as it chooses to use as premise for its argument a feature which (undeniably) is unique and peculiar to humans—but why cannot one select in turn a feature about the lion which can be said to be unique and peculiar to the lion, or about the oak which can be said to be unique and peculiar to the oak to justify that they too are intrinsically valuable? Moreover, it also turns the argument into a tautology: humans are said to be intrinsically valuable because they possess such a type of consciousness, but why does the possession of such consciousness be said to be a ground for attributing intrinsic value? Ultimately, it is because one has defined "intrinsic value" in terms of the peculiarity of human consciousness.

The failure of the first claim to survive critical scrutiny means that the second conclusion also falls by the wayside. In turn, the third can be shown to be a *non-sequitur*—in terms of strict logic, one is equally permitted to draw the very opposite of that conclusion, namely, that humans, far from being entitled to ride roughshod over non-human nature in the name of human superiority, have duties and responsibilities to it.[6] The possession of such unique consciousness renders humans alone aware that other non-human natural entities have "goods" of their own, lives to lead, trajectories of their own; it also makes humans capable of morality, to take into account such goods, lives and trajectories which have value of their own, independent of the fact that they could be of use to human projects.

To recognise the uniqueness of human consciousness in the way just mentioned is to admit that humans are indeed intrinsically valuable—a sense which shall soon be made clear—without, however, denying that non-human entities could also be said to be valuable in other possible senses of the term.

Humans are intrinsically valuable "in themselves" for, as we have already seen, they are beings whose consciousness involves language which allows them to articulate abstract concepts and thoughts, to construct theories (whether religious, metaphysical, scientific). Human language goes beyond the expressive ("ouch!") and the signalling ("fire!") levels of language use to the descriptive and the argumentative levels. While some of the higher animals may be able to communicate at the first two levels, their type of consciousness does not encompass the other two. As a result, they are not as powerful and as successful as humans in ensuring their own personal survival and reproduction of their species. Today, humans via their scientific theorising, have induced increasingly more powerful technologies to transform and manipulate the non-human world, according to their conception of promoting human well-being, freedom or self-realisation. Some conceptions, like Hitler's, are partial, being confined only to certain human groups; but others are universal, and preach equality, justice and freedom for all. Whatever the differences, humans alone are capable of acting for evil as well as for good. Such beings—saints and sinners alike—are undeniably intrinsically valuable "in themselves."

The command of language at an abstract, symbolic and conceptual level enables humans to articulate their values in a self-conscious, and even critical, manner. In this way, they are at once the source as well as the locus of recognized-articulated values. Humans can articulate their own conceptions of what is good or what is evil, and of that value which is bound up with the ability to arrive at such conceptions. In this way, they are both the source and locus of such values, which is also to say that they are intrinsically valuable "in themselves."

While humans can articulate not only their own intrinsic value, and can also recognise and articulate that other natural non-human beings, too, have a good or, in Aristotelian language, a *telos* each of its own, these other beings themselves cannot articulate such a good; they can only enact it in their very existence and their activities. They are, therefore, both the loci and the source of what may be called mutely-enacted values, which render them intrinsically valuable "for themselves."

Thus the necessarily anthropogenic recognition and articulation of nonanthropocentric, mutely-enacted values shows that it is false to maintain that human consciousness is the loci of all values, and misleading to say it is the source of all values *tout court*.

As we know, humans did not emerge on Earth till recently (judging by geological time). In other words, once upon a time, there were only mutely-enacted values on Earth. Like other species, humankind may one day go extinct and, should no other being evolve with an equivalent type of consciousness in such a world of the future, there will also be no recognized-articulated values, only mutely-enacted ones (assuming that life in general continues). However, in a world where there are both humans and other non-human organisms, there are both sets of values, and there are beings which are both intrinsically valuable "in themselves" and "for themselves" (that is humans) as well as beings which are only intrinsically valuable "for themselves."

The Aristotelian concept of *telos* should not, and need not, be burdened with the conceptual baggage of human consciousness. The oak, unlike the hog or the lion, even lacks consciousness of any kind; yet its life history and its associated activities could be understood as one of being goal-oriented or "teleonomic." What it does is not haphazard and random, but has a coherence and a unity which is to be understood in terms of its need to survive and to reproduce. To capture this coherence and wholeness, one can say, then, that the individual organism exists "for itself" and that it is valuable "for itself." It is a mistake to infer from the fact that it fails to be valuable "in itself" that it is also not valuable "for itself"—such a mistake is a categorical one, as it judges something to be inferior or defective by a standard which is not appropriate to invoke in the first instance. It is as if to claim that a cat is defective (as a dog) because it does not bark, like a dog, but only meows. In other words, oaks and lions are just not humans and humans are not oaks or lions. The intrinsic value of each being must be judged in terms of concepts appropriate to the kind of being it is. Non-human biotic beings, nevertheless, possess a value in their own beings, which have nothing to do with their also being instrumentally valuable in certain contexts to humans.

It is vital, then, to re-affirm that the *telos* of the individual organism is independent of the intentions, goals and projects of humans. For this reason, one is wrong to conclude that they exist only to serve human ends, that is to say, that they only have instrumental value for humans who alone are intrinsically valuable. Aristotle is often said to be the source of this mistaken understanding. That might be so, but his account of the organism's *telos* is capable of a more sophisticated interpretation than the one usually assigned to him. For Aristotle, what is primary is the unfolding of that *telos* within the context of its own life and existence—in other words, the oak does not give forth acorns in order to feed the hog which, when fattened, would be slaughtered to feed the human. The oak produces acorns for the goal of reproducing itself; it does not do so in order to serve either hog or human ends. Of course, the hog does find the acorns useful for its own end of survival, and in turn, the human considers the hog useful as food. However, the hog does not eat the acorns in order to be served up as roast pork at the hunter's table. The oak and the hog, each does what it does, in accordance with its own *telos*, solely to further its own good. That it may be found to be useful to other non-humans or humans is a secondary matter—a contingency and a happenstance.

In other words, in this context, one should distinguish between at least two different theses of teleology: external teleology and intrinsic/immanent teleology.[7] The former simply embodies a strident anthropocentrism as it holds that the world has come into existence for a very specific purpose, namely, ultimately to serve humans, a view which is not uniformly sanctioned and championed by contemporary (Christian) theological thought, nor is it compatible with scientific thought since Darwin, whose theory of natural selection precisely dispenses with such a thesis—natural evolution in accordance with natural selection does not require the intervention of extraneous ends or designs, whether supernatural or human. Instead, the mechanism of natural selection is consonant with and requires the thesis of intrinsic/immanent teleology.

Each organism is simply concerned with maintaining its own functioning integrity in order to survive and to reproduce itself; those individuals which happen to have certain traits (in their genetic make-up) which are favourable to survival and reproduction in a certain environment are able to leave behind offspring (or more offspring); those who do not, either die before they reach maturity to reproduce, simply fail to reproduce, or their offspring themselves die as soon as they are born or fail to thrive and to survive to sexual maturity. In other words, the theory of natural selection in natural evolution, while denying external teleology, presupposes intrinsic/immanent teleology if it is to be intelligible. To hold anthropocentrism based on external teleology is, therefore, to fly in the face of the scientific consensus which has grown up in the last hundred and fifty years.

On the other hand, Aristotle may be said to hold both theses; but it would be a travesty of this thought to ignore his implication that intrinsic/immanent teleology is prior to external teleology. As a result, his anthropocentrism is more nuanced than the modern version, which simply upholds external teleology, while rejecting intrinsic/immanent teleology.

3

This section explores in outline the arguments that we should respect the abiotic even in a context which excludes the biotic altogether, such as the greater part of Antarctica or the Sun at the centre of our Solar system. (Mars might not be such a good example as some experts hold that it might once have had life on it and might still do—hence the missions to that planet to determine the validity of such a thesis.) On Earth, the biotic and the abiotic are often inextricably intertwined—soil is a perfect example. As already remarked earlier, in conceiving an environmental philosophy, which is wholly Earth-bound, one does not need to worry unduly about providing a defence of the abiotic per se. One can minimally assume that the abiotic has instrumental value for the biotic, which can more readily be argued to be intrinsically valuable. Such a strategy of economy saves one from the more difficult philosophical task of justifying respect for the abiotic when it occurs independently of the biotic, as on the other planets of our Solar system as well as on the Sun itself, which as far as one knows does not contain life.

Unlike the biotic, the abiotic is "teleomatic," as it lacks not only consciousness (like the oak), but also because its behaviour is not goal-oriented (unlike the oak). It cannot be said to be autopoietic, to strive to maintain its own functioning integrity. It is simply subjected to the natural (geo-hydro-chemical-bio) processes at work, which determine its trajectory. It cannot and does not take steps (at least there are no mechanisms one knows of) to avoid certain outcomes or to attain others. A lake comes into existence, often as the result of glacial action in the past; eventually, it will cease to be a lake and become a meadow. Lots of lakes have come and gone in this way in Earth's history. Some abiotic entities are more enduring than others as geological forms—even very large lakes are relatively ephemeral compared to mountains, and so are volcanoes.

Obviously, abiotic entities are not intrinsically valuable either "in themselves" like humans or "for themselves" like plants and animals. However, they may be said to possess a value which is bound up with the fact that they have a trajectory all of their own—they come into existence, continue to exist, and eventually go out of existence entirely as the result of certain natural processes at work which fashion, sustain and ultimately annihilate them, in the absence, in principle, of human manipulation and interference of any kind. (Of course, humans, with their increasingly powerful technology, which enables them to some extent to manipulate abiotic nature, may choose to do so, especially when the abiotic is not in general thought to be morally considerable.) To mark these differences, let us move away from the term "intrinsic value" itself, to a deeper one which may be called "independent value."

Abiotic entities may be said to possess independent value, to be valuable "by themselves." This sense of value—the existential sense—embraces both abiotic nature as well as non-human biotic nature. Organisms as well as abiotic entities are valuable "by themselves" although organisms are also valuable "for themselves." Organisms, too, have come into existence, continue to exist and go out of existence, entirely because of certain other natural processes at work, and do so, in principle, independent of human manipulation and control.[8] To be valuable "by itself" may then be said to be deeper than to be valuable "for itself"—the naturally-occurring organism first exists in this independent manner before it acts autopoietically.

This existential value is at the same time an ontological one. It marks out a space which is not that occupied by the products of human manipulation and control, which does not embody human intentionality. Unlike human artefacts which are designed to serve human ends, and therefore, are the instruments of humans, biotic and abiotic nature (in the wild) belong to a totally different ontological domain. In principle (in general), they do not owe their existence and their being to humans. They come and go according to their own trajectories.

We have already seen that the thesis of external teleology is scientifically false in the case of the biotic (on Earth). It is even more obviously false and patently implausible in the case of the Sun itself and its other planets in the system. Humans at least exist on Earth and are part of natural evolution of life on it. But the Sun and its other planets have and had no humans. Today, (some) humans would like one day eventually to transform Mars, possibly also Venus (if not the Sun, Mercury or Jupiter) to make it fit for human habitation. It is deliberately to turn these planets into objects to serve our own ends. But this possibility of use for humans does not undermine, in any way, the assertion that their *raison d'être* has nothing to do with human projects. They exist "by themselves", no more and no less. It is both an illusion and a fallacy to hold that whatever we humans happen to find useful in the light of our technology must have come into existence just for the purpose of serving our ends, or that their existence is to be justified solely in terms of such a purpose. Similarly, the undeniable fact that life (overwhelmingly) on Earth, including human life, is impossible without the Sun, should not be understood to mean that the Sun came into existence and/or continues to exist for the purpose of sustaining human life. Whether the Sun and its other planets do or do not sustain human life (and life in general) has nothing to do with the fact that they are valuable "by themselves," that they may be said to possess independent value.

Furthermore, their genesis long antedated the appearance of humans. Their continuing existence also has nothing to do with humans. What happened or happens to them, or on them, is totally independent of us (barring the fact that we have already significantly manipulated and modified Earth's surface). For instance, Mars might once have had water or indeed even bacterial life (as is being speculated), but today it is said to be waterless and lifeless. But neither state of affairs has been caused by human effort or design. Left on its own, Mars might or might not eventually have water again, and might or might not support life (and what sort of life) again in the absence of human intervention and manipulation. The Sun, its planets, the Solar System itself, each has its own trajectory.

Something else, too, must not be forgotten. Human survival and flourishing are dependent on Earth's atmosphere and its biosphere, which depend on Earth's diurnal rotation around its own axis and its year-long revolution around the Sun, which, in turn, depend on the Sun staying still while the other planets rotate around it, exerting gravitational pull on one another in certain ways. So while the existence of humans depends on the existence of the Sun and the other planets, the existence of the latter would not be affected should humans, as a species on earth, become extinct. Causally and cosmologically speaking, the extinction of *Homo sapiens* would not be significant. Of course, the extinction of humans would involve a loss of value, but not all values—the value peculiar and unique to humans, namely, being intrinsically valuable "in themselves" will be gone, but in so far as biotic life goes on, beings which are intrinsically valuable "for themselves" will continue to exist, and beings which are valuable "by themselves" (both abiotic and biotic entities) will continue to exist. Such values will simply be mutely-enacted values; they will exist as such, just as they have existed in the past before the emergence on Earth of beings with a very peculiar type of consciousness enabling them to recognise and articulate the values embodied in other biotic and abiotic things.

The above should provide some reasons which may be considered to be compelling for us to admit that abiotic nature as represented by the rest of the Solar system (and on Earth when divorced from biotic nature) has a value which is independent of humans altogether, and, therefore, is not simply the result of mere human presence and the projection of human consciousness upon (non-human) nature. The thesis of Humean projectivism is not unchallengeable *tout court*.

4

In the light of the fore-going sections, let me very briefly refer back to the notion of "trajectory" in order to anticipate and clear up a possible query. Critics could well point out that the notion may be understood agnostically. By this is meant, that it could be used indifferently in connection with any entity whether the entity is an oak, a lion or a teapot. Each of these entities has its own trajectory. The teapot has come into existence at a particular point in time, continues to exist for a certain period of time and will go out of existence at a future point in time. However, the point to emphasise here is the context in which their respective trajectories take place—while the trajectories of oak and the lion in the wild occur in the absence of deliberate human manipulation and control, in other words, in the absence of human design, the trajectory of the teapot occurs specifically in the presence of explicit human design, and is the direct product of such design and manipulation.

While the thesis of intrinsic/immanent teleology holds true in the case of the oak and the lion in the wild, the thesis of what may be called extrinsic/imposed teleology obtains in the case of the teapot. Human artefacts like teapots, cars or houses embody the latter thesis of teleology: their "essence," so to speak, is defined in terms of the intention, the end or goal of the person(s) who design(s) and fabricate(s) them. A teapot comes into existence only because we humans engage in the activity of making and drinking tea; as an artefact, it cannot come into existence independently of human design and fabrication. (We might of course come across a large shell, let us say, which could do duty as a teapot; by appropriating it—what has come into existence independently of our intention and control—for the purpose we have in mind, we have turned into something which is of instrumental use for us. However, this is not the usual context one bears in mind when talking about artefacts). Unlike the oak and the lion in the wild, *ex hypothesi*, the teapot embodies no other *telos* than the one which we, humans, have given it and imposed on it. As an artefact, the teapot is nothing more than the material embodiment of our intentionality.

To conclude on this point: it is not so much that the term "trajectory" cannot be used in connection with a teapot or any other human artefact, but the sense which is emphasised here arises in a context where entities are valuable "for themselves" and "by themselves" (as in the case of oaks and lions in the wild), or are valuable "by themselves" (as in the case of mountains and rivers); it arises in a context when the thesis of intrinsic/immanent teleology obtains (in the case of the oak and the lion in the wild) or when the thesis of extrinsic/imposed teleology fails to obtain (as in the case of the oak, the lion in the wild as well as mountains and rivers).

5

The absence of extrinsic/imposed teleology in contexts where entities exist "by themselves" forms the basis for saying that such entities have independent value. The term "independent value," obviously, refers to "independence of human intention, design, manipulation, control and fabrication." So again, the critic might say: you are simply attributing value to whatever arises spontaneously in nature. Yes, why not? The short answer to this is simply to admit the charge because in one crucial sense, it is not damaging. The charge only makes sense from the anthropocentric standpoint; that is to say, it assumes only two categories of values—humans which are intrinsically valuable "in themselves" and other entities which are instrumentally valuable to humans. No other values obtain, whether these be values which reside in entities which are only intrinsically valuable "for themselves" as well as "by themselves" or the value which resides in entities which are only valuable "by themselves." The charge begs the very issue at question, whether anthropocentrism is correct and whether anthropocentric values exhaust all the values that there may be.

6

I hope what I have set out in this chapter does not amount to a distortion of Rolston's wide-ranging thoughts. Although it is true that his defence of the abiotic *per se*, in terms of output, is not great, it is not obvious, at least to me, that what I have to say on the matter is necessarily incompatible with his general respect for the abiotic. It is in this sense, that I see my thoughts as complementing his, rather than in opposition to them, in spite of the fact that the kind of environmental philosophy I advocate is more radical than his. However, it would be presumptuous of me to imply that Rolston would necessarily agree with what I have written in every detail or, indeed, even in principle.

NOTES

[1] I only know of one publication in which Rolston has explicitly devoted himself to abiotic nature, completely outside of the biotic context. See "The Preservation of Natural Value in the Solar System" in *Beyond Spaceship Earth: Environmental Ethics and the Solar System*, edited by Eugene C. Hargrove (San Francisco, CA: Sierra Club, 1986) where he defends the thesis that the Sun is worthy of environmental respect.

[2] See "Environmental ethics in Antarctica" in *Environmental Ethics* 24 (No.2, 2002): 115–134. See also his "Environmental ethics on Antarctic Ice" in *Polar Record* 36 (No. 199, October 2000): 289–290.

[3] For the sake of brevity, I will draw attention to only one of Rolston's numerous works as representative of his environmental philosophy: *Conserving* Natural Values (New York: Columbia University Press, 1994).

[4] For a fully worked out attempt, see Keekok Lee, *The Natural and the Artefactual: The Implications of Deep Science and Deep Technology for Environmental Philosophy* (Lanham: Lexington Books, 1999).

[5] This is not the place to sort out the contribution of other leading environmental philosophers, such as that of Baird Callicott, to this issue about the source and locus of intrinsic value. For a detailed account of my comments on the Rolston/Callicott debate, see Keekok Lee, "The source and locus of intrinsic values–a re-examination," *Environmental Ethics* 18 (1996): 297–309. Although I disagree with both Callicott and Rolston in this discussion, nevertheless, I have argued that Rolston, on the whole, is more right than wrong, while Callicott is more wrong than right.

[6] Rolston does use the term "human superiority" himself, but he does not mean that from it one could derive anthropocentrism. Far from it, he argues explicitly that a duty of respect follows from such "superiority." He is using the term in a different sense, presumably to refer to the fact that human consciousness is much more developed and advanced than any other known form of consciousness in naturally evolved non-human beings.

[7] Later in the chapter, a third thesis of teleology will be introduced.

[8] Today, the biotic appears to stand in greater danger than the abiotic of being undermined by human technology. In particular, biotechnology (since the 1970s) has for the first time in human history succeeded in transcending the barriers established within natural evolution for reproduction—transgenic organisms can now be created, crossing not simply the species but also the Kingdoms barriers. This means that humans can now create organisms which are no longer valuable "by themselves," as their very coming into existence is the outcome of human manipulation and control. The subversion of their ontological status as beings, which are valuable "by themselves" has, at the same time, also subverted their status as beings which are valuable "for themselves"—human technology has simply hi-jacked their autopoietic capabilities to sustain their own functioning integrity to serve, not their own, but human, ends. The transgenic cow like the ordinary cow produces milk, in accordance with its mechanisms of reproductive physiology, but its milk is no longer cow's milk as it contains a human protein—the transgenic cow has become a drug-making biotic machine.

CHRISTOPHER J. PRESTON

3. REFINING ROLSTON: A NATURAL ONTOLOGICAL ATTITUDE TOWARDS NATURAL VALUES

It is hard to overstate the influence that Holmes Rolston, III has had on this engaging field of inquiry we call environmental ethics. From the first paper in a major philosophy journal ("Is There an Ecological Ethic?" (1975)),[1] to the co-founding of the now leading journal in the field (*Environmental Ethics* (1979)), to one of the first book length collections of essays in environmental philosophy (*Philosophy Gone Wild* (1986)), to his role in establishing the International Society for Environmental Ethics (1990), to the Gifford Lectures (1997), to the Templeton Prize (2003). Rolston is probably the only philosopher to have given lectures on all 7 continents and he packs a busy schedule as a keynote speaker at conferences and as a visiting lecturer at universities throughout the world. Important to add to this greatly abbreviated list of public accomplishments is the personal level of care and attention that Rolston continually offers those in the field around him. Over the years he has provided an enormous amount of intellectual nurturing and support to the environmental ethics community at large by reviewing work, writing recommendations, putting people in touch with each other, and generally doing what he can to promote the philosophical interests of others. The field of environmental philosophy owes him a great debt.

Over the nearly thirty years that Rolston has been writing environmental philosophy he has carved out for himself a signature position that, at least in the early days of the discipline, also became the position most readily identified with the field as a whole. This position is characterized by the claim that nature possesses objective intrinsic value, value that exists independently of any human valuers. This strong environmental position provides some seemingly solid foundations for environmentalists who had previously struggled to find a non-subjective and non-relativistic reason for why they should protect nature. Rolston insists that the moral reason for the environment's protection in no way depends upon any human preferences since "what counts as value in nature is not just brought to or imposed on the ecosystem; it is discovered there."[2] The arguments Rolston uses to support his claims about the intrinsic value of nature are multiple and varied but they include a good deal of detailed scientific description of biological and ecological processes, close attention to what we mean when we say something is valuable, an in depth consideration of the creativity and biological complexity that evolution makes possible, and an investigation of the relationship between normativity and narrative.[3] With lyrical writing, word plays, and seductive argumentation, Rolston makes a compelling case that nature in itself is morally considerable. His position rests on the central claim that "systemic nature is valuable intrinsically, as a projective system . . . for its capacity to throw forward (pro-ject) all the storied natural history."[4]

C.J. Preston and W. Ouderkirk (eds.), Nature, Value, Duty, 29–44.

As a result of both its prominence and its philosophical boldness, this signature position of Rolston's has been heavily criticized from a number of quarters. These criticisms include that Rolston offers no argument for value in nature just Delphic-sounding assertions; that Rolston commits a naturalistic fallacy by moving from scientific description to ethical prescription; that the ethic is too wedded to Kantian ideas about rationally generated duties towards some property of nature; and that an environmental ethic that rests upon claims about the intrinsic value of nature fails on both pragmatic and epistemological grounds. It is the alleged failure on pragmatic and epistemological grounds, failures that are often claimed to be linked to each other, that I am going to discuss in this chapter.

TWO INITIAL PRAGMATIST CRITICISMS

Environmental pragmatists have spent a good deal of time since the nineteen-nineties suggesting that it was an error to make questions about the intrinsic value of nature so central to environmental philosophy.[5] This complaint has been made on several grounds. The first is that Rolston made a mistake in centralizing a rather abstract debate in axiology that apparently does little for environmental activists and policy-makers who urgently need to do something to stop environmental destruction.[6] It is a tragedy, claims Bryan Norton, that environmental ethicists should engage in philosophical hairsplitting about differences between intrinsic, instrumental, and inherent values when there is important practical work to be done. Norton fears a future in which environmental philosophers "continue to dance with Cartesian ghosts even while tropical forests, both ancient and modern, burn."[7] Andrew Light, equally frustrated with this aspect of environmental philosophy but a little more sympathetic to at least some abstract work, recommends a "meta-philosophical pragmatist" approach to environmental ethics in which environmental philosophers maintain some sharp lines between their public and professional personae, keeping the arcane debates about value theory for their conferences and their private conversations while adopting a much more straightforward and robust policy-oriented approach to environmental problems when talking with the public.[8] There are at least two voices discernable in this Norton/Light line of criticism. The first is the voice of the activist or policy maker, the person that actually wants to get stuff done and who finds all this philosophy nonsense rather unhelpful to the task. The second is the voice of the Deweyan pragmatist, the one that is skeptical of ever being able to establish intrinsic properties of anything and who finds it much more helpful to frame ethical problems in terms of the resolution of problems that exist in relation to perceived needs.

There are a couple of quick and easy responses available to Rolston in the face of these claims about the unhelpfulness of his work. The first is to question whether the complaint is a fair one to make of a philosopher in the first place. What is it that an environmental ethicist is supposed to be doing in their job? Should they be making policy and campaigning on behalf of certain issues or should they be working through complex theoretical puzzles? If Rolston is primarily an ethical theorist—and his sustained defense of the concept of objective, intrinsic value in nature looks more like ethical theory than it does any other branch of academic work—then he is surely under little if any obligation to supply solutions to social and political problems. He is articulating and defending a particular axiology and not trying to be an activist or politician. Philosophers rarely get directly in involved in political issues. Blaming Rolston for not being very helpful in changing environmental politics would be rather like blaming A.J Ayer for not doing enough on the ground to prevent the rise of fascism in Europe in the nineteen-thirties. It was not really his job.

A second response to the pragmatist complaint, one that does not let environmental philosophers of the hook quite so easily is to bite the bullet on the question of helpfulness and directly challenge the claim that intrinsic value theory has not been helpful to policy-makers over the thirty years of its existence. There is a good case to make for quite the contrary conclusion.[9] Since its spread through environmental ethics, the theory of the intrinsic value of nature has not only become common currency across the reaches of philosophy but it has also made considerable inroads into both national and international political discourse. The influence of the discussion initiated by philosophers like Rolston is evident in a number of places. The 1982 World Charter for Nature, for example, states that "[E]very form of life is unique warranting respect regardless of its worth to man." Not too many years after the publication of that document, George H. Bush's EPA director Bill Reilly claimed, "natural ecosystems have intrinsic values independently of human uses." The 1993 Global Convention on Biodiversity affirms the participating states' consciousness of "the intrinsic value of biological diversity." The Earth Charter adopted in an advisory capacity at the Rio + Ten meetings in Johannesburg in 2002 asks we "recognize that all beings are interdependent and every form of life has value regardless of its worth to human beings." While Rolston can not claim to have coined the term "intrinsic value," he has certainly been a major player in giving it this degree of prominence in public debates about environmental issues. Former radical environmental activist Dave Foreman has identified professional environmental ethics (including the debates about intrinsic value theory) as one of four major forces shaping the conservation movement in the 1990's.[10] Only the animal liberation movement inspired by Peter Singer's 1975 book has shown a similarly rapid and influential conveyance of a new philosophical idea into public discourse.[11] So perhaps a better way to respond to the first version of the pragmatist critique is to simply spell out the considerable political potency of Rolston's work.

A second version of the pragmatist critique—one with some epistemological shades to it—is the complaint that, even if intrinsic value were a helpful notion in principle, unless one had a formula for determining just how these intrinsic values are to be pitched against each other in cases of conflict then the whole concept of intrinsic value would remain more or less unworkable. The critique has epistemological shades to it because this critic wonders how we are able to know intrinsic values in such a way that they can be useful to us. How can we use such a murky notion to solve, for example, management dilemmas over wolf and elk? If wolves and elk both have intrinsic value and the elk population is stressed by a combination of wolf depredation and the lack of available habitat, then is the notion of intrinsic value helpful in telling us whether we should allow hunting of the wolves? The complaint against intrinsic value here seems to be a consequence of the epistemological difficulty of ever being able to know intrinsic values in such a way that one could quantify them for the purpose of policy decisions. The epistemological difficulty is well illustrated by the acrobatics of those that have tried to solve it. J. Baird Callicott has made the mistake of trying to provide a complex framework of first order and second order considerations for inherent values. The resulting prioritizing is not only confusing to adjudicate but ends up prioritizing human values over natural values and taking us back down an undesirable path to anthropocentrism.[12]

There are a number of better ways to respond to this pragmatic/epistemological complaint. A first response is to use the very same strategy as above and suggest that is it not the ethical theorist's job to solve policy dilemmas. While this evasion of the criticism again has some plausibility, such evasiveness is even more unsatisfactory here than it was the first time around. Rolston himself feels an obligation to show that his theory has policy consequences. It would be very unsatisfactory indeed if the idea of intrinsic value were completely unusable in practice.

A more purposive strategy might be to first highlight the extraordinary difficulties philosophers face when trying to extract complex policy guidelines from the conceptual territory that is the business of philosophy. Much work in philosophy is done deliberately in abstraction from the messy contingencies of social, political, and personal environments. The goals of universality and certainty seem to demand it. To set the bar too high for environmental philosophers, given complexities such as ecological uncertainty and the multiple interests of different environmental stakeholders, appears to be grossly unfair to the environmental philosopher.

Once this has been said, it could then be noted that Rolston has in fact on several occasions written in some detail on how to work with his intrinsic value theory in policy making, most notably in the numerous rules and guidelines articulated throughout *Conserving Natural Value. Conserving Natural Value* is a book written by Rolston after the United Nations Conference on Environment and Development specifically on environmental policy. It keeps the policies articulated in close contact with the axiological commitments that ground them. Almost every chapter of this book contains a list of policy guidelines based upon Rolston's commitments to instrumental, intrinsic, and systemic values in nature. For example, in the first chapter on the question of how to balance natural and cultural values Rolston recommends such policy guidelines as "emphasize non-rival natural and cultural values" and "protect minority values."[13] In a later chapter concerning international law, Rolston uses his theory to suggest to policymakers that "[c]ommon natural resources are more fundamental than national or private resources" and that policymakers should "avoid global irreversible change" (229, 231). In an earlier paper commissioned by the U.S. Forest Service Rolston offered ten policy guidelines on how public forestlands should be managed in the United States.[14] To say that Rolston's account of intrinsic values supplies no guidelines for environmental policy is simply false. While intrinsic natural values cannot be known mathematically or experimentally nor can they be precisely weighted on every occasion, Rolston shows that it is possible to make general policy guidelines based on intrinsic value. In what we should acknowledge to be a very tricky area for moral philosophy in general, Rolston's work, particularly in *Conserving Natural Value*, at times looks like a great example of how to move from ethical theory to applied ethics.

THE PROBLEMATIC ONTOLOGY OF INTRINSIC VALUES

There is a third version of the pragmatist/epistemological critique, one that moves away from the policy emphasis of the first two criticisms and back into more traditional philosophical territory.[15] This critique centers upon the accusation leveled at Rolston that his position relies upon outdated, modernist, epistemological staples such as foundationalism, representationalism, and master narratives. Critics contend that Rolston's commitment to the real existence of objective, intrinsic values in nature commits him to a metaphysical and epistemological picture that is no longer supportable. Over a decade ago, Bryan Norton and Baird Callicott made several criticisms of Rolston on this score.[16] They accused him of subscribing to a simplistic Cartesian metaphysics in which knower and known are kept separate and of coupling this metaphysics with an unsophisticated epistemology in which knowledge amounts to the representation in the mind of objective facts in the external world. Norton suggested that if Rolston got beyond his outdated Cartesianism he would see how his strenuous exertions to prove the existence of objective intrinsic values are "a classic example of a philosophical pseudo-problem, a bewitchment of ossified language."[17] As it stands, Norton claims, Rolston relies upon a theory that "cannot escape skeptical collapse."[18] J. Baird Callicott allied himself with Norton on the issue, suggesting that Rolston employs an outmoded Cartesian metaphysics that in turn demands an impossible epistemology, rendering his arguments for objective values a "magic show, brought off with smoke and mirrors."[19]

The exact nature of the criticisms Norton and Callicott offer requires some teasing out. There are both ontological and epistemological issues at play and both critics mix them in a way that sometimes makes it hard to know exactly what they are trying to pin on Rolston. Very loosely speaking, it is possible to characterize both of their criticisms as rooted in a cluster of post-modern epistemological developments of the second half of the twentieth century. This cluster compels us to reject the kind of foundationalism in epistemology that Rolston allegedly requires in order to claim we know anything objectively about nature. The cluster includes the following post-modern theses: the theory dependence of observation thesis, the mix of analytic and synthetic components in every belief, the critique of what has been called the "Myth of the Given" in empiricism, the web-like nature of our systems of belief, the value biases present in epistemic claims, the fallibility of perception, understanding scientific empiricism as interventionalist rather than representationalist, and the operation of language as a system.[20] According to Norton and Callicott, it is Rolston's failure to appreciate any of these epistemological developments that makes his position old-fashioned and untenable.

To his credit, Rolston has in fact invested a considerable amount of energy in answering these criticisms in at least two recent papers.[21] Contrary to the critiques, it seems to me that Rolston's position is anything but simplistic on these issues. This recent work confirms how he has all along been well aware of post-modern epistemological slipperiness. "The constructions we see always depend upon the instructions with which we look" is one of the elegant ways that Rolston acknowledged the theory dependence of observation thesis in 1982 before Norton or Callicott had levied any critique.[22] But while recognizing the human influence on our knowledge claims, Rolston still insists upon realism about objects, realism about natural values, and objectivism about our ability to know the both of them. It is the human ability to get out of our skins and say something for certain about nature for real that Rolston believes is part of the genius of *Homo sapiens*. "There is a hawk in the spruce tree beside that granite boulder," Rolston sensibly wants to assert with confidence.

But Rolston may be trying to squirm off the hook just a little bit too easily here. Of course there is a hawk. Few would doubt that. But in what sense can the philosopher confidently say we *know* the hawk given his or her relentless pursuit of certainty? Can we be sure of an exact correspondence between the idea of the hawk that we have in our minds and the physical hawk perched on that Rocky Mountain hillside on the spruce bough? Environmentalists are almost by the nature of their inquiry necessarily committed to post-modern epistemological views such as the belief that humans are always ecologically, evolutionarily, biologically, perceptually, and culturally situated. With this in mind, it must be still worth asking—in the vein of Norton and Callicott—if Rolston's claim about human genius isn't just a bit to hasty.

ROLSTON'S CRITICAL REALISM

A fuller version of the sophisticated position Rolston takes on this issue goes something like this: "Natural science," he states, "is a primary place where humans know nature for real."[23] The sharpness of the wolf's fang is as real for the biologist as it is for the elk calf. This is a real fact about the world that humans can come to know with certainty even though they will likely never have to experience it for themselves. Trees really do use chlorophyll to photosynthesize. The Krebs cycle, DNA, and the Permian/Cretaceous extinction are for real. The world remains how it is, independent of how different beings come to experience it and it contains these things that humans describe when they do their science properly. When humans make a claim about the sharpness of the wolf's fang or the Krebs cycle, they get nature right. The biologist's realist ontology and objectivist epistemology both have the armor to resist the philosopher's epistemological uncertainty.

At the same time as insisting on this ontological realism in science, Rolston continues to show a handy awareness of some of the tag lines of the opposition. He consistently rehearses the line taken by constructivists, even those neo-Kantians that would insist on a kind of co-construction between subject and object in which scientific claims are "interactive activities between humans and a nature out there that we know only through the lenses, theories, and equipment that we humans have constructed."[24] Rolston is well aware of the fact that scientific theories are not necessarily pure, that instrumentation transforms nature as it records it, that environments are describable only relative to particular organisms, and that wilderness is alleged to be a Euro-American construct. He knows in other words, about claims that there is no given in experience, that the analytic components of our claims cannot be separated from the synthetic.

But even in the face of these epistemological cautions, Rolston presses on with his realism and objectivism, suggesting that science allows us to cut through all the fuzziness. He warns us not to confuse our philosophical "-ologies." "What is getting contaminated conceptually," he claims, "is epistemological making up the world with ontological making up the world, the order of knowing with the order of being."[25] He wants to grant the epistemological limitations of our conceptual schemes and embodied situations, but deny that these limitations prevent us from being realists about the objects described and objectivist about our knowledge of them. One of his favorite strategies to reinforce the realism is to give those wooly-headed post-modernists some temporal and spatial perspective on their claims. He draws their attention to the famous photographs of earth from space, the ones that show our home to be a blue-green pearl floating in the front of a sea of blackness. He makes the historical point that human cultures, and in particular those that labor under post-modern epistemic difficulties, are only late arrivals on this planetary scene. After approximately 4 billion years of geo-chemical and biological activity on earth, only in the last five thousand or so since humans developed a written language have there been beings that even possess the wherewithal to worry about epistemology. Only in the last fifty or so have those worries become debilitating.

The strategy Rolston uses here is a well-tested and effective one for the environmentalist. Putting our current situation into some evolutionary and geological context has proven to be a good way for environmentalists to cast a shadow on anthropocentric approaches to ethics that revolve too much around humans. The strategy is supposed to work the same way for the post-modern epistemologies that would also be too anthropocentric. In Rolston's view, the addition of some perspective should be enough to persuade post-modern epistemologists that the objects of our studies in the natural sciences are really there, existing before we started to get caught up in our epistemic worries about them. Rolston notes that knowledge claims are themselves subject to selective pressures and observes how our concepts get tried out and sifted over millennia. Those that prevail, if not actually true, are at least more nearly true than those that are rejected. "All those persons that do not think that 'lion' refers to a real predator lurking in the grass," he dryly observes, "are extinct."[26]

The key part of the rhetorical strategy here, one even more centrally employed in the ethical debate, is to threaten doubters with the accusation of hubris. Rolston points out how "arrogant and myopic" it would be for us late arrivals to subordinate the grandeur of what is out there in nature into the conceptual prison of what is stuck inside our heads. Reaffirming the wilderness idea, Rolston insists that "[c]ontrary to Nash, wilderness is not a state of mind; it is what existed before there were states of mind."[27] While different people can mean different things when they use particular labels, the objects out there to which these labels refer can still be the same objects. Studies across cultures by anthropologists inform us that even wildly different cultures existing within their dramatically different conceptual frameworks tend to broadly agree about how to classify many species.

One final and novel check on our hubris that the environmental philosopher can offer is to draw attention to the epistemic situation of our non-human relatives and kin. In contrast to the arguments offered by those that get stuck within the play of words that exist in human conversations, the conversations that take us out of our skins are not just conversations that we have with "mankind," they are conversations that "find significance in others in their ecologies too."[28] The ecologies being referred to here are the ecologies of non-human others. We have the resource of numerous species-to-species checks and balances on our claims to free us from our conceptual prisons. The venom of the taipan snake is equally toxic in the blood of a human as it is in the blood of an antelope. If we see an antelope avoiding the snake we have a confirmation from a conceptual scheme radically different from our own that tells us of the venom's toxicity. Some have even used this possibility of inter-species epistemic checks and balances as supplementary reasons why non-human animals should be of ethical concern. N. Katherine Hayles, for example, suggests that "[t]o sacrifice animals or exterminate species in this model, directly reduces the sum total of knowledge about the world, for it removes from the chorus of experience some of the voices articulating its richness and variety."[29] Trips into the wilderness help to dispel the myth of humans trapped in the prisons of their minds. It is for this reason that poet Gary Snyder brusquely suggests that post-modern philosophers should get out and wade around in the streams a bit more. It would be a useful antidote to their post-modern listlessness. Epistemic doubt seems altogether too nit-picky during a hike in the woods or, even more so, an environmental crisis.

The real world perspective that the environmentalist can put at the feet of the post-modern epistemologist is certainly appropriate for those that are too out of touch with common sense. We should be all for philosophy that is worldly, realistic, and motivating. But it would be good to remember at this point that when David Hume recommended in the seventeenth century not getting too caught up in the epistemological problems he had already detected in modernism it was not because he thought his skeptical conclusions were wrong but because he thought they were too distracting for day-to-day living. Hume recognized that we need to develop certain mechanisms and psychological habits of mind in order to cope with reality but this was necessary precisely because epistemological uncertainty was so trenchant. Rolston and Snyder are right that some types of philosophical worry are rather distracting for day-to-day environmentalism. But even if we can go backpacking to walk off our post-modern queasiness does that adequately deal with the epistemological problem? Can we transfer *tout court* into the philosophical realm the kind of commitment to the actual existence of mountain lakes, photosynthesis, and hawks in spruce trees that we might employ on a backpacking trip? If we are really going to take epistemology seriously, can we afford to get out and wade around in the stream just yet?

I believe that we cannot and that Rolston does in fact have a serious epistemological problem with his position. His insistence on the realism of natural values and on the possibility of us knowing them objectively is problematic given post-modern epistemology. It is particularly problematic for Rolston because he himself admits many of the post-modern epistemological issues that the critics want to level against his position. We do not have infallible access to reality, he admits, and he says that we see constructions that are shaped by "the instructions with which we look." And yet he thinks we can "fight a way through how we know what we know . . . in order to reach the state of affairs in the real world and to be able to defend the existence of value there."[30] Rolston has articulated the realist/constructivist dilemma and yet impales himself on the realist horn regardless of its problems. This would be troubling enough if he were simply being a realist about the existence of physical objects but it is double trouble since he also wishes to be a realist about natural values. A J. L.Mackie type of argument about the queerness of values already casts doubt on Rolston's value realism even before post-modern epistemological considerations are added to the mix.[31]

But at the same time, Rolston is no doubt correct to recognize that the constructivist side of the dilemma is at least as unappealing as the realist side. It is arguably even worse for an environmentalist to be a constructivist about natural values than it is for a different epistemologist to be a constructivist about tables and chairs. In the latter case, the tables and chairs don't actually disappear when admitting that our accounts of them contain an irremovable human component and a measure of doubt. But in the case of intrinsic natural values, the very reason for positing them in the first place—to provide a non-anthropocentric and objective reason for protecting nature—does in fact disappear if the values have too much of a human component. The realist/constructivist dilemma is one upon which Rolston really cannot afford to be impaled. So how can he get off these horns? The rest of this paper is devoted to tentatively articulating an approach to natural values that may give Rolston what he wants, an escape from the either/or of a realism that is impossible to maintain or a constructivism that evacuates his position of its power.

A DIGNIFIED SURRENDER OF REALISM

We start this approach to natural values a long way from the territory traditionally occupied by environmental philosophers by turning to a debate in the philosophy of science. This is a debate about the ontological status of unobservables in science, entities such as quarks, gluons, or laws of gravitation that the scientist does not see, but whose existence are inferred from what the scientist can see. These remarks about unobservables will stand us in good stead later on when we begin to talk again about natural values.

In his discussion of how to mediate realist and anti-realist positions on unobservables in the philosophy of science, Arthur Fine starts with a claim about a core position that both realist and anti-realist share. He makes the very general and relatively uncontestable observation that whatever ontological status they grant to unobservables both realist and anti-realist end up accepting the entities into their life in more or less the same way that they accept everyday objects of the senses like tables and chairs.[32] Realists and anti-realists both think, in other words, that claims about unobservables are incorporated into our belief system at face value. As a result, realists and anti-realists make appropriate changes in behavior and attitude (both theoretical and practical) just as they do when they accept the reality of tables, chairs, and cracks in the pavement. This shared belief constitutes what Fine calls the "core position."

Realists and anti-realists part company according to Fine when it comes time for each to add to this shared core position some interpretation of what their position means when they each state they "accept" the unobservables. Anti-realists accept the unobservables science posits because they believe that these entities can variously do such things as "save the phenomena," "cohere with our web of belief," or show themselves to be "instrumentally useful in helping us secure our ends." Realists accept claims about unobservables because they believe that these entities describe what is really real underlying our scientific claims. The realist, Fine suggests, in fact wants to add "a desk-thumping, foot-stamping shout of 'Really!' " to any description of an unobservable in the world.[33] Nothing less will do when we make an existence claim about an unobservable. The anti-realist explanations are a bit less assertive but they too add what Fine refers to as an "essence" to their existence claim that captures what they have mind when they make an ontological commitment.

Both interpretations, according to Fine, attempt to do too much. They both try to supplement the core acceptance of the entity with something unhelpfully metaphysical. And in so doing, they create epistemological problems for themselves. The realist, believes Fine, simply deludes him or herself when claiming to be able to "stand outside the arena watching the ongoing game, and then . . . to judge (from this external point of view) what the point is."[34] Getting outside of the game is something a human being that is always physically, culturally, linguistically, ecologically, and psychologically already immersed in the world simply cannot do. The anti-realist position according to Fine includes equally damaging faults such as inconsistency, question begging, and a lack of robustness.[35] The lesson Fine wants us to draw from what he sees as both realism's and anti-realism's excesses is that sometimes, when you are interested in making an ontological commitment, less can be more.

We also find this invitation to do more with less, when we look at the work of Richard Rorty.[36] Rorty is important to the current discussion because, unlike Fine, Rorty is a philosopher with whom Rolston specifically engages on these epistemological issues. In a similar fashion to Fine, Rorty thinks that realist dogmatism about correspondence between ideas and the world is pointless. This is not because Rorty feels paralyzed or intimidated by epistemological uncertainty nor because he has any question that there is an actual world out there that creates specific physical demands on us and occasionally, in the case of wildcats and cancers, can end up doing us real harm. He eschews realism because he has a different idea about what it is that language—including the language of science—actually does for us. Rorty asks us to make a distinction between the claim that the world is out there and the claim that truth is out there. He insists that,

> [T]o say that the world is out there, that it is not our creation, is to say, with common sense, that most things in space and time are the effects of causes, which do not include human mental states. To say that truth is not out there is simply to say that where there are no sentences there is no truth, that sentences are elements of human languages, and that human languages are human creations. Truth cannot be out there—cannot exist independently of the human mind—because sentences cannot so exist, or be out there. The world is out there but descriptions of the world are not.[37]

Having established that language is not the kind of thing that should ever really be said to map or mirror the world—"the fact that Newton's vocabulary lets us predict the world more easily than Aristotle's does not mean that the world speaks Newtonian"[38]—Rorty shows at length how realists insisting on correspondence theories of truth are really engaged in a very strange game, a game that our intellectual inheritance makes look very plausible but one our epistemic situation renders impossible to complete. Like Fine, but for different reasons, Rorty is no more impressed with anti-realists than realists. Anti-realists, very roughly speaking, look inside their social communities rather than outside for legitimation of their claims, which is a strategy as attributable to a misguided intellectual inheritance as the realist errors were in the first place. Rorty observes that this intellectual inheritance has tended to create "an unfortunate . . . pendulum swinging between realism and idealism."[39] And while this pendulum creates one type of difficulty for the philosopher of science, it creates equally enduring difficulties for Rolston when he demands realism about natural values.

Realism about values is certainly different from realism about physical objects but there seems to be a general lesson here. If the see-saw or pendulum that you are riding between realism and anti-realism (or realism and constructivism) is beginning to make you feel nauseous, you might want to think about how to get off. One reason Rolston needs to get off the realist/constructivist see-saw is as a matter of epistemological consistency. To give up the kind of realism upon which he insists would be more consistent with the contemporary epistemological observations that he already admits. When under fire from the constructivists about his realism, Rolston shot back that his critics were mistakenly carrying their epistemological queasiness over into their ontology. But it seems that he is wrong about this. If your epistemology incorporates uncertainty, it is hard to know what kind of ontological realist you can be.[40] When Rolston claims that to contaminate your ontological reality with your epistemological uncertainty shows a failure of nerve, it is just possible that he has it the wrong way around. To buy into the epistemological uncertainty as he does and yet to insist on the ontological reality for the entities we have described is to retreat from taking implications of epistemological uncertainty as far as they need to be taken. Perhaps it is Rolston who shows the failure of nerve and Rolston who seems to be frightened of relaxing his grip on a picture of knowledge as correspondence with the real.

The failure of nerve is not surprising. There are many psychological reasons to hold on hard to realism. If beliefs are not evaluated for whether they correspond to the world, we find ourselves in a very large and occasionally dangerous place in which the internal map of the world we had crafted turns out to be not the document we thought it was.[41] The claim of "correspondence with reality" plays a welcome legitimizing role. Environmental ethicists in particular, setting out to make what was always going to be a very challenging case about natural values, will struggle hard for this legitimacy. But having the nerve to give up the kind of realism to which he subscribes will make Rolston's position more internally consistent. Those who are pleased that Rolston appreciates how our situation is one of always "looking out from where we are, using our senses and our brains, from an anthropocentric perspective"[42] will be relieved when his ontology releases its take-no-prisoners realism. If he amends this realism, the rather crude criticisms leveled by Callicott and Norton in the early nineteen nineties about Rolston's "Cartesian metaphysics" and "modernist epistemology" will start to lose their traction. At the same time, for his own sake he will need to remain secure in the conviction that the world does not float away as soon as this grip is relaxed.

But there is more to say beyond the issue of epistemological consistency for why Rolston should amend his realism. Doing so, ironically, might give him a much better ground from which to defend the claim that there are natural values in nature, values that exist independent of human valuers. Rolston can learn some more helpful lessons here from Fine. We heard already about Fine's beef with realists and anti-realists. But where did this ultimately lead him?

A NATURAL ONTOLOGICAL ATTITUDE

Recall first that Fine was criticizing realists and anti-realists for what they added to their statements about unobservables. Both assumed that science is only legitimated if there is some kind of metaphysical explanation about the essence of truth in addition to the practice of science itself that gives scientific claims their authority.[43] Since both of the authorizing moves contain problems and, in Fine's view, neither adds anything substantial to what the business of science already does very effectively on its own, Fine suggests that both realist and anti-realist metaphysical additions can be dropped. Scientific claims could then be considered in an entirely new, less metaphysically demanding light. In effect, Fine wants us to relax a bit and "[t]ry to take science on its own terms, and try not to read things into science. If one adopts this attitude, then the global interpretations, the 'isms' of scientific philosophies, appear as idle overlays to science: not necessary, not warranted and, in the end, probably not even intelligible."[44]

Fine offered as his minimalist solution a view he whimsically refers to as "California natural—no additives please."[45] What he called before the "core position" should be left to stand on its own. So our claims about unobservables in science should all be taken at face value, with the reality of the objects described in theoretical science treated on a par with the reality of objects of everyday experience like tables and chairs. It is a stripped down approach to ontology already accepted by both realists and anti-realists, each relying on the practices and methods internal to science to make it reasonable to believe in something. Science does not need any additional interpretation about what those claims about unobservables mean because its history and its practices are rich enough to give its results all the context they need. A claim about an unobservable is true if it meets current standards of scientific practice. This truth doesn't have to be inflated by adding something like "what this means is that the belief corresponds with reality" or "saves the phenomena." Rather felicitously for our current discussion about environmental values, he calls this whole approach "the natural ontological attitude (NOA)."

Fine's view looks like it might be useful to Rolston because the way Fine would like us to talk about unobservables sounds very much like the way Rolston already talks about natural values. (Values, we might note in passing, are good candidates for unobservable entities). It is a suitable approach for Rolston to adopt because once we have used all reasonable evidence to posit the existence of natural values we won't need to stamp the foot and insist on their reality (as the realist and Rolston would have us do) nor would we have to say that they exist in the sense of "cohering with our web of belief" or being "empirically adequate" (as the anti-realist would have us do). We have only to say that it is as reasonable to make the same commitment to these unobservables as we do to everyday objects like tables and chairs. "NOA involves a trusting attitude," enthused Fine. "It trusts the overall good sense of science, and it trusts our overall good sense as well."[46] I take it that this is roughly how Rolston wants us to commit to natural values. He tries to make values emerge as a matter of good sense. As you read a Rolston text you learn that you are expected to accept natural values in much the same way as you accept wolves and DNA. Just as there are lions lurking in the grass ready to eat antelopes or passing humans, so are there normative sets of lion genes that these individual lions subconsciously go about preserving. Just as there are Amazonian forest canopies in Brazil absorbing sunlight and converting CO_2 into oxygen and sugars, so are there systemic values in those canopies that support the lives of the species that live within them.

As mentioned above, Fine contends that the best reasons to believe in the existence of unobservable entities are reasons taken from the methods and practices of science itself. Science provides its own, non-metaphysical and local context for legitimating its claims. We should note as a point of considerable difference from the ethical case that it is unlikely that these scientific methods and practices are on their own going to support the existence of intrinsic natural values. Values are not immediately revealed by the methods and practices of science. However, it does not seem one would have to amend Fine's position very far in order that the natural ontological attitude could also be the attitude we take towards natural values. Instead of the standards of science, one would only have to consider the standards that moral philosophy gives us to think about those things that we deem to be morally considerable.

What exactly are these standards? We could start with claims like the following: Moral philosophy tells us that objects of a certain kind (such as humans) possess things we variously call values, rights, or dignity. Moral philosophy also suggests that some beings are properly the foci for our care. It tells us that these objects with moral standing are things that deserve a certain kind of treatment, perhaps require respect, perhaps should be treated equally by different moral agents, and certainly should motivate particular kinds of behavior. The qualities found in humans that warrant this behavior are qualities that some people believe we can intuit, others think emerge alongside certain of our capacities—for example, that we feel pain or that we can reason—and still others think emerge out of the narrative context of our lives. These are admittedly somewhat vague remarks about the standards provided by moral philosophy. But I don't think it matters too much that we cannot definitively specify our moral ontology. A benefit of drawing the parallel to the discussion of unobservable entities in science comes when we note that scientists too sometimes struggle to specify their ontologies. That, after all, is what the realist/anti-realist debate in science is all about. Fine cautioned us only against looking for something extra-scientific to legitimate our scientific claims about what exists. Rolston would similarly have to stay within our existing discourses about science and values, in the process surrendering his own quest for a metaphysical additive outside of his claims about values. Since there would be nothing added in this new, purer Rolstonian attitude towards environmental values, we should probably call his position "Colorado natural—no additives please."

To build the case for a natural ontological attitude to values just a bit more, we can draw the parallel between the scientific and ethical discussion a little tighter. To do this it helps to remember how Rolston takes on the naturalistic fallacy in his own work. Rolston always insists on keeping the philosophy as close as he can to the natural science. His argument is one of the most naturalistically informed of all ethics. It is the truths of science, in particular of evolution, ecology, and biology that lead us directly towards his conclusions about natural values. The way that Rolston argues towards the existence of natural values is to use everyday accounts of observation and accepted beliefs in science to infer something reasonable about values. Given what we know about ecology and evolution, given what we know about DNA transferring mechanisms of survival to offspring, what we know about species lines, and what we know about the struggle of individuals to survive and to propagate, we infer from this scientific knowledge that there are also objective natural values outside of the human realm. Rolston argues that genetic sets are propositional sets that become normative sets. "The ought is not so much derived from the is as emerges simultaneously with it" Rolston contends.[47] The natural values are postulated only when we are confident that we have got the science right. He doesn't think he is going out on much of a limb here. The natural science provides 95% of the grounding for the natural values.

The fact that Rolston's case for values relies so heavily on natural science is an advantage in the current discussion. It is the emergence through evolution of diverse and complex life forms that brings forth the intrinsic values Rolston describes. A natural ontological attitude to values would therefore accept the existence of values in our moral ontology in a broadly similar way to how it accepts the existence of DNA in our scientific ontology. The unobservable values emerge almost simultaneously alongside the science with the help of a few additional insights drawn from our moral practices.

And so the environmental philosopher positing natural values could now proceed largely free from those epistemological hang-ups. He or she could still investigate both the world of nature and the world of natural values without the complications of searching for appropriate realist or anti-realist metaphysical additives. Adopting a natural ontological attitude, it is perfectly legitimate to make claims about the world that include claims about unobservables and, if one is prepared to think more expansively, about natural values. The natural ontological attitude, Fine points out, remains flexible on its commitments. "It has only an attitude to recommend: namely, to look and see as openly as one can what it is reasonable to believe in, and then to go with the belief and commitment that emerges."[48] There is no metaphysical reason not to accept claims about unobservable values in the same way that we accept claims about unobservable entities. And Fine core position suggests that we should accept unobservable entities in the same way that we accept the objects of everyday experience.

The tough pill to swallow is that all this has to be done without saying things about correspondence with reality. Fine admits that he sometimes regrets the loss of the realist fantasy and the comfort it used to provide. But the boldness of letting it go yields a position more consistent with contemporary epistemology and more in keeping with the way that knowledge claims appear to operate. These might be better ontological grounds for Rolston to rest his compelling arguments for the existence of natural values. Rolston might even end up quoting Fine on his natural ontological attitude: "Look and see as openly as one can what it is reasonable to believe in, and then go with the belief and commitment that emerges."

NOTES

1 Holmes Rolston III, "Is there an Ecological Ethic?" *Ethics* 85 (1975): 93–109.

2 Holmes Rolston III, *Environmental Ethics: Duties to and Values in the Natural World* (Philadelphia: Temple University Press, 1988), 231.

3 See my "Intrinsic value and care: making connections through ecological narratives," *Environmental Values*, 10 (2001): 243–263 and Jim Cheney's "Post-modern environmental ethics: ethics as bioregional narrative" *Environmental Ethics* 11 (1989): 117–134 for elucidations of this narrative aspect of Rolston's work.

4 Rolston, Environmental Ethics, 1988, 198.

5 See, for example, the essays collected in Light and Katz, *Environmental Pragmatism* (New York: Routledge, 1996).

6 See the contribution to this volume by John Lemons.

7 Bryan Norton, "Epistemology and environmental values" *Monist* (1992), 224.

8 Andrew Light, "Taking environmental ethics public," in *Environmental Ethics: What Really Matters? What Really Works?* eds. D. Schmidtz and E. Willott (Oxford: Oxford University Press, 2002), 556–566.

9 See Callicott, J. Baird, "The pragmatic power and promise of theoretical environmental ethics: forging a new discourse," *Environmental Values* 11(2002):3–25 for additional support for this claim.

10 Dave Foreman, "The new conservation movement," *Wild Earth*, vol. 1, #2 (1991): 6–12.

11 Peter Singer, *Animal Liberation* (New York: Harper Collins, 1975).

12 J. Baird Callicott, "Holistic environmental ethics and the problem of ecofascism" reprinted in *Beyond the Land Ethic: More Essays in Environmental* Philosophy (Albany: SUNY Press, 1999): 59–76.

13 Rolston, *Conserving Natural Value*, 26, 28.

14 Rolston, "Valuing wildlands," *Environmental Ethics* 7 (1985): 23–47.

15 My 1998 paper "Epistemology and intrinsic values: Norton and Callicott's critiques of Rolston" *Environmental Ethics* 20 (1998): 409–428 was a preliminary discussion of these epistemological issues. While those remarks were mostly defensive of Rolston, the remarks here go further and offer some critique.

16 Bryan G. Norton, review of Holmes Rolston, III, *Conserving Natural Value* (New York: Columbia University Press, 1994), *Environmental Ethics* 18 (1996): 209–214. Norton, "Epistemology and environmental values," 208–226. J. Baird Callicott, "Intrinsic values, quantum theory, and environmental

ethics," *Environmental Ethics* 7 (1985): 357–375 and "Rolston on intrinsic value: a deconstruction" *Environmental Ethics* 14 (1992): 129–143. See also Norton's APA criticism of Rolston December 1998, Washington DC.

17 Norton, "Epistemology and environmental values," 217.

18 Norton, Review of *Conserving Natural Value*, (1996), 213.

19 Callicott, "Rolston on intrinsic value" (1992), 138.

20 For these developments we thank epistemologists and philosophers of science such as Norwood Hanson, Willard Quine, Paul Feyerabend, Wilfred Sellars, Ludwig Wittgenstein, Sandra Harding, Lorraine Code, and Ian Hacking.

21 Holmes Rolston, III, "F/Actual Knowing: Putting Facts and Values in Place" in *Ethics and the Environment* 10 (2)(Special Issue) (Winter 2005): 137–174, edited by Christopher J. Preston, and "Nature for real: is nature a social construct?" in *The Philosophy of the Environment*, edited by Timothy Chappell (Edinburgh: Edinburgh University Press, 2001): 38–64.

22 Holmes Rolston III, "Are values in nature subjective or objective?" *Environmental Ethics* 4 (1982): 125–151.

23 Rolston, "Nature for Real?" 38.

24 Ibid., 38–39.

25 Ibid., 54.

26 Ibid., 42.

27 Ibid., 49.

28 Ibid., 45.

29 N. Katherine Hayles, "Searching for common ground," in *Reinventing Nature*, edited by G. Lease and M. Soulé (Washington DC: Island Press, 1995), 58.

30 Holmes Rolston III, *Philosophy Gone Wild* (Buffalo: Prometheus Books, 1986), 92.

31 J.L. Mackie, *Ethics: Inventing Right and Wrong* (New York: Penguin Books, 1977).

32 Arthur Fine, "The natural ontological attitude," in *Philosophy of Science*, edited by David Papineau (Oxford: Oxford University Press, 1996), 37.

33 Ibid.

34 Ibid., 39.

35 Arthur Fine, "And Not Anti-realism Either," *Nous* 18 (1) (1984): 51–65.

36 The pairing of Fine with Rorty is suggested by philosopher of science Joseph Rouse who puts Rorty and Fine in the same wing of the post-modern philosophy of science camp for the way they both attempt to undercut the whole realist/anti-realist debate. Rouse, "The politics of post-modern philosophy of science," *Philosophy of Science* 58 (1991): 607–627.

37 Richard Rorty, *Contingency, Irony, and Solidarity* (Cambridge: Cambridge University Press, 1989), 5.

38 Ibid., 6.

39 Richard Rorty, *Philosophy and the Mirror of Nature*, (Princeton, NJ: Princeton University Press, 1981), 325.

40 One definition of scientific realism is that claims about the world accurately describe objects to which they *refer*. For there to be *reference* there must exist objects *about which we can know something*. What this definition entails is that our ontology and our epistemology cannot always, as Rolston suggests, be kept separate.

41 Perhaps we can console ourselves with the thought that rabbits, snakes, and other non-linguistic organisms are in the same map-less boat. They tend not to do so badly most of the time.

42 Rolston, "F/Actual Knowing," 140.

43 Fine, "And Not Anti-Realism Either," 51–65.

44 Ibid., 62.

45 Arthur Fine, "Unnatural attitudes: realist and instrumentalist attachments to science," *Mind* 95 (378)(1986), 177.

46 Ibid.

47 Holmes Rolston III, *Environmental Ethics*, 232.

48 Fine, "Unnatural Attitudes," 176.

MARK WYNN

4. IN ROLSTON'S FOOTSTEPS: HUMAN EMOTIONS AND VALUES IN NATURE

It is a commonplace of recent discussion in the ethics of personal relationships that the emotions can serve as agents of discovery, alerting us to values that might otherwise pass unobserved, or that might otherwise fail to be accorded sufficient weight.[1] It is also commonly supposed that even where we are capable of registering the presence of some value independently of any emotional response, there will remain some deficiency in the mode of our understanding if the emotions are not engaged.[2] Given this larger intellectual context, we might expect an environmental ethicist to take our emotional responses to nature as one central clue to the values that are to be found there. And we would form such an expectation above all where an environmental ethicist treats the natural world as charged with something like a personal presence.[3] Holmes Rolston is a most distinguished environmental ethicist, and he thinks of the natural world as a kind of partner in conversation, but even so, he seems to give relatively little explicit attention to the emotions as agents in the discovery of natural values.[4] In this paper, I shall argue that notwithstanding the relative neglect of the emotions in the bulk of his oeuvre, Rolston does offer a highly suggestive account of their contribution to the discernment of natural values in some of his more personal essays, where he is describing his own experiences of nature. So taking Rolston's firsthand descriptions of nature as our starting point, we shall re-trace his footsteps across a variety of landscapes, and thereby we shall also re-trace his footsteps intellectually, by examining the thoughts to which he is moved in these encounters with nature. To bring out the force of his account, I shall also introduce a number of recent theories of the emotions, and argue that their perspective is highly consonant with the understanding implied in Rolston's personal reflections.

THE EMOTIONS AS VEHICLES OF UNDERSTANDING

In this paper, I shall be primarily concerned with the essays which fall within Section IV of Rolston's work *Philosophy Gone Wild*. This section is entitled "Nature in Experience", and it is here, where his voice is personal and autobiographical rather than formal and professorial, that Rolston broaches most directly the question of the emotions' contribution to our appreciation of the natural world.

In these reflections, Rolston acknowledges that his own intellectual formation in philosophy and the sciences has in some degree predisposed him to understate the importance of the emotions for an informed appreciation of the values inherent in the natural order. For instance, in "Meditation at the Precambrian Contact" he

C.J. Preston and W. Ouderkirk (eds.), Nature, Value, Duty, 45–61.

writes that:

> The lifeless granite, this fossil conglomerate inescapably now become charged with a numinous mystique. They address me, alas, they enchant me—the gods have been here—and I succumb again to the naïve emoting from which I so deeply disciplined in scientific criticism and positivist analysis, had thought myself to have escaped.[5]

Significantly, Rolston goes on to repudiate the thought that philosophy and the sciences might properly require the suppression of such responses: "Could it be that modernity's ken ignores and suppresses an authentic, primitive organon that now quickens within me? Need I apologize for my wonder? I think I shall rather boast of it."[6] So when writing as it were confessionally, Rolston clearly recognises the role of the emotions in sensitising us to natural values; but he is also conscious of the difficulty in articulating this contribution in the language supplied by "modern" philosophy and the sciences. I want to argue now that recent developments in the philosophy of the emotions have made it easier to record in a philosophical idiom the kind of insight that Rolston is enunciating in these remarks.

Much twentieth century philosophical thinking about the emotions took them to have two constituent parts, a thought component and an affective component, where the thought component was said to account for whatever cognitive content could be assigned to the emotion, and the affective component was construed as a kind of sensation, a raw feel, albeit one that was occasioned by a thought rather than contact with the sensory world.[7] On this sort of account, embarrassment for example should be construed as the thought that I find myself in circumstances which might lower my standing in the eyes of others, where this thought gives rise to a feeling of discomfort. A number of recent accounts of the emotions have taken issue with this model, and sought to assign feelings an intentional significance in their own right. If these models are correct, then we should no longer assume, on philosophical grounds, that in so far as the emotions are revelatory of the nature of things, this is solely on account of their status as "thoughts", where thoughts are considered as distinct from feelings. On the contrary, it may be that feelings themselves give us access to the nature of things. And if that is so, then "naïve emoting" may after all have a part to play in disclosing natural values, where these values are taken, in Rolstonian fashion, as part of the fabric of the world.

Briefly, I shall set out three recent accounts of the emotions which share this commitment to the thought-like character, or intrinsic intentionality, of feeling. These accounts are all self-consciously set against the traditional model that I have just described, at any rate if that model is taken to offer a comprehensive account of the emotions. Consider first of all John Deigh's proposal that certain "primitive" emotions reveal the character of the world independently of the mediating influence of any discursive conceptualisation. For example he writes:

> Roughly speaking, one feels fear at what is scary, horror at what is gruesome, and disgust at what is foul. These properties characterise the way

> things look, sound, taste, and smell. . . . the important point is that the scary differs from the dangerous in being at least sometimes a true or direct property of the way something looks and sounds. Something that looks dangerous is something one can infer is dangerous from the way it looks, whereas one need make no inference to see that something looks scary.[8]

So on Deigh's account, our affective responses can reveal the world to be scary, gruesome or foul, independently of the mediating influence of any affectively neutral thought or affectively neutral perception. (Contrast the recognition that something is dangerous. Here too I may feel fear, but this kind of fear is more readily analysable in the terms supplied by the traditional model of the emotions: by way of an affectively disengaged inference, I judge that the thing poses a threat, and this judgement causes me to feel fear.) Deigh's account evidently involves a claim about the phenomenology of our experience: on his view, the recognition that something is scary (or gruesome or foul) is (typically) embedded in our felt responses, rather than requiring (whether in the form of an initial perception or some inference from such a perception) an affectively neutral assessment of our circumstances. Hence such responses may be called "primitive" in the sense that they operate independently of any discursive appreciation of the character of our surroundings. However, they are not therefore impervious to reflection. For instance, the fearfulness which reveals an object to be scary may be diffused as I come to understand the object better. It is also true that reflection may deepen or elaborate upon, rather than cancel or subvert, such primitive responses. For instance, romantic love depends in part for its energy upon the primitive emotion of lust, but is also conditioned by the linguistic and other resources supplied by culture.

In a further recent discussion of the emotions, Peter Goldie has defended a similar thesis, only he is concerned not so much with the possibility of affective responses picking out some feature of the world independently of any (affect-free) conceptualisation, but with the case where affects build on what we know by verbal means, but offer a deeper kind of intentionality than is available by those means. He gives the example of how a person may have a purely intellectual grasp of the dangers of falling on ice, but then acquires a new, affectively-toned appreciation of these dangers once they have actually fallen.

> Coming to think of it in this new way is not to be understood as consisting of thinking of it in the old way, plus some new added-on phenomenal ingredient—feeling perhaps; rather, the whole way of experiencing, or being conscious of, the world is new . . . The difference between thinking of X as Y without feeling and thinking of X as Y with feeling will not just comprise a different attitude towards the same content—a thinking which earlier was without feeling and now is with feeling. The difference also lies in the content, although it might be that this difference cannot be captured in words.[9]

Here Goldie, like Deigh, argues that affects can be the bearers of understanding in their own right. But he also introduces a further and more radical thought, when suggesting that the insights disclosed by feeling may be such as cannot be articulated by verbal means.

Lastly, I note briefly the views of Geoffrey Maddell, another recent commentator on the emotions. Maddell is also opposed to the "add-on" construal of the contribution of feelings to emotions. He proposes that we should think of feelings as "feelings-towards" (that is, as having intrinsic intentional significance), and in support of this thought, he cites our musical experience:

> hearing the dominant seventh evokes a desire, and sometimes something akin to a longing, for its resolution. That is a state of consciousness directed to an intentional object; it is also an affective state of consciousness. It is not the entertaining of an evaluation which (magically) leads to certain bodily disturbances. One may, if one is so disposed, regard the desire for the tonic resolution as ground for the evaluation that such a resolution would be "a good thing", but it would be a total distortion to suppose that the desire, or the longing, is an evaluation, one which inexplicably leads to certain physical effects. It is a mode of "feeling towards" its intentional object.[10]

So with Deigh and Goldie, Maddell assigns feeling a role that is in some measure independent of that of "thought" or "evaluation": with them, he proposes that feelings are capable of directing us towards an intentional content in their own right. Maddell thinks that musical experience offers a particularly clear example of this possibility, but he also supposes that it is characteristic of the emotions generally that they should be comprised of an affective component, where this component is not just occasioned by thought, but itself thought-ful. Hence the fear that the stock market will fall is to be understood as constituted of the thought of such a fall and a fearful feeling-towards that thought, where the feeling takes the thought as its object, rather than just being occasioned by the thought, rather as a sensation of pain might be occasioned by a pinprick.

So in recent philosophical writing, we find a broadly-based tendency to re-conceive the significance of feelings and their contribution to the emotions. Specifically, writers like Goldie, Maddell and Deigh all agree that feelings are capable of picking out the nature of things in their own right, rather than simply by way of the mediating influence of affectively neutral thoughts or perceptions. If that is so, then we have the sort of conceptual framework that will allow us to articulate the thought that affective responses are integral to our appreciation of values, and cannot be regarded as mere add-ons in a theory of value recognition. This is a thought that Rolston implicitly endorses in the passages I quoted at the head of this section, where he maintains that "naïve emoting" has a part to play in our appreciation of nature (where "naïve emoting" may be glossed as "feeling" taken independently of discursive reflection). Next I want to draw on Rolston's writings in rather more detail, together with the models of Deigh, Goldie and Maddell, to consider more carefully how such an understanding of feelings may inform an account of natural value recognition.

PRIMITIVE EMOTIONS AND THE RECOGNITION OF NATURAL VALUE

I shall seek to apply each of the three models of the emotions we have just discussed to the question of the epistemology of natural value recognition, beginning with Deigh's model. In each case, I shall turn to Rolston's work, and specifically the "Nature in Experience" section of *Philosophy Gone Wild*, to frame the discussion and provide pointers to the route to be followed.

Rolston allows that certain primitive emotions, of an apparently culture-transcendent kind, provide a mode of access to the values in nature. For instance, in his "Meditation at the Precambrian Contact" he writes that:

> One steps into the abandoned tunnels enroute here, lower in the gorge, with an initial shudder. He enters the stone bowels of the Earth as though they were haunted with the jinn of Hades. The darkness is lonesome and alien. Intuitions of the savage persist, modern as I am. But the shudder passes, and, as is the case with one's initial encounter with the sea, there follows a fascination born of the intuition of connection, or reconnection.[11]

In support of the possibility of this kind of value recognition, we can refer once more to Deigh's description of the primitive affects of fear, disgust and horror. It might be thought (by a "modern" mind) that such responses belong to a "savage", untutored assessment of the world, and should be set aside in favour of a scientifically and philosophically informed appreciation of the world's meaning. But to dismiss such responses as lacking in any enduring significance seems to me a mistake. To see why, we may recall William James' account of the role that may be played by such primal responses in the development of the conceptual edifices of philosophy and religion. He writes: "I do believe that feeling is the deeper source of religion, and that philosophic and theological formulas are secondary products, like translations of a text into another tongue. . . . When I call theological formulas secondary products, I mean that in a world in which no feeling had ever existed, I doubt whether any philosophic theology could ever have been framed."[12] Here James seems to be suggesting that our emotional responses provide an initial, pre-theoretical value orientation in the world, one that can then be deepened by way of conceptual reflection. Such responses may be "primitive", in the sense of lacking the conceptual sophistication of a philosophical or theological system, but they provide, James implies, a necessary impetus for our more refined enquiries, which would not have begun but for some such stimulus.

Some will no doubt draw the conclusion that this goes to show that religion is after all a kind of superstition, grounded in primitive, pre-rational (even irrational) responses to the world. But this rejoinder is again too quick, for certain ventures of indisputable cognitive standing also seem to have their roots in affectively toned responses to the world. After all, one important contribution of the emotions to the economy of human life is to establish pre-reflectively the salience of certain features of our surroundings, and thereby to provide a starting point for any investigation of those surroundings. (Compare Michael Stocker's comment that anger, self-pity and other such states can "seek out and collect, even create, sustaining or concordant facts (or "facts"), which they then use to justify and sustain that emotion, which then leads to further seeking, collecting, creating and colouring".)[13] And to the extent that they have this character, then the emotions can play a role akin to that of scientific paradigms. As Ronald de Sousa puts the point, "paying attention to certain things [as we do when our emotions are engaged] is a source of reasons, but comes before them. Similarly, scientific paradigms, in Kuhn's sense, are better at stimulating research than at finding compelling and fair reasons for their own adoption. They are too "deep" for that, too unlike specific, easily formulated beliefs."[14] Accordingly, the basic, pre-theoretical ("deep") emotions that are engaged in primal apprehensions of natural value are arguably not so much irrational as supra-rational. Rather than simply generating rationalisations of themselves, they may help to constitute a research programme (by analogy with the role played by paradigms in science), one that is at least prima facie innocent, in so far as we need some such initial orientation if our enquiries are to get started. Of course, once these more specific enquiries get under way, the original affect may be diffused (compare our earlier example of feeling scared); but equally it may be in some fashion deepened or re-channelled (compare the example of romantic love). Rolston himself provides an example of the latter case in his further remarks in the passage from which I have just quoted:

> Earth's carbonate and apatite have graced me with the carbon, calcium, and phosphate that support my frame. The iron of hornblende and augite is the iron of the blood that courses through my life. Those stains of limonite and hematite now coloring this weathered cut will tomorrow be the haemoglobin that flushes my face with red. So now would I, this rock parasite, return to praise my natural parents. Ephemeral, anomalous, if so I am, erudite, conscious, proud, I can no longer suppress, but yield to, rejoice in, and humbly confess yet another primitive intuition, only enriched by my intellectual sophistication. Here is my cradle. My soul is hidden in this cleft of rock.[15]

Here is a wonderful example of a "primitive" emotion (that of connectedness) being deepened by its association with a complex, scientifically literate understanding of the nature of our origins. But the initial, affectively charged value revelation does not thereby lose its significance, surely. Instead, it forms part of the affective complex that Rolston describes with such power here, a complex that includes scientific insight, but where this insight is woven through (rather than just causally conjoined to) the affective dimension of our response to our natural context.[16]

There is one further point at which James's reflections on the epistemic priority of the emotions vis-à-vis philosophical-scientific reflection intersect with Rolston's thinking. Rolston notes the possibility of various global emotions in our dealings with nature. And he comments on the contribution of intellectual systems in fostering a variety of such emotions:

> Modernity began when Descartes divorced mind from matter, and the self was already lonesome in Newton's mechanistic universe. In Darwin's jungle that alienation greatly increased, under the variant dualism of the organism struggling against its world. . . . As had Darwin, Marx found that man rises up out of nature to be set in dialectical struggle with it by means of his labouring . . . For Freud too, the self evolves out of nature and is set against it.[17]

Rolston goes on to note that, in virtue of the prevalence of such theories, the predominant "modern" emotions towards the natural world are those of "loneliness" and "pathos". These emotions have their proper part in human life, he suggests, but "unless there are other emotions to relieve the tragic sense, it alone makes us ill".[18] Here Rolston is hinting at another role for the emotions. We have seen already how certain "primitive" emotions, such as that of connectedness, may give impetus to our appreciation of the natural world, and be woven through a more considered grasp of its significance. But here the emotions bear on our enquiries in a rather different way, by ensuring that certain world-views are from an existential-affective point of view self-defeating (in so far as they make us "ill"). Once again, this is a view defended by James. Consider for instance this passage:

> A philosophy may be unimpeachable in other respects, but either of two defects will be fatal to its universal acceptance. First, its ultimate principle must not be one that essentially baffles and disappoints our dearest desires and most cherished powers. A pessimistic principle like Schopenhauer's . . . or Hartmann's . . . will perpetually call forth essays at other philosophies. . . . But a second and worse defect in a philosophy than that of contradicting our active propensities is to give them no object whatever to press against. A philosophy whose principle is so incommensurate with our most intimate powers as to deny them all relevancy in universal affairs . . . will be more unpopular than pessimism. Better face the enemy than the eternal Void![19]

Here James is arguing that a world-view must not only satisfy certain abstract intellectual criteria (consistency with the data, freedom from self-contradiction, and so on). If it is to represent a natural stopping point for enquiry, a theory of the nature of things must also meet certain affective-existential needs of human beings, by representing the world as in some fashion receptive to our most deep-seated concerns and aspirations. James goes on to argue that materialism fails when measured against this requirement (in particular, it exhibits the second of the two deficiencies he lists in this passage). Rolston is evidently of a somewhat similar mind on this question. He too thinks that unless we feel in some sense "at home" in the world, then we cannot really flourish; and if that is so, then any world-view that fails to show how we can be at home in the world must be deemed an unsatisfactory terminus of enquiry. Making the same sort of point in a further way, he writes:

> What was earlier a healthy, composing fear of nature seems, under theories that overexplain the offensiveness and underexplain any attractiveness in nature, to have gotten us lost on a darkling plain. Lostness in the wild is, by all accounts, an intensely emotional state which breeds irrationality and disorganization, and in which we become our own worst enemies.[20]

To cast this point in terms of primitive emotions, we might say that there are certain primitive emotions focussed on the need to feel at home in the world (compare Rolston's allusion to the "ever dormant yearning to return to the womb of Mother Earth")[21] that need to be satisfied if we are to be able to function effectively in the world.[22]

So here too, we find that Rolston's reflections provide the raw materials for an account of the emotions and their bearing on our appreciation of natural value. We have seen that such an account will encompass both negative and positive attitudes to nature (it will include alienation and lostness, and also feelings of connectedness and wonder). On the one side, the role of such emotions is to provide an initial value orientation in the world of the kind that will help to stimulate our particular enquiries. Such emotions may then come to infuse our developing theoretical appreciation of our natural surroundings (compare Rolston's example of how theory may contribute to the deepening of a primitive sense of connectedness). On the other side, the emotions can set a check on our theories, where those theories yield an account of the world that is unreceptive to the primitive emotional need to find in the world a "home". In this way, basic affects can serve as a stimulus to our enquiries both positively and negatively: they can lead us to elaborate a theoretical perspective (compare the example of connectedness again) and they can give impetus to an enquiry which is directed against the established paradigm (when that paradigm frustrates our need to feel at home in the world).

"SOPHISTICATED" EMOTIONS AND THE RECOGNITION OF NATURAL VALUE

We have been discussing the role of primitive emotions in forging a sense of natural value, and specifically in sustaining the thought that we human beings are "connected" to the natural world in ways that make it our "home". In support of the possibility of such "primitive" or conceptually unarticulated emotions, I appealed to the work of John Deigh. I turn now to the work of Peter Goldie. Here again I shall be concerned with the possibility that an account of the emotions can contribute towards a theory of natural value.

Like Deigh, Goldie subscribes to the intrinsic intentionality of feeling thesis. However, Goldie is more interested in the case where feelings deepen a purely verbal appreciation of some state of affairs in ways that may escape verbalisation. (Think again of his example of falling on ice.) If Goldie is right on this point, then feelings may have a non-eliminable role to play in the identification of natural values (in so far as words provide no substitute for the value recognition embodied in feeling). In the heading of this section, I have called the kinds of emotions which conform to Goldie's model "sophisticated", to mark the fact that rather than setting the scene for verbal understanding, they represent a deepening of such understanding. Once more, Rolston's descriptions of his encounters with nature will provide a way into our topic.

In his essay "The Pasqueflower", Holmes Rolston first examines the pasqueflower from a biological perspective, before supplementing this perspective with another, more affectively resonant description:

> The brilliance of this pasqueflower has its simplest explanations in mechanisms for flowering so soon at the winter's end. It must have petals (or, as the botanists prefer, petallike sepals) large enough to attract the few insects that are out so early. The downy surface of transparent hairs on its palmate leaves and stem insulates and also, as do those of pussy willows, allows a radiation heating to temperatures high enough for development, providing a miniature greenhouse effect.

Rolston continues in this vein for another few lines, and then comments:

> but the pasqueflower helps me to glimpse something more, the skill of art superimposed on the science of survival. This is exuberance in the fundamental, etymological sense of being more than expectedly luxuriant. Does not such an encouraging beauty speak of that face of nature that overleaps the merest hanging on to life to bear the winds of the storm with vigorous, adorning beauty? Nor is it just the grand petals of delicate purple whorled about the yellow stamens and pistils, for the fingered involucre frames the flower so well, and the villous coat has a sheen that, seen backlighted by the sun, gives a lustrous aura to complement the gentle leafy green.[23]

Here we find two kinds of vocabulary juxtaposed: the vocabulary of biological science and an aesthetic vocabulary (the latter involving such terms as "grand", "delicate" and "lustrous"). And as Rolston says, the aesthetic perspective goes beyond (while still being informed by) the scientific, since the pasqueflower proves to be "more than expectedly luxuriant" (the expectations here are, I take it, those befitting a purely scientific account). So here Rolston acknowledges that a scientific appreciation of the pasqueflower, however accurate and exhaustive, may fail to disclose certain important truths concerning its character. Elsewhere, Rolston has commented that "a science-based landscape aesthetics is urgent, but it must also be a science-transcending aesthetic of participatory experience."[24] This suggests that it is only on the strength of firsthand experience that we can grasp the aesthetic truths he records here in full, for only so will we be able to appreciate the particular quality of the plant's delicate colouring and lustre.

This sort of point is familiar from standard philosophical discussions of the incommunicability of the phenomenological feel of experience by verbal means alone.

(Compare Nagel's celebrated question: what is it like to be a bat?) But Rolston speaks here of an "encouraging beauty" and finds in this beauty a clue to the meaning of the natural order, a meaning that runs beyond mere survival. Here again we are in the realm of the emotions; and the implication of the passage is therefore that an appreciation of the value of the pasqueflower (an appreciation of the particular character of its exuberant beauty, and the correlative encouragement that it offers) depends in significant degree upon the quality of our affective response to it. To this extent, Rolston's remarks bear out Goldie's proposal that feelings may build upon and surpass the kinds of value insight that are communicable by verbal means alone.

This reading of Rolston suggests a parallel with Raimond Gaita's claim that understanding the nature of our moral relationships with other human beings requires some reference to the quality of our felt responses to them. In part, Gaita is led in the direction of this claim by his sense that the standard moral theories cannot give an adequate account of the moral standing of certain "afflicted" human beings. For instance, if we take a capacity for rational, autonomous decision-making as the key to moral value, or again if we take "happiness" as the key while allowing for grades of happiness and ranking intellectual above other kinds of fulfilment, then we will be unable to articulate the thought that human beings suffering from a cognitive disability are fully members of the moral community, and properly the objects of an "undiminished moral response".[25] On Gaita's account, seeing someone as fully a member of the moral community requires seeing them as intelligibly the object of someone's love; and where the afflicted are concerned, this requires the example of the "saints" (where the term bears no necessary religious connotation), who can show love even to such people.

Gaita makes this point very clearly by means of an example. Here he is describing his experience of working at a psychiatric hospital. He has noted that the psychiatrists at the hospital subscribed to the view that their patients were fully their equals, as he did himself. But he comes to see this verbal profession of their equality as flawed when a nun visits the ward:

> In her middle years, only her vivacity made an impression on me until she talked to the patients. Then everything in her demeanour towards them—the way she spoke to them, her facial expressions, the inflexions of her body—contrasted with and showed up the behaviour of those noble psychiatrists. She showed that they were, despite their best efforts, condescending, as I too had been. She thereby revealed that even such patients were, as the psychiatrists and I had genuinely professed, the equals of those who wanted to help them; but she also revealed that in our hearts we did not believe this.[26]

The nun's behaviour, Gaita implies, flows from something she believes "in her heart". And it is this affective response to the patients that grounds her non-condescending behaviour towards them. By contrast, the psychiatrists, while they may have the vocabulary of human dignity, equality of respect, and the rest, and may use this vocabulary in all sincerity, still fail to register what the nun recognises "in her heart." Given our typology of the emotions, it is worth adding that on Gaita's view, this recognition is not embedded in some "primitive" emotion, but dependent upon a specific linguistic tradition, in particular the language of prayer and divine love.[27]

Gaita goes on to claim that it is only in the light of such an impartial, saintly love (and the quality of affective response that it embodies) that we can appreciate the true worth of the patients. (Compare Goldie's thought that it is only in the light of certain affects that we can grasp fully the values or disvalues which arise in our dealings with ice.) He writes that: "if someone were now to ask me what informs my sense that [the patients] are rightly the objects of such treatment [as that evident in the nun's behaviour], I can appeal only to the purity of her love. For me, the purity of the love proved the reality of what it revealed. . . . From the point of view of the speculative intelligence, however, I am going round in ever darkening circles, because I allow for no independent justification of her attitude."[28] In other words, Gaita believes that we cannot fully explicate the character of our moral relations with others (where these moral relations extend equally to all human beings, just by virtue of their "humanity") without reference to the quality of our affective response to others. To be sure, you and I can grasp that the patients in Gaita's example are deserving of an undiminished moral response without feeling what the nun feels; but our insight is dependent upon her example (which reveals that they are intelligibly the objects of a non-condescending regard), and her example is in turn dependent upon the quality of her affective response (it is dependent on what she believes "in her heart").

This is not the place to consider the plausibility of Gaita's account, but rehearsing this view here does serve two goals. First, it allows us to see that the reading of Rolston that I offered above would place his understanding of affective response and its relationship to value recognition into a broader stream of ethical reflection. Secondly, it invites us to put to Rolston's account (on my reading of it) a question that Gaita puts to himself: if quality of affective response is presupposed in this way in at least some of our value insights, then does that not suggest that those values are a projection? Gaita's judgement is that: "The contrast between inventing or making and discovering cannot be applied in any simple way here."[29] Similarly, he speaks in this context of the "interdependence of object and response."[30] For the most part, Rolston is concerned to uphold a robustly objective account of values in nature (and it is partly for this reason no doubt that in the bulk of his writing, he prefers to ground the idea of natural value in scientific insights, rather than by reference to the emotions).[31] But he does sometimes display a sympathy for Gaita's view. Consider for example this comment: "How lovely is this crystal lake beneath Mt. Alice! And this aesthetic experience is neither my invention nor simply my discovery; it emerges in relational encounter."[32]

However exactly we choose to make this sort of point (regarding the contribution of human response to the values in nature), Gaita's kind of perspective leads on to the thought that our affective responses do not only provide evidence for our value claims: rather, such responses are also partly constitutive of the meaning of those claims. (For instance, on Gaita's view, our concept of "humanity", where this expression is morally charged, and signifies more than simply membership of the species *Homo sapiens*, cannot be fully explicated without reference to quality of felt response, and what is revealed in such responses.) This position is compatible, I suggest, with a commitment to the "objectivity" of value; but it does mark a difference between at least some of our value concepts and the concepts of the natural sciences, in so far as the latter, while they may reflect human interests of various kinds, are not so evidently conditioned by our particular, human sensibility.

To take stock, we have been considering the accounts of value recognition that are at least implied in the writings of Goldie, Rolston and Gaita. On Goldie's view, my felt response to falling on ice may provide a new and deeper understanding of the dangers (or disvalues) of ice than any which can be communicated by verbal means alone. Rolston's description of the pasqueflower suggests the possibility of a similar sort of deepening: the language of aesthetics offers a deepening of the kind of understanding that is available in the language of science, and we might add that this is partly because the language of aesthetics is more directly tied to our particular, human sensibility; and the aesthetic qualities of the pasqueflower can in turn be understood most fully in the light of firsthand, affectively resonant experience of the flower. Lastly, Gaita suggests that our recognition of the moral significance of certain human beings may rest upon the quality of affective response towards such people that is exhibited in the lives of the saints. These writers are not making precisely the same point. In particular, whereas Goldie sees the quality of our affective responses as taking us beyond what can be expressed verbally, Gaita sees them as contributing to the meaning of our terms. But together these authors all point us in the direction of the thought that affects are not just caused by "thoughts", where it is thoughts that carry cognitive, value-recognising significance; rather, it is through, and non-eliminably through, our affective responses that at least certain values are most fully communicated.

Having broached the question of whether values in nature are subjective or objective, it is worth adding that, uncontroversially, the emotions are constitutive of certain values at least in part. Consider for instance Graham Nerlich's comments on grief, and its relationship to our valuing of other human beings:

> Grief can do no obvious service to the dead. It pays its due to the deep and complex array of threads that tie one to a beloved. It is a human debt, paid because we are, quite contingently to personhood, creatures who need a time for recovery from the tearing-away of those manifold connections. But it is also personal, in the sense that many of these inarticulately forged connections and judgements constitute the value placed on the person for whom one grieves and were a part, perhaps quite dimly perceived, of the process of coming to value her.[33]

We can make this same sort of point by noting that the value I place on another person may relate in part to their irreplaceability in my life—to the fact that I cannot simply substitute other life projects for those that are bound up with my relationship to them. And this irreplaceability may in turn be grounded in my emotional attachment to the person, for it is in virtue of this attachment that I feel grief at their death, with the consequence that for a time I am unable to cultivate new interests, but must instead endure the "tearing away" of the bonds that united us. Extending this example into the sphere of natural values, we may say that the value of a certain landscape, for example, is, in part, constituted by our attachment to it, since this attachment means that we would be in some degree disabled by the landscape's destruction. But this is not to say that the value of the landscape is reducible to our emotional attachment; after all, such attachments are formed, at least in part, in response to features of the natural world that are antecedently deemed to be valuable.[34]

A NATURAL THEOLOGICAL CODA

I have been considering the contribution that may be made by "primitive" and "sophisticated" emotions to our recognition of natural values, and in the process endeavouring to follow in the footsteps of Rolston. But any treatment of natural values which is genuinely following in his footsteps must make some larger reference than I have so far to the natural theological dimension of these issues. (Compare the role of Rolston's essay on the pasqueflower in the *Philosophy Gone Wild* collection: this essay, the most theologically engaged of the book, falls last.) So I conclude with a theological excursus on the character of natural values. Again, I shall take a passage from Rolston as my starting point, and again I shall make a connection between Rolston's observations and the perspective of a recent commentator on the emotions.

We have taken note already of Rolston's reference to the pasqueflower's "encouraging beauty". He goes on to propose that the pasqueflower owes its particular beauty (and time of flowering, as recorded in its name) to the rigours of winter: "This pasqueflower springs forth in its particular form of early beauty as much because of winter as to spite it; it buds and blossoms because it is blasted."[35] In this way, the pasqueflower represents in miniature the natural order in general, and specifically it records this more general truth: "Life is pressed by the storms, but it is pressed on by the storms, and environmental necessity is the mother of invention in life".[36] And this is also the basic truth of human life considered in its natural context: for "we owe all culture to the hostility of nature, provided only that we keep in tension with this the support of nature that is truer still, the one the warp, the other the woof, in the weaving of what we have become".[37] In concluding this paper, and the book, Rolston offers the obvious Christian reading of all these observations: "The way of nature is, in this deep though earthen sense, the Way of the Cross. Light shines in the darkness that does not overcome it. This noble flower is a poignant sacrament of this"[38]

In these remarks, Rolston sets out the main lineaments of his own theory of natural value, namely, that over the long haul, natural processes work to generate life not merely in spite of but through death. This view can of course be expounded in scientific terms, by noting the trend of evolutionary processes to produce new and more complex forms of life, notwithstanding periodic mass extinctions. But here the truth is grasped more poetically, and in ways that are affectively resonant. Indeed, Rolston seems to suggest that this most basic of truths concerning the character of natural value can be apprehended in something like a primitive emotional response. For instance, he writes that: "We are born to die, but it is life rather than death which is the principal mystery that comes out of nature, and our emotions are stirred proportionately."[39] This remark could be taken to mean that we grasp by way of discursive thought that it is life rather than death that is nature's last word, and this thought gives rise to an affective response. But in keeping with the theories of the emotions we have been examining, we might also suppose that the affective component of the emotion itself registers the prevailing tendency of natural processes. In support of this reading, it is worth noting that Rolston is clearly talking about a trans-cultural pattern of emotional response, and hence one that is not dependent on the thinking of particular, culturally specific thoughts. For the passage I have just quoted is immediately prefaced by these words: "we are not born clean of nature, and in any cultural education we do ill to neglect those emotions that are native to this birth." Similarly, in his essay on the pasqueflower, Rolston recalls the example of the Shanidar Cave, where a Neanderthal corpse was found strewn with the remains of flowers, and he comments:

> If the flower has for fifty thousand years served as an emblem of resolution in the face of death, then my thoughts run steady in a natural track as perennial as the springs since Neanderthal times. The flower is a very powerful symbol, it has had a psychologically elevating effect in every culture, and if anyone cares to say that this is not scientific, but romantic, that does not make it any less real. Our recent "flower children" knew this impact when they hung flowers in protest in the guns of destruction.[40]

Here again we seem to be dealing with an affective response that is pre-reflective, or in the terms we have been using "primitive". This suggests that such responses may reveal the most basic natural value of all, concerning the regenerative powers of the natural order, and its capacity to bring life from death.

We have been considering the pasqueflower in particular and flowers more generally as disclosing a general truth about the character of the natural order and human experience within its natural context. As Rolston notes, these truths are of obvious theological resonance in so far as they deal with the ultimate destiny of human life and the cosmos itself. But they are also theologically significant in a further respect, in virtue of the connection between the character of the world as a whole and truths concerning the divine nature. Hence Mircea Eliade can write that: "The great paradox common to all religions is that God in showing Himself to mankind is free to take any form whatsoever but that, by this very assertion of His freedom, He "limits Himself" and reduces Himself to a mere fragment of the whole which he represents."[41] This same sort of point can be made in Christian terms by saying that in Jesus, the Logos (who is both God and the decisive clue to the meaning of the world as a whole) is revealed as self-sacrificial, radically regenerative love. In this sort of way, truths concerning the character of the world as a whole can also be taken as revelatory of the divine nature.

As we have seen, Rolston finds that this Christian sense of the world's cruciform meaning is borne out in the natural order, in so far as nature conforms to the law of new life through suffering. This suggests the possibility of a new kind of natural theology, one that can be formulated in the language of science, with reference to the long-range tendency of evolutionary systems,[42] or alternatively, in an idiom that is more poetic and affectively resonant (if we recall the ways in which a flower may symbolise these larger truths). If we take this latter approach, then we may suppose that just as the real meaning of our fellow human beings (their worth and the sense in which they set a limit to our will) is revealed (on Gaita's account) in forms of felt response, so the world's meaning, its Logos, may be revealed by means of an affectively-constituted insight.

Claiming to discern the meaning of the world implies a claim about its future and not just its past, of course. And Rolston observes as much in the closing line of "The Pasqueflower": "This noble flower is a poignant sacrament of this [the fact that the darkness does not overcome light], and to chance to find it in earliest spring, and to pause at that meeting, is to find a moment of truth, a moment of memory and promise. Let winters come, life will flower on as long as Earth shall last." Here Rolston is still speaking from his natural theological perspective. A more radical, specifically Christian view would be that at the eschaton, the cosmos itself will be transformed, in the life of the resurrection. This suggests a parallel with Maddell's account of the emotions.

Remember that on his view, our affective responses to a piece of music can provide an anticipatory glimpse of its resolution, independently of any discursive thought or judgement. Analogously, in a mystical vein, we might suppose it is at least possible that in feeling our minds can be cast forward to the character of the world's resolution at the eschaton: just as feeling may carry the mind from a certain musical input to an anticipation of musical climax, so it may carry the mind from a certain sensory-affective input, concerning the life-giving character of the world, to a foretaste of cosmic resolution. Of course, this is to move some way from what is claimed in Rolston's essay. But it represents a natural, Christianising extrapolation of his view that affects can reveal the world's meaning.

I have been arguing that natural values can be disclosed in our emotional responses to the world, where those responses may be "primitive" or "sophisticated". And these values, I have suggested, may in turn be folded into theological values, concerning the ultimate meaning of our world's history. If all of this is so, then we can agree with Rolston that "our encounter with nature is as passionate as it is cognitive".[43] And drawing on his own insights, we may add that in this context, passion and cognition are not to be understood antithetically, nor even as simply distinct, but in certain cases at least as aspects of one and the same movement of understanding.

NOTES

1 See for example Lawrence Blum, *Moral Perception and Particularity* (Cambridge: Cambridge University Press, 1994), p. 35.

2 Compare Nancy Sherman, *The Fabric of Character. Aristotle's Theory of Virtue* (Oxford: Clarendon Press, 1989), p. 47.

3 Compare Holmes Rolston, III, *Philosophy Gone Wild* (Buffalo, NY: Prometheus Books, 1989), pp. 223–4.

4 This sort of measure needs to be supplemented by closer reading, but even so, it is significant that there are only four page references under "emotions" in the index of *Environmental Ethics. Duties to and Values in the Natural World* (Philadelphia: Temple University Press, 1988), and no such references in *Conserving Natural Value* (New York: Columbia University Press, 1994). In Rolston's most recent

book, *Genes, Genesis and God. Values and Their Origins in Natural and Human History* (Cambridge: Cambridge University Press, 1999) there are eight page references for "emotions", in a work whose text runs to 370 pages. In his more recent work, however, Rolston does speak of "care" for the environment, a term which clearly suggests affective engagement, not least because of its usage in feminist writing. See for instance Holmes Rolston, "From Beauty to Duty: Aesthetics of Nature and Environmental Ethics", in Arnold Berleant (ed.), *Environment and the Arts: Perspectives on Environmental Aesthetics* (Aldershot: Ashgate Publishing, 2002), pp. 127–41.

5 Rolston, *Philosophy Gone Wild*, pp. 239–40.

6 *Ibid.*, p. 240.

7 Summarising this view, Geoffrey Maddell writes: "A dominant view of emotion has been that emotions are essentially evaluations which cause certain bodily changes, and it is our awareness of these which constitutes feeling": "What Music Teaches About Emotion", *Philosophy*, 71 (1996), 76.

8 John Deigh, "Cognitivism in the Theory of Emotions", *Ethics*, 104 (1994), 842.

9 Peter Goldie, *The Emotions: A Philosophical Exploration* (Oxford: Clarendon Press, 2000), pp. 59–60, emphasis in the original.

10 Geoffrey Maddell, "What Music Teaches About Emotion", 78, emphasis in the original.

11 Rolston, *Philosophy Gone Wild*, p. 233.

12 William James, *Varieties of Religious Experience* (Harmondsworth: Penguin, 1982), p. 431.

13 Michacl Stockcr with Elizabeth Hegeman, *Valuing Emotions* (Cambridge: Cambridge University Press, 1996), p. 94.

14 Ronald de Sousa, "The Rationality of Emotions", in Amélie Rorty (ed.), *Explaining Emotions* (Berkeley: University of California Press,1980), p. 139.

15 Rolston, *Philosophy Gone Wild*, pp. 233–4.

16 Rolston cites some research which gives further substance to the thought that certain basic attitudes to nature can provide a stimulus for our enquiries. Drawing on the work of Edith Cobb, he notes that geniuses typically report a period of close identification with natural processes during their middle childhood. This suggests that there may be deep-seated connections between human creativity and attitudes to nature, and specifically feelings of connectedness to nature: Rolston, *Philosophy Gone Wild*, pp. 253–4.

17 *Ibid.*, p. 251.

18 *Ibid.*, p. 252.

19 William James, "The Sentiment of Rationality", in William James, *Essays in Pragmatism* (New York: Hafner Press, 1948), p. 17.

20 Rolston, *Philosophy Gone Wild*, p. 252.

21 *Ibid.*, p. 233.

22 The biophilia hypothesis sheds an interesting, scientifically informed light on this thought. The hypothesis maintains that we are fitted for evolutionary reasons to appreciate certain kinds of landscape, notably those marked by water and spatial openness. See for instance Roger Ulrich, "Biophilia, Biophobia, and Natural Landscapes", in Stephen Kellert and Edward Wilson (eds), *The Biophilia Hypothesis* (Washington D.C.: Island Press, 1993), pp. 89–92.

23 Rolston, *Philosophy Gone Wild*, p. 257.

24 Holmes Rolston, "Does Aesthetic Appreciation of Landscapes Need to Be Science-Based?", *British Journal of Aesthetics*, 35 (1995), p. 384. The role of firsthand experience is also discussed in *Environmental Ethics*, p. 352.

25 *Good & Evil. An Absolute Conception* (Basingstoke: Macmillan, 1991), p. 196.

26 Raimond Gaita, *A Common Humanity. Thinking About Love & Truth & Justice* (Melbourne: Text Publishing, 1999), pp. 18–19.

27 *Ibid.*, p. 22.

28 *Ibid.* pp. 21–22.

29 *Good & Evil*, p. 125.

30 *Ibid.*, p. 166.

31 See for instance his essay "Are Values in Nature Subjective or Objective?", Rolston, *Philosophy Gone Wild*, Ch. 6. Here he makes this prefatory observation: "I keep the whole discussion as close to science

as I can, while demanding a full-blooded, no-nonsense account of the phenomenon of value in, and valuing of, the natural world": p. 92. In general, Rolston's position is that: "There is value wherever there is creativity": *Conserving Natural Value*, p. 195.

[32] Rolston, *Philosophy Gone Wild*, p. 231. It is no doubt significant that here Rolston is talking about the values revealed in aesthetic experience, and not those revealed in science.

[33] Graham Nerlich, *Values and Valuing. Speculations on the Ethical Lives of Persons* (Oxford: Clarendon Press, 1989), p. 164.

[34] More controversially, it might be said that through our attachments to others, we contribute to the mattering or meaning of their lives. And if it is a good that their lives should matter, then my emotional attachments are constitutive of value in this further respect. Analogously we might argue that our emotional attachments to the natural world contribute to its mattering, and that this constitutes a value. For further development of this thought, see Mark Wynn, "Valuing the World. The Emotions as Data for the Philosophy of Religion", *International Journal for Philosophy of Religion*, 52 (2002), 103–5.

[35] Rolston, *Philosophy Gone Wild*, p. 261.

[36] *Ibid.*, p. 260.

[37] *Ibid.*, p. 261.

[38] *Ibid.*, p. 261. Rolston develops the theme that our natural history is cruciform at greater length in his essay "Does Nature Need to Be Redeemed?", *Zygon: Journal of Religion and Science*, 29 (1994), 205–229.

[39] Rolston, *Philosophy Gone Wild*, p. 255.

[40] *Ibid.*, p. 260.

[41] Mircea Eliade, "Divinities: Art and the Divine", reprinted in Diane Apostolos-Cappadona (ed.), *Mircea Eliade: Symbolism, the Sacred and the Arts* (New York: Crossroad, 1985), p. 56.

[42] See Rolston's version of the argument from design in *Science and Religion. A Critical Survey* (Philadephia: Temple University Press, 1987), pp. 107–22.

[43] Rolston, *Philosophy Gone Wild*, p. 248.

NED HETTINGER

5. RELIGION IN ROLSTON'S ENVIRONMENTAL ETHICS

Is there religion deep down in Holmes Rolston's environmental ethics? As anyone who has read Rolston's work or heard him speak will find obvious, a profound love of earthen nature and a desire to experience, understand, appreciate, and protect it animates his work. Rolston is as articulate and insightful a champion of wild nature as we have on the contemporary scene. Rolston's defense of nature is multi-disciplinary: It spans ethics, aesthetics, and religion, and all of these are informed by a philosophy of nature tightly hitched to science.[1]

With Rolston's winning of the Templeton Prize, it has become clear that others value Rolston at least as much for his religious insight as for the scientifically-informed love of wild nature that drew me to his work. Religion is at the core of Rolston's worldview. When accepting the Templeton prize he said: "I've spent my life in a lover's quarrel . . . with the two disciplines I love: science and religion" (Ostling, 2003). Any attempt to get to the bottom of Rolston's philosophy of nature will have to determine what role religion plays in his account.

In this chapter, I consider the role religion pays in Rolston's understanding, evaluation, and defense of nature, particularly its role in his environmental ethics. How central is religion to Rolston's defense of nature? What sort of religion, if any, fits with—or is required by—his environmental ethics? My conclusion will be that while an immanent nature spirituality is a congenial part of his environmental ethics, a transcendent deity is not. I suggest that Rolston's appeal to such a deity in his account of natural history undermines important aspects of his environmental ethics.

IS RELIGION NEEDED FOR ROLSTON'S ENVIRONMENTAL VALUE THEORY?

One of the most striking features of Rolston's philosophy is how intensely valuable he finds nature (both aesthetically and ethically) and how forcefully he argues for such value in light of possible evidence to the contrary. In the aesthetics of nature, Rolston embraces positive aesthetics. Quoting John Muir, he writes: "None of Nature's landscapes are ugly so long as they are wild" (1988: 237). Rolston acknowledges that if one looks at particular aspects of nature one will find ugliness: "If hikers come upon the rotting carcass of an elk, full of maggots, they find it revolting" (1988: 238). But this "momentary ugliness is only a still shot in an ongoing motion picture" (1988: 239). Rolston argues that a scientifically-informed approach will note with appreciation that "The rotting elk returns to the humus, its nutrients recycled; the maggots become flies, which become food for the birds; natural selection results

C.J. Preston and W. Ouderkirk (eds.), Nature, Value, Duty, 63–76.

in a better-adapted elk for the next generation" (1988: 239). Although for Rolston, virgin nature is not "invariably aesthetically positive in immediate detail" (1988: 245–46), "these ugly events . . . (are) anomalies challenging the general paradigm that nature's landscapes almost without fail have an essential beauty" (1988: 243). With regard to ethically important non-instrumental values, Rolston locates them throughout nature: In humans, animals, plants, microbes, abiotic natural wonders, natural systems and processes, and especially in the creative, prolife, systemic nature that drives natural history on earth: "The projective system is fundamentally the most valuable phenomenon of all" (1988: 225). Rolston finds nature so spectacular that he characterizes it as a "wonderland" and "sacred gift."

Not all commentators find this intense positive valuing of nature in Rolston. In a helpful article that examines the provocative fact that Rolston, more than any other environmental philosopher, worries about possible defects in nature (i.e., the "disvalues in nature"), Wayne Ouderkirk describes Rolston's overall evaluation of nature thus: "I think it is also fair to say that he considers the whole system, *including the disvalues*, as valuable. . . . I believe that Rolston is right that thus far we can say that the system has gone *mostly* in the positive direction" (Ouderkirk, 1999: 144–45). But Rolston's appraisal of the value of earthen nature is hardly lukewarm. One might think it is because Rolston is a master at expressing the case against nature's value. For example, Rolston writes: "The wilderness teems with kinds but is a vast graveyard with hundreds of species laid waste for one or two that survive. . . . Wildness is a gigantic food pyramid, and this sets value in a grim, deathbound jungle. All is a slaughterhouse, with life a miasma rising over the stench" (1983: 193). But when Annie Dillard appraises evolutionary history as an odious scene of suffering and violence ("I came from the world, I crawled out of a sea of amino acids, and now I must whirl around and shake my fist at that sea and cry shame"), Rolston is compelled to respond: "If I were Aphrodite, rising from the sea, I think I would turn back, reflect on that event, and rather raise both hands and cheer. And if I came to realize that my rising out of the misty seas involved a long struggle of life renewed in the midst of its perpetual perishing, I might well fall to my knees in praise" (1994b: 58).

Ouderkirk is quite right that Rolston's focus on and account of disvalues in nature parallels the theodicy provided by theists attempting to make sense of evil in a world created by a beneficent deity. Rolston (1992) argues that the so-called disvalues in nature—predation, suffering, death, parasitism, randomness, disasters, selfishness, blindness, indifference, waste, and struggle—are necessary for the far greater positive values evidenced in natural history—viz., life, sentience, mind, and culture. These disvalues systemically drive nature's value achievements. Rolston writes: "It is a sometimes tragic view of life, but one in which tragedy is the shadow of prolific creativity. . . . A world without blood would be poorer, but a world without bloodshed would be poorer too, both less rich in biodiversity and less divine" (2003: 85).

Rolston's strenuous defense of positive aesthetics and of the intensely positive overall value of natural history invites a religious interpretation. A supreme deity would not make an ugly world filled with evil. Thus, on this interpretation, Rolston accounts for these negative dimensions of nature in order to provide a theodicy.

However, a secular rationale is equally available: An unattractive, disvaluable nature where ugliness and evil compete closely with beauty and goodness is not one that we ought to worry much about preserving for its own sake, and this goal is at the heart of Rolston's project in environmental philosophy.

One might argue that religion is implicated by the way Rolston defends nature's positive value rather than simply because he does so. Rolston conceives of natural processes as "sacrificial" in the sense that good is resurrected out of evil, life out of death. Rolston interprets the fundamental character of the biological world as involving struggling and suffering in the achievement of something higher. The evolutionary advancements Rolston sees as the trend of natural history—from matter to microbes to plants to animals, from reflex to instinct to learning, and from nature to culture—have been earned at the price of extinction, death, and suffering: Millions of species and billions upon billions of individuals have been sacrificed to attain the marvelous life forms we find in nature today. Rolston routinely describes nature as "cruciform" (2003: 84) and he interprets "the way of nature as the way of the cross" (1987: 144). He argues that "the capacity to suffer through to joy is a supreme emergent and an essence of Christianity. Yet, the whole evolutionary upslope is a lesser calling of this kind, in which renewed life comes by blasting the old" (2003: 84). Thus, one might argue that Rolston's defense of nature's value implicates not just religion but Christianity in particular.

While there can be no doubt that Rolston's account has been influenced by Christian theology, whether that account is a successful defense of nature's value is independent of the truth of the related theological claims. Nature may well have the character of struggling through to something higher and have an essential goodness that is dependent on death and suffering whether or not Christian theology provides an accurate metaphysic.

In one of the few treatments in the environmental ethics literature of the role of religion in Rolston's environmental philosophy, Francisco Benzoni argues that "The theological dimension of Rolston's work . . . undergirds his entire ethic" and that "his ethic is . . . finally a theological ethic" (1996: 339, 349). Benzoni claims that the objective intrinsic value theory central to Rolston's environmental ethics is "finally grounded . . . in his theology and thus in the divine" (1996: 339). On Benzoni's account of Rolston's views, the creative prolific earth would not be objectively good if there were no God. Benzoni writes:

> The intrinsic value of creation is based in the divine creator. The creator is the basis and foundation of reality who declares that the earth is good. In other words, it has real, objective intrinsic value that is utterly independent of human valuation, and indeed human existence. Ultimate reality has declared creation to be good. This is entirely different than a value theory in which human loving or valuing nature for its own sake endows it with intrinsic worth. . . . Because God declares that creation is good and God loves creation for its own sake, quite simply, creation is good and is intrinsically valuable (1996: 348).

Benzoni is right that Rolston wishes to avoid a Callicott-style account whereby nature's intrinsic value results from humans valuing nature for its own sake. But unless one accepts the view that intrinsic value is a kind of valuing and thus requires a valuer, there is no need to posit a God who intrinsically values nature in order for nature to have intrinsic value independent of human valuing. On Benzoni's account of Rolston, we don't get a nature with objective value, but rather one that has subjective value conferred by God. This is clearly at odds with Rolston's own insistence on the objective intrinsic value of nature.

One might interpret Benzoni as claiming that Rolston needs God to bestow objective, non-instrumental value on nature which it then has on its own and not because God or anyone else values it.[2] But if one is suspicious about the possibility of objective intrinsic value, it is not clear how positing a God as bestowing such value clears up any mystery. Further, far from finding it mysterious, Rolston is, if anything, too sanguine about the reality of objective intrinsic value in nature. Perhaps Benzoni thinks Rolston's ethic is "theological" because Rolston posits God as accounting for certain valuable dimensions of nature (as we shall see below). On this interpretation, Rolston's value theory depends on God because his account of natural history is ultimately theological. In short, because his metaphysic of nature is theological, so is his value theory. But even if God created significant aspects of nature and is thus responsible for them (and thus for their value), they need not be valuable because God created them rather than because of their own value-adding features. I see no evidence in Rolston writings that suggests he thinks that nature is valuable not because of the kind of thing it is, but because it has a causal origin in the hand of God. As I show below, Rolston does understand certain valuable aspects of nature (e.g., nature's "enthralling creativity" and its prolific, prolife fertility) as manifestations of divine activity. When we value these aspects of nature, we are—on his account—valuing God's presence in nature. But here again Rolston's theological appeal serves an explanatory and not evaluative role: He argues that the earth's prolife creativity can not be explained without God, not that its value somehow depends on being a manifestation of God. In the last section of this chapter, I argue that this explanatory appeal to a transcendent deity is in serious tension with fundamental features of Rolston's environmental ethics rather than being a centrally important part of it—as Benzoni would have it.

While it is clear that Rolston's conceptualization of nature's value has been greatly influenced by religion, fundamentally his environmental value theory does not depend on supernaturalist, religious metaphysics. Thus I think it is a mistake to characterize Rolston's environmental ethic as a theological one.

RELIGION IN ROLSTON'S ACCOUNT OF NATURAL HISTORY

I have argued that Rolston's environmental value theory stands by itself, largely free of any reliance on transcendental religion. In the next section of this chapter, I argue that an immanent nature spirituality, while not strictly required for his environmental ethics, fits harmoniously with it. In this section, I explore the role a transcendent deity plays in Rolston's account of the origin of natural history. Rolston's explicitly theological writings provide a prominent place for a transcendent deity in our explanation of the natural world.

There are times in these writings when Rolston downplays the need for a transcendent religious account of valuable nature and suggests that such an account is open to doubt. He writes:

> Whatever you make of God, biological creativity is indisputable; There is creation, whether or not there is a Creator . . . Some biologists decline to speak of creation, because they fear a Creator lurking beneath. Well, at least there is genesis, whether or not there is a Genitor. . . . Biologists may doubt whether there is a Creator, but no biologist can doubt genesis (1994b: 58).

But Rolston clearly thinks that our scientific understanding of natural history leaves room for religious interpretation and explanation. "Biological explanation," he says, "is modestly incomplete" (1998: 429). "God," he writes, "is an explanatory dimension (a cause in the Aristotelian, though not the scientific sense) for which contemporary biology leaves ample space" (1999: 368).

Further, Rolston argues that religious explanation is not just possible but is required for the fullest account of the phenomena. "Laws are important," he writes, "but natural law is not the complete explanatory category for nature" (1994b: 56). "This phenomenal world, studied by science, urges us on a spiritual quest" (2003: 82). "The history of Earth . . . is a story of the achievement, conservation, and sharing of values. Earth is a fertile planet . . . This creative systemic process is profoundly but partially described by evolutionary theory . . . Such fecundity is not finally understood until seen as divine creativity" (1994b: 57). Rolston argues that when the need for an explanation of nature's creativity is acknowledge by the naturalistically minded, the "usual turn here is simply to conclude that nature is self-organizing (autopoiesis)" (1998: 429). But this is to respond to "the mysterious genesis of more out of less" with a label that looks scientific but is "really a sort of mystical chant over a miraculous universe" (1998: 429). Purely naturalistic accounts give you a nature with "overwhelming mystery" and insufficient explanation.

One might think that Rolston limits religion's role to explaining the meaning of what has happened in natural history: "God gives meaning to the world, which science is incompetent to evaluate" (1999: 368).[3] But Rolston also suggests that religion plays a causal explanatory and (perhaps even) a predictive role in the account of natural history. Although he grants that we don't "need God to do biology" (1998: 433), he points out that "the only forces biology is competent to detect are natural ones" (1998: 433) and he writes about both "biological and theological forces producing" forms of life (1994b: 49). He does say that "God does not intervene as a causal force in the world," but then immediately takes it back by saying "not at least of such kind as science can detect" (1999: 368). God is "a countercurrent to entropy, a sort of biogravity that lures life upward" (1998: 430). Although the role Rolston provides for God may seem to be a traditional "god of the gaps" position—whereby God is posited to fill in the gaps in our scientific knowledge of the world—Rolston appears to craft a causal role for God that is not open to falsification by future scientific discoveries.

Rolston argues that God helps to explain numerous dimensions of natural history. God is "the divine wellspring from which matter-energy bubbles up" (2003: 84) suggesting that Rolston sees God as the creator of matter-energy and thus as providing an answer to the question of why there is something rather than nothing. In accordance with "the anthropic principle," Rolston argues that the set up of the universe was "spectacularly fine-tuned for life. Hundreds of microphysical and astronomical phenomena . . . have to be almost exactly what they are if life is to be possible" (1994b: 52). For example, "the charges of electrons and protons, the strengths of the four binding forces . . . the expansion rate of universe, the proportion of hydrogen and helium" (1994b: 52). Rolston concludes that "Even before there is life we already get a pro-life universe" (1994b: 52). Rolston argues that the earth also has a set up that makes it a "prolife planet." It is the right distance from sun and has liquid water, a suitable atmosphere, the right elements, an energy source deep in the earth, and so on. He writes that "the Earth-system is a kind of cooking pot sufficient to make life probable, even inevitable" and claims that "God is somehow behind that set up" (1998: 430).

Additionally, Rolston argues that the diversity of life that has issued forth on this planet is so stupendous and progressive that it is indicative of God. "The story goes from zero to five million species in five billion years, passing though perhaps one billion species en route" (1994b: 53). Not only have there been an incredible number of kinds, but there has been advancement toward complexity:

> Once there was no smelling, swimming, hiding, defending a territory, gambling, making mistakes, or outsmarting a competitor. Once there were no eggs hatching, no mothers nursing young. Once there was no instinct, no conditioned learning. Once there was no pleasure, no pain. But all these phenomena appear, gradually, but eventually . . . Natural history suggests a creative genesis of life transmitted across long-continuing turnover of kinds, shared across a long history of struggling toward more diverse and complex forms of life (1998: 418, 422).

In short, Rolston sees Earth's biodiversity as "miraculous" and requiring explanation. "It would be a rather anomalous result if there had appeared novel kinds steadily over many millennia but only by drifting into them" (1998: 422). He concludes that "when such construction of valuable biodiversity has gone on for millennia, the epic suggests mysterious powers that signal the divine presence" (1998: 432).

More generally, Rolston thinks God helps to explain the "remarkable negentropic, cybernetic self-organizing that characterizes the life story on Earth" (1998: 432). Rolston asks "where the information comes from by which matter and energy becomes steadily so informed that there is, across evolutionary history, this river of life that flows uphill, this brilliant output from a beginning in mindless chaos" (1998: 433)? Quoting Dan Dennett, he wonders how "out of next to nothing the world we know and love created itself" (1998: 433). Without the posit of divine intervention we get "information floating in from nowhere" and "appearing ex 'nihilo' " (1998: 429, 430).

Rolston thinks science has not explained—and perhaps never will be able to explain or predict—certain dimensions of natural history. The miraculous, complex biodiverse earth is not even in the "possibility space" of the original elements: "The know-how, so to speak, to make salt is already in the sodium and chlorine, but the know-how to make hemoglobin molecules and lemurs is not secretly coded in carbon, hydrogen, and nitrogen" (1998: 428). "Given chemistry as premise, there is no deductive or inductive logic by which biology follows as a conclusion. . . . There is no covering law (such as natural selection), plus initial conditions (such as trilobites), from which one can deduce primates" (1994b: 53, 56). He argues that "Looking at a pool of amino acids and seeing dinosaurs or Homo sapiens in them is something like looking at a pile of alphabetical letters and seeing Hamlet. In fact, Hamlet is not lurking around a pile of A-Z's; such a play is not within their possibility space—not until Shakespeare comes around" (1998: 427). This gap in the scientific account of biodiversity's origins can be "congenially" filled by a divine author.[4] For Rolston, God can be seen as having input at crucial turning points in natural history:

> Anyone who takes the divine inspiration seriously will have to posit occasions . . . during which God provides information in the world . . . by some inspiration that first animates matter and energy into life, or launches replication and genetic coding . . . or moves life onto land, or invents animal societies or acquired learning, or endows life with mind, and inspires culture, ethics, religions, science (1998: 433).

Rolston explains the "mechanism" of God's intervention as follows: "Monotheists who take genesis seriously do not suppose a Deux ex machina that lifts organisms out of their environment, redesigns them, and reinserts them with an upgraded design. Rather they find a divine creativity that leads and lures along available routes of Earth history" (1994b: 55). Rolston argues that the openness and randomness inherent in nature allow for divine intervention without trace:

> God could also be in the details. That might be difficult to know, especially if God operated with the resolve to maximize the creaturely autonomy and integrity, to prompt rather than to command. Still, God could be slipping information into the world. That will not be detectable as any gap in or perforation of the natural order . . . Chance is an effective mask for divine action (1998: 431).

It is noteworthy that Rolston frequently sums up his view of God's role in explaining natural history somewhat equivocally: "That there is a divine presence underneath natural history becomes as plausible as that there is not" (1998: 434).

IMMANENT NATURE SPIRITUALITY AND ROLSTON'S ENVIRONMENTAL ETHICS

It is clear that a transcendent deity figures prominently in Rolston's account of natural history. I now return to the issue of the relation between Rolston's environmental ethics and religion. I argue that while an immanent nature spirituality fits harmoniously with his environmental ethics, the appeal to a supernatural and transcendent deity, in many ways, takes us away from it.

Nature spirituality is a religious attitude toward earthen nature that is naturalistic in the sense that it is the natural world itself that is the appropriate object of religious attitudes.[5] It is not committed to supernatural beings and has a thoroughly immanent conception of the godly. While apart from human culture (and rudimentary animal culture), nature is all there is, this spirituality finds that nature is precious, awesome, even enchanting. For nature spirituality, salvation is to be found in an altered understanding of and relation to the natural world, not in getting in touch with something beyond this world.

Naturalistic spirituality sees earth as a holy place and claims that if anything is sacred and worthy of reverence and devotion, it is this miraculous earthen community of life. It views the Earth as a magnificent, majestic place that should elicit our love, thanks, and humble respect. Because the earth produced us and sustains us and every other living being, it treats the earth as our creator and the ground of our being. For nature spirituality, earth warrants profound parental-like respect and should be cherished, revered and defended.

Nature spirituality fits well with Rolston's writings in environmental ethics. In fact, as I argue below, its fits far better than the more explicit transcendent religious approach found in his theological writings. Consider Rolston's outrage at the idea of humans managing planet earth. In writing as powerful as any in the Rolston corpus, he rebukes the "planetary managers" paradigm of the human relation to nature as a kind of sacrilege and idolatry that involves a misplaced trust in humans rather than in nature. The idea that the human "relationship to nature (is) one of engineering it for the better" involves, he says, "a danger of false gods, and an overweening trust in 'Science, Technology, and Industry' (that) may result in too little trust in 'Mother Earth' after all" (1994a: 226–27). Rolston asks:

> Is man the engineer in an un-engineered world? . . . There is ample inventive and engineering power in nature, which has built Earth and several billion species, keeping the whole machinery running with these species coming and going for billions of years. Who built the engineers, with their clever brains and hands, with which they propose now to manage the planet? . . . Hands are for managing, but hands are also for holding in loving care (1994a: 227–28).

This appeal for humility about humans' place in nature and the suggestion that we trust earthen nature given its multi-billion-year record of success fit nicely with nature spirituality and do not suggest—much less require—a transcendent deity.

Rolston argues that we are earthlings first and foremost and that our responsibility to the earth is the most fundamental of all. "What is principally to be protected (is) the value of life arising as a creative process on Earth" (1994a: 234). The earth's natural processes, he says "are the ground of your being, and we all owe the Earth system far more than we owe obedience to civic laws, the national history, or even the heritage of our cultural system" (1994a: 236). Rolston repeatedly claims that "Nature is grace (whatever more grace may also be)" (2003: 81). Grace, he explains, is a goodness that one has no cause to expect, a favor that one does not deserve. Nature has this type of surprising and uncalled for goodness. Life on earth is a spectacular gift and such a gift calls for ultimate responsibility.

While Rolston allows that we can doubt the appropriateness of directing our religious faculties toward the transcendent, he claims there can be no doubt that we should take a religious attitude toward the earth: "Perhaps there is a God above, and this marvelous earth creation may witness to that God, but meanwhile what cannot be doubted is that on this enthralling earth we live and move and have our being. . . . If there is any holy ground, this is it" (1994a: 236).

Rolston argues—as does Wendell Berry (1992: 103)—that what has happened on earth is far more "miraculous" than the religious miracles typically mentioned. The story of the development of life on earth is a story far more marvelous and spectacular, far more deserving of praise and wonder, far more of an account of a holy event than are stories about—say—Jesus walking on water. If the parting of the seas is a miracle that should elicit religious response, what about the existence of the seas in the first place?

> Moses thought that the burning bush, not consumed, was quite a miracle. We hardly believe any more in that sort of supernatural miracle; science has made such stories incredible. But what has it left instead? A self-organizing photosynthesis driving a life synthesis that has burned for millennia, life as a strange fire that outlasts the sticks that feed it. . . . This is hardly a phenomenon less marvelous even if we no longer want to say that it is miraculous. . . . To go back to the miracle that Moses saw, a bush that burned briefly without being consumed, would be to return to something several orders of magnitude less spectacular (1996: 398–99).

Rolston thinks the idea that "science chases out the holy" is superficial (1996: 387) and he rejects the secular/sacred contrast "as though anything sacred must be of some other, supernatural realm, not of this present world" (1996: 387). He asks, "What if the secular world proves to be pretty spectacular stuff? What if we lose our confidence in the supernatural, only to find it replaced by increasing confidence that nature is super, superb, mysteriously animated and inspirited?" (1996: 387). Rolston argues that biology and religion have "increasingly joined in recent years . . . in admiration for this marvelous planet that we inhabit" (1996: 410). "That respect," he says, "sooner or later passes over to a reverence. . . . *Homo sapiens* reaches a responsibility that assumes spiritual dimensions" (1996: 410). Such reverence, he suggests, is not dependent on a transcendent deity. He writes: "And if one cannot get clear about God, there is ample and urgent call to reverence the Earth" (1996: 408).

Rolston's writings in environmental ethics make a powerful case that a religious attitude toward nature as a sacred and holy place is important for proper acknowledgment of the intense value of the earth, for underlining the seriousness of our obligations to preserve it, for an appropriate condemnation of its destruction (as a type of sacrilege), and more generally for developing the proper virtues (such as humility) in our relationship with it. Rolston's environmental ethics lends significant support to an immanent nature spirituality and is itself most consistently interpreted as involving such a religious dimension.

Some will object to naturalistic spirituality because they think the religious attitudes involved require an intentional or conscious being as an appropriate object, something nature spirituality purports to do without. Talk of the earth as our creator, or of being thankful to the earth, would seem to treat the earth as a conscious intentional being. If life on earth is a gift, it might be argued, this requires a giver to whom we can be thankful. How can one revere and be devoted to something that is not kindly or lovingly disposed towards us?

Some religious attitudes are, however, perfectly legitimate when directed at non-intentional, non-conscious entities. That this spectacular planet and the natural history that underlies it were unplanned and not created by an intelligent designer does not lessen our ability or duty to love it or defend it. Nor does it entail that we should be any less thankful. It is truc that such attitudes will differ in some of their details when they are not directed at a personal, loving deity. Reverence for earth is not likely to be the identical type of attitude as reverence for a personal God. Love and thankfulness when directed at beings that cannot be aware of these attitudes are different than are love and thankfulness directed at a person. In the latter case, one expects some kind of response; one expects that these attitudes will make a difference to their object. The earth will not respond to those who love it and are thankful for it any differently than to those who fear and denigrate it. Nevertheless, for nature spirituality, these are appropriate attitudes and they make a difference in the lives of those who are so disposed and in how they respond to and treat the earth.

Rolston distinguishes between nature spirituality—a view that is close to what he calls "soft naturalism" in *Science and Religion* (1987: 253)—and the supernaturalism he thinks is required for an adequate explanation of natural history. At one point he describes how a transcendent religion differs from immanent religions thus: "In contrast with the surrounding faiths from which biblical faith emerged, the natural world is disenchanted; it is neither God, nor is it full of gods, but it remains sacred, a sacrament of God. Although nature is an incomplete revelation of God's presence, it remains a mysterious sign of divine power" (1991: 5). Although he has provided powerful arguments and language supporting religious naturalism (what I have called "nature spirituality), Rolston thinks it inadequate: "My problem with that is you get a nature with overwhelming mystery and minuscule sense of explanation" (2004). When he criticizes soft naturalism in *Science and Religion*, he claims it is unable "to give much account of the central creativity" in nature and becomes a "mystic chant over an unintelligible Universe" (1987: 256, 257).

TROUBLES WITH TRANSCENDENCE

I have argued that while Rolston's environmental ethics can stand on its own apart from religion, it is congenially interpreted as involving an immanent nature spirituality.

I now turn to my concern that Rolston's posit of a transcendent deity to account for certain aspects of natural history undermines some of the ideas and attitudes his environmental ethic has so powerfully defended.

At the most general level, I worry that the appeal to a transcendent God is likely to take the focus off valuing the earth and place it instead on valuing the transcendent creator. Why revere the proximate cause when we can revere the ultimate cause, especially when only the latter appreciates our attitude? If there is a God above and this marvelous earth witnesses that God, then should we not be valuing the earth instrumentally as witness to the transcendent—at least in the first instance? While it is true that we could intrinsically value both the earth and its creator, the transcendent creator account seems in tension with Rolston's insistence on the fundamental importance of nature's intrinsic value.

In addition to shifting the focus off valuing of the earth to valuing God, a religious attitude toward earthen nature is threatened by Rolston's appeal to transcendent deity. If the holiness of earth is a function of its manifestation of a divine presence, then veneration of the earth would seem a kind of idolatry. If—as Benzoni (1996: 347) interprets Rolston— "God is the divine parent, coaxing forth ever more diversity and complexity through the available pathways of nature," then fundamentally the glory is God's rather than the creation's. This does not fit happily with Rolston's insistence on the incredibly intense, fundamental, and ultimately religious value of the earth itself.

The transcendent account of natural history would also seem to diminish our appreciation of it. By explaining the mysteries he identifies in natural history, Rolston takes away from its miraculous nature. Earthen nature's prolific, prolife story is more incredible if it exists without conscious design or "orchestration" (2004), to use a word Rolston prefers. If God planned the course of natural history, "leading and luring it upward," then its origin and development becomes less astounding and less awe inspiring. A central part of why the natural world is so amazing is that it has not been intentionally produced. Gene Hargrove makes the same point about natural beauty: "Our aesthetic admiration and appreciation for natural beauty is an appreciation of the achievement of complex form that is entirely unplanned. It is in fact because it is unplanned and independent of human involvement that the achievement is so amazing, wonderful and delightful" (1994: 183). On the transcendent God account, the richness of the world we have inherited becomes less of a fortuitous and uncalled for gift and more of an expected result. One would expect this much from a transcendent creator, at least one who has the traditional characteristics of a monotheistic god. In fact, one might expect more.[6]

Furthermore, the transcendent creator account of earthen nature is in serious tension with Rolston's emphasis on valuing wild, spontaneous nature. To the extent that systemic nature is a product of mind and purpose, to that extent it is artifactual. In his theological writings, Rolston has argued that in crucial ways and in the dimensions that science hasn't or can't explain, natural history has been "orchestrated." God directs the play and makes sure the story is ultimately a good one. Thus, in fundamental ways, nature is more of a divine artifact, than a spontaneous, unmanaged performance.

Additionally, the value of nature as other is compromised. While it is true that a divinely "inspired" nature remains other in the sense of having a profoundly nonhuman origin and fundamental reality, there is an important respect in which, on this account, nature becomes more familiar and less other. In many ways, a transcendent deity is more like humans than is nonhuman nature. Like humans—but unlike nonhuman natural processes—such a god is purposeful, conscious, and personal. Just as human presence in wild nature diminishes its wild otherness, so too, perhaps, would God's presence: There is another person there, or at least evidence of—or the workings of—another person.

Rolston has resources with which to respond to these objections. Most importantly, Rolston's account of how God interacts with nature allows for significant spontaneity, autonomy, and creativity in the creation and its creatures. Rolston describes God's involvement with nature as "leading," "luring," and "urging" the creation onward and as "prompting" rather than commanding. He explicitly avoids conceptualizing God's creative activity toward the world as one of design (1999: 369) and argues that God's involvement is an "empowering permission that places productive autonomy in the creation" (1999: 370). Rolston argues that "there is enormous self-creativity in the creatures, but the context and ambience of that is the possibility space opened up by divine inspiration" (2004). "In Earth's wildness," he says there is "a complex mixture of authority and autonomy, a divine imperative that there be communities (ecosystems) of spontaneous and autonomous ('wild') creatures" (1991: 7). He accepts a "loose teleology, a soft concept of creation, and yet one that permits genuine, though not ultimate, integrity and autonomy in the creatures" (1998: 431).

Rolston might appeal to a parent-child analogy to argue that divine origin does not lessen the value of the natural world. That parents produce and then empower their children to lead productive lives neither diminishes the children's autonomy nor devalues their accomplishments. So too, God's production and empowerment of the creation and its creatures should not undermine their autonomy or devalue their accomplishments. Nor should it weaken our appreciation or diminish our valuation of them.[7] Because Rolston sees God's involvement in natural history as empowerment of a genuinely autonomous creation, his account does not take away all the wild spontaneity in nature and it leaves much that we can intensely value for its own sake. For example, tigers are no less intrinsically valuable or awe inspiring (as tigers) because God "opened up the possibility space" for such a being to evolve.

Nevertheless, Rolston's account of a transcendent deity's activity in natural history changes the fundamental character of nature from an amazing, spontaneous, progressive diversity that is of fundamental value to a nature whose substantial achievements have been orchestrated by something beyond nature that is the fundamental value. While there remains localized spontaneity and achievement in nature to value for its own sake, the fundamental creativity of the system, the "systemic value" of nature that Rolston has argued is most to be cherished, is due to the activity of the deity. The most miraculous facts of natural history—that life persists in the midst of its perpetual perishing, that life fights entropy and increases in diversity and complexity, that there is matter-energy when there might not have been, and that the universe is suited for life—are all now manifestations of God's grace. I do think this shifts the focus from valuing nature to valuing God, places the ultimate value in God rather than nature, takes away the amazement that purposeless spontaneity could have produced such valuable products, and significantly shifts our conception of nature from a wild otherness to an artificed creation, lovingly given.

CONCLUSION

If I am right about these concerns, then Rolston's appeal to a transcendent deity in his theological writings is not only unneeded for his environmental ethic, but is in serious tension with it. In describing his seminary and religious studies training, Rolston writes: "The trouble was I had to fight theology to love nature" (1991: 2). I guess I am here suggesting that he continue that fight. Rolston has provided strong support for an immanent nature spirituality and such a spirituality—though it is not needed for a Rolstonian environmental ethic and defense of nature—fits well with it. The posit of supernatural agency to account for what our science of natural history leaves unexplained not only raises significant explanatory problems of its own, but causes trouble for what I take to be the most compelling environmental ethic in the field.[8]

NOTES

[1] No environmental philosopher knows the natural sciences better than does Rolston. Scientific description pervades Rolston's writing, and appropriately so, for we can't properly appreciate, protect and love what we don't understand, and the natural sciences are central to our understanding of nature. Rolston significantly "scientizes" his aesthetics, ethics and religion of nature. The title of one of Rolston's papers (1995) asks: "Does the Aesthetic Appreciation of Landscapes Need To Be Science-Based?" Rolston's answer is "yes," at least for the deepest sort of aesthetic appreciation. In the end, Rolston also thinks that the deepest types of environmental ethic and natural theology also need to be science-based.

[2] Footnote #1 in Benzoni's paper suggests such an alternative.

[3] At times, Rolston (1987: 311–313) pushes hard on the idea that religion explains the meaning of events whereas science explains the causes.

[4] Rolston's argument here raises a host of questions: Does he reject causal determinism? Is he saying that even an all knowing being could not predict this biologically rich world given full knowledge of initial conditions and the relevant laws, perhaps because the free will of God is an additional factor? Is this why natural selection is not a sufficient explanation for how we get primates from trilobites? Does Rolston reject the possibility that there can be emergent properties in the course of natural history, at least none without a causal origin in God?

[5] It need not be naturalistic in the sense of embracing a reductionist account of religion in terms of the categories of the natural (or even social) sciences.

[6] Thus a theodicy for natural evil becomes paramount. Rolston clearly takes up this challenge, as Ouderkirk shows.

[7] While there is much analogous here, there is also significant disanalogy. Typically (and ultimately) parents let go of their children, who really do then lead their own lives. In contrast, while God lets his autonomous creatures make their choices, God continually empowers, inspires and orchestrates the context of such choices.

[8] I thank Phil Cafaro for the chance to speak at an American Philosophical Association sponsored session on Rolston's work. I also thank Holmes Rolston for comments on a draft of this chapter.

REFERENCES

Benzoni, Francisco. 1996. "Rolston's theological ethic," *Environmental Ethics* 18 (4), pp. 339–52.

Berry, Wendell. 1992. "Christianity and the survival of creation," in *Sex, Economy, Freedom & Community*. New York: Pantheon Books, pp. 93–116.

Hargrove, Eugene. 1994. "The paradox of humanity: two views of biodiversity and landscapes," in Ke Chung Kim and Robert D. Weaver, eds., *Biodiversity and Landscapes*. New York: Cambridge University Press, pp. 173–86.

Ostling, Richard. 2003. "Colorado pioneer in environmental ethics wins religion prize worth more than $1 Million," Associated Press (March 19).

Ouderkirk, Wayne. 1999. "Can nature be evil? Rolston, disvalue, and theodicy," *Environmental Ethics* 21 (2), pp. 135–50.

Rolston, Holmes, III. 1983. "Values gone wild," *Inquiry* 26, pp. 181–207.

Rolston, Holmes, III. 1987. *Science and Religion: A Critical Survey*. New York: Random House.

Rolston Holmes, III. 1988. *Environmental Ethics*. Philadelphia, PA: Temple University Press.

Rolston, Holmes, III. 1991. "Respect for life: Christians, creation, and environmental ethics," *CTNS Bulletin: The Center for Theology and the Natural Sciences*. 11 (2), pp. 1–8.

Rolston, Holmes, III. 1992. "Disvalues in nature," *Monist* 75 (2), pp. 250–278.

Rolston, Holmes, III. 1994a. *Conserving Natural Value*. New York: Columbia University Press.

Rolston, Holmes, III. 1994b. "Creation: God and endangered species," in Ke Chung Kim and Robert D. Weaver, eds., *Biodiversity and Landscapes*. New York: Cambridge University Press, pp. 47–59.

Rolston, Holmes, III. 1995. "Does Aesthetic Appreciation of Landscapes Need to be Science-Based?" *British Journal of Aesthetics* 35 (4), pp. 374–386.

Rolston, Holmes, III. 1996. "Scientific inquiry" (Secular scientific spirituality) in Peter H. Van Ness, ed., *Spirituality and the Secular Quest*. New York: Crossroad, pp. 387–413.

Rolston, Holmes III. 1998. "Evolutionary history and divine presence," *Theology Today* 55, pp. 415–34.

Rolston, Holmes III. 1999. *Genes, Genesis, and God*. New York: Cambridge University Press.

Rolston, Holmes, III. 2003. "Naturalizing and systematizing evil," in Willem B. Drees, ed., *Is Nature Every Evil? Religion, Science and Value*. London: Routledge, pp. 67–86.

Rolston, Holmes, III 2004. Rolston's comments on a draft of this chapter.

LISA SIDERIS

6. WRITING STRAIGHT WITH CROOKED LINES: HOLMES ROLSTON'S ECOLOGICAL THEOLOGY AND THEODICY

> Improvement makes straight roads, but the crooked roads, without Improvement, are roads of Genius.—William Blake

In Barbara Kingsolver's novel *The Poisonwood Bible*, a southern Baptist missionary named Nathan Price brings his family to the politically volatile Belgian Congo of the 1960s, expecting to convert the Congolese to the unimpeachable logic of Christianity. Soon after their arrival, Nathan and his young daughter Leah survey their newly planted vegetable garden. Intended as a model for their neighbors to emulate, the garden is the product of good Kentucky seeds, honest Christian labor, and strange African soil. The vines stretch endlessly toward the hot sun, apparently thriving, but day after day they refuse to set fruit. As they ponder this horticultural mystery, their conversation turns to the size of heaven, and whether it is capacious enough to accommodate all who are saved. The minister confidently assures his daughter that there will always be room for the righteous, whom God delivers from every affliction. "But you know, Leah," he goes on to say, "sometimes He doesn't deliver us out of our hardships but through them." Leah has never considered the possibility that God's plan might include the suffering of the righteous and she finds this new piece of theology as disorienting as their funereal garden. It is her first inkling that hard work and sacrifice are not necessarily compensated, at least not in the ways we expect.

Christians have always struggled to find redemptive value in sacrifice and suffering, and ecological theologians are no exception. The need to affirm that deliverance from suffering, perhaps even a heavenly reward, awaits all righteous beings—human and nonhuman alike—is sometimes very strong in ecotheology. In much of this literature, as in Kingsolver's story, a desire to see all suffering compensated is intertwined with a longing to see nature itself improved or recreated according to an imported and transplanted ideal, to bring it into closer conformity with deeply held convictions about the way the world ought to work. The ecological imperatives that issue from such expectations sometimes fail to find purchase in the natural world we actually inhabit. When our expectations are not met, whom should we blame—nature? God? ourselves?

Just as the minister's remarks seem to Leah an insufficient response to the problem of suffering and the meaning of sacrifice, Holmes Rolston's natural theology and theodicy may be unsettling to many ecotheologians. The God of Rolston's nature does not liberate organisms from conditions of pain and deprivation, nor does God compensate individual loss. The (very real) existence of these "disvalues" is not an argument against God, in Rolston's view. He does not believe that wholesale, eschatological redemption of nature is necessary in order to realign nature with the "true"

C.J. Preston and W. Ouderkirk (eds.), Nature, Value, Duty, 77–101.

or "ultimate" plan of God. The natural world is godly—cruciform and deiform—Rolston claims, precisely because of struggle and suffering, not in spite of it. Rolston seriously considers the possibility that the natural world, in its given ordering, rather than its past or future perfection, may reflect the will and ordering of God, even while elements of it remain troubling, or frustrating, or even improvident, from our perspective.

In fact, as Mark Wynn has noted, Rolston's theology of nature is open to criticism from theologians who may object that a "thoroughgoing affirmation of the natural world is inconsistent with Christian teaching, because it does not take seriously enough the doctrine of the Fall, or does not leave open the possibility that the natural world will be subject to radical improvement at the eschaton."[1] Convinced that God is essentially good and praiseworthy, many ecotheologians find fault instead with natural processes. Restoring nature to its original, pre-Fall goodness brings it back in line with the intentions of the creator and aligns us more closely to the will of God.

Rolston explicitly rejects this sort of redemptive environmentalism and he is wary of ecological ethics that are inordinately preoccupied with "saving" (theologically or biologically) individual organisms, over and above concern for natural systems and natural processes. As Wynn suggests, Rolston's biological realism may not sit well with the soteriological and eschatological orientation of much of Christian ecotheology. Yet, from the standpoint of some Darwinians—and perhaps Darwin himself, were he to encounter it—Rolston's natural theology might well appear too sanguine, offering more comfort than is warranted in light of nature's cruelty, randomness, and wastefulness.

Darwin, like ecotheologians, was troubled by the harshness of natural processes and evolution's apparent indifference to the fate of individuals. But, in general, his response was not so much to critique nature but to critique God. In reasoning through the conflict between widespread natural suffering and belief in a benevolent God, Darwin took some initial, tentative steps in the direction of Rolston's theology of nature, before retreating once and for all to atheism.[2] In the end, Darwin retained his commitment to natural selection and turned against God: "disbelief crept over me at a very slow rate but was at last complete," he reports in his autobiography. "I can indeed hardly see how anyone ought to wish Christianity to be true."[3] In the end, as Donald Fleming has argued, Darwin "resolved not to be an accomplice in the evil of the world by assenting to God's dominion." Darwin yearned for a "better God than God," but never found him in nature.[4] In light of their understanding of God, on the other hand, many Christian ecotheologians appear to have turned against nature, longing for a "better" natural world than the one we have.

Whether or not they would accept it, Rolston's gift to both Darwinians and Christians—and especially to Christian Darwinians—is his refusal to censure nature or God. Rolston's account of systemic value and cruciform nature addresses many of the concerns and objections raised by Darwin and ecotheologians. Though not perfect in every respect, Rolston's reconciliation of evolutionary science, religion, and environmental ethics remains the most comprehensive and convincing proposal offered thus far. As I will argue, Rolston's key to harmonizing natural selection and

Christianity—while still offering a realistic environmental ethic—lies in part in his uncompromising stance on what has proven to be a particularly, and perennially, thorny issue: the status of the individual organism, and its suffering, within the overarching processes of evolution. Rolston faces this fact of nature squarely without repudiating God or resorting to eschatological improvement of nature. In order to appreciate Rolston's accomplishment, we need to consider a few alternative solutions to the problem of reconciling evolution and theism. We begin by examining the views of the best spokesperson for a Darwinian worldview, Charles Darwin himself.

DARWIN'S CRITIQUE OF GOD

The apparent expendability of individual organisms in natural processes has been one of the most intractable problems for theistic evolutionists. This brutal fact of nature was a compelling reason behind Darwin's creeping atheism, but it was also one of the best pieces of evidence for his theory of natural selection. Portions of Darwin's critique of Christianity were for many years omitted from his autobiography at the request of his wife Emma, who deemed certain passages too "raw" for public consumption.[5] Darwin somewhat wistfully remembers his naively "orthodox" views of God's creation while a young naturalist on board the Beagle. "The old argument of design in nature, as given by Paley," he recalls, sufficed in those days to explain the natural ordering.[6] But with time and careful observation, he came to believe that there is "no more design in the variability of organic beings and in the action of natural selection, than in the course which the wind blows."[7] Nature's randomness in and of itself casts some doubt on the existence of divine providence, but the deal-breaker for Darwin was the presence of so much suffering in the animal world. Such suffering makes sense in light of natural selection, Darwin argues, which after all is "not perfect in its action, but tends only to render each species as successful as possible in the battle for life with other species, in wonderfully complex and changing circumstances."[8] But suffering was not compatible with belief in the Christian God because it "revolts our understanding to suppose that his benevolence is not unbounded, for what advantage can there be in the sufferings of millions of the lower animals throughout almost endless time?"[9]

Darwin was of course aware of theodicies that posit "moral improvement" or "character building" as a partial compensation for suffering. But these accounts ring hollow in the nonhuman world where we find "sentient beings [who] often suffer greatly without any moral improvement."[10] The prevalence of seemingly pointless suffering in nature as evidence "against the existence of an intelligent first cause" is too strong to ignore, Darwin concludes.[11]

Though Darwin presents his findings as an open-and-shut case against God, some of his insights suggest that he probed the issue more deeply, and it is here that he adumbrates aspects of Rolston's proposal. Darwin and Rolston, as well as myriad other evolutionists, have wrestled with the problem of "counterproductive pain" in evolution—pain that seems excessive, serves no obvious purpose, or is otherwise maladaptive. Darwin's recognition of this category of counterproductive or dysfunctional pain suggests that there is also pain and suffering that is in some sense productive or functional. Indeed, despite Darwin's rejection of theism on the grounds of widespread natural suffering, he suggests that counterproductive pain is in fact anomalous, the exception that proves natural selection's power, in general, to utilize and even streamline pain. Here Darwin discerns an opening toward an evolutionary theodicy that he never completely steps into. His take on the role of pain (and pleasure) in natural selection is as follows.

> If all the individuals of any species were habitually to suffer to an extreme degree they would neglect to propagate their kind; but we have no reason to believe that this has ever or at least often occurred. . . . Now an animal may be led to pursue that course of action which is most beneficial to the species by suffering, such as pain, hunger, thirst, and fear—or by pleasure, as in eating and drinking and in the propagation of the species, &c. or by both means combined, as in the search for food. But pain and suffering of any kind, if long continued, causes depression and lessens the power of action; yet is well adapted to make a creature guard itself against any great or sudden evil.[12]

Darwin's reflections are familiar terrain to anyone who has thought at length about the role of pain in nature, but it was relatively uncharted territory in his day. As he notes, pain may be functional on different levels. If not so extreme or continuous as to cause "depression," it can serve as a warning system that motivates the organism to act, to "guard itself" against harms. While it may be inaccurate to say that the organism undergoes moral improvement through pain, it may very well learn how better to respond to it, or avoid it in the future. Benefits may accrue to the species as a whole of which the suffering individual is a member. Those individuals who respond "well" to pain (whatever that entails in the particular environment) may survive and pass on their genes. Sometimes only the species benefits: it is improved by weeding out of less fit members, even while no benefits accrue to the individual (which dies and/or does not "propagate its kind.") Neither of these cases is dysfunctional, though the second case is clearly more tragic from the standpoint of the individual organism. Yet the existence of predation, usually considered a primary source of suffering, may also mitigate it. For organisms weakened by injury, hunger, or disease, death from predation may be preferable to a lingering death and we may thus "console ourselves," as Darwin notes in the *Origin of Species*, that "death is generally prompt" in the war of nature.[13] Darwin also recognizes that pain and pleasure are intertwined: the unpleasant sensation of hunger drives the organism to procure food which is pleasant to consume (and in turn, painful, even if only briefly, for sentient prey organisms). But rarely if at all, Darwin argues, do we find pain that stands alone, with no resultant or commingled pleasure or benefit; pain that is dysfunctional in the sense that no advantage or improvement is evident at either the individual or the species level is rare-to-nonexistent. Over time, suffering that constantly afflicts "all the individuals of any species" is not favored in the evolutionary process.

Darwin would seem to have largely answered his own question of what possible "advantage" there can be in the suffering of millions of animals (though apparently that answer is not sufficient for reconciling natural selection with the existence of God). In fact, he argues that we find in nature a "generally beneficent arrangement" which "harmonises well with effects which we might expect from natural selection."[14] But if nature is generally beneficent, as Darwin claims, then why is atheism alone compatible with natural selection? The sticking point for Darwin seems to be that this generally good system is nevertheless sustained at the expense of the suffering individual who may receive none of its benefits. Even in those cases (which he feels constitute the majority) where suffering produces some positive benefit somewhere, the problem remains that a God whose goodness is "not unbounded" could surely have devised a better scheme. Jay McDaniel, an ecotheologian who has given some thought to this issue, puts it this way: "God's love for nature cannot be limited to the whole of nature at the expense of nature's parts . . . If the love of God is anything like that of an all-loving parent, God would want to redeem the parts as well as the whole."[15] Darwin does not deny that certain "improvements" emerge from suffering but he remains uneasy with their inequitable and impersonal distribution. Add to that the weight of those anomalously rare but deeply troubling cases of truly dysfunctional pain, and the scales tip toward atheism.

Still, the natural world ought not to be faulted—it is after all merely the result of "fixed laws" that, however blindly and clumsily, manage to produce "endlessly beautiful adaptations."[16] As Darwin concludes in the famous, final passage of the *Origin of Species*, there is undeniable "grandeur" in an evolutionary view of life. But the same cannot be said of evolution's creator.

PEACEABLE KINGDOMS AND COMMUNITY ETHICS

Ironically, this preoccupation with uncompensated individual suffering is one that Darwin shares with many ecotheologians who otherwise tend to ignore the implications of Darwinism. Rather than critique God, they view nature's apparent cruelty as a symptom of the Fall which corrupted nature's original goodness.[17] As I have argued elsewhere, the environmental ethic they propose remains incompatible with natural processes, as it seeks to guide nature toward conditions that are biologically impossible.[18]

Newcomers to the field might be surprised to discover just how rampant the belief in nature's inadequacy—its fallenness—is in ecotheology, particularly given recent efforts by many Christian environmentalists to attain greater scientific accuracy in their writing on nature.[19] For example, the motif of the Peaceable Kingdom, in which predator and prey exist in harmony and nature is once more infused with a tranquil abundance, recurs throughout ecotheological literature. Above all, ecotheologians want to join the good of the individual and the good of the natural "community" in a way that dissolves all conflicts between parts and wholes. The "problem" of predation is the most obvious, though not the only, form of conflict between the individual and the natural system in which it is embedded.

For some of the most prominent ecotheologians, Biblical stories depicting paradisiacal nature serve as a touchstone for environmental objectives and obligations toward nonhuman life. Rosemary Radford Ruether hopes for a final restoration of nature that will usher in "right relations" among all creatures, thus "healing nature's enmity." The book of Isaiah, she reminds us, promises that "even the carnivorous conflict between animals will be overcome in the Peaceable Kingdom."[20] Sallie McFague offers a blueprint for Christian environmentalism—a "subject-subjects model"—that draws inspiration from Biblical stories "in which the lion and the lamb, the child and the snake, lie down together; where there is food for all; where neither people nor animals are destroying one another."[21] Jürgen Moltmann awaits the time of *creatio nova* when the spirit of God "drives out the forces of the negative, and therefore also banishes fear and the struggle for existence from creation."[22] A "peaceable kingdom of shalom and ecological harmony" in which "predatorial behavior will no longer characterize human and non-human relations" captures Michael Northcott's vision of God's will for creation.[23] Charles Birch also anticipates a time when "paradise is regained, and everyone not only goes back to a non-meat diet, but the friendliest relations subsist between all species. The wolf shall dwell with the lamb, the leopard shall lie down with the kid, and a little child shall lead them!"[24] Even secular environmental ethicists such as Tom Regan have at times advocated animal rights as an important step in the "journey back (or forward) to Eden [and] God's original hopes for and plans in creation," though here the issue is more one of humans eating other animals than nonhuman animals eating each other.[25]

Such censuring of nature, and the attendant hope for its transformation, is motivated by ecotheology's discomfort with natural communities that seem altogether "too casual about pain, suffering and death."[26] As ecotheologian Larry Rasmussen observes, predation in particular is "not a pattern of morality we praise and advocate" for our own communities—at least, he quips, "not on our better days."[27] If, as many Christian environmentalists argue, we ought to extend to nature (or, alternatively, derive from it) a community ethic, what are we to make of these conflicts? Surely God would have us facilitate the flourishing of individuals as well as their communities. In nature, as in culture, we should dismantle destructive hierarchies, banish all forms of oppression and domination of the weak by the strong. Physical health must be recognized as a "basic right" in natural communities, just as it is in human communities.[28] Yet a natural community of this sort could come into existence only if natural selection were somehow to cease. And in fact, as we shall see, some theologians call for nothing less than an end to the whole process of evolution.

Since, in many cases, these basic requirements of health and wholeness are not being met in nature (presumably due to its corrupted state), it falls to ecotheology to ensure that they are. In taking on this responsibility, some ecotheologians suggest, we not only meet the needs of our fellow natural "neighbors" but also bring about the sort of life in community intended by God. Ecotheologians discern God's intentions in Biblical stories such as the Garden of Eden and the book of Isaiah, but also in the example of Jesus. Extending Jesus' radically inclusive community ethic to our nonhuman neighbors implies caring for the "new poor"—nature's rejected, despised, and oppressed beings. It means "healing the wounds of nature and feeding its starving creatures" just as a Christian community would focus on "feeding and healing its needy human beings."[29] However, far less attention is given to the possibility that we might discern God's will in the natural ordering and processes—including natural selection—evident in the world we actually inhabit.

This possibility receives sustained attention in Rolston's ecotheology. His discussion of the desire to redeem nature's suffering reads as a direct rebuttal to many ecotheologians' longing for a world remade; at the same time, his account of cruciform nature (augmented in helpful ways by Robin Attfield, as I will suggest) locates possibilities for rapprochement between evolution and theism that may have eluded Darwin and other evolutionists. These features of Rolston's thought are held together by a general theory of systemic and intrinsic value that gives priority to natural processes, to collective entities such as ecosystems, and to the species line which is the lifeform (form of life) of which the individual is a valuable but subordinate representative. With ecotheologians' call for a restored, paradisiacal creation still echoing in our ears, we turn now to Rolston's response.

REDEMPTIVE ENVIRONMENTALISM

Rolston argues that the real source of conflict between biology and the Bible is not, as is often assumed, the creation story of six days. Rather, the problem lies in the idea of a Fall when a "once-paradisiacal nature becomes recalcitrant as punishment for human sin."[30] Rolston is cognizant, as was Darwin, that the Fall and the account of evil it assumes, makes no sense in light of natural history which suggests that struggle and suffering were features extant in nature long before humans evolved. Biblical passages implying that nature is fallen and in need of restoration ought not to be "taken to suggest that existing wildlands are fallen, nor can they be interpreted in terms of redemptive wildlands management."[31] Such hopes for nature can easily distort our understanding of what it means to restore nature. A "peaceable natural kingdom, where the lion lies down with the lamb, is sometimes used as a symbol of fulfillment in the Promised Land," Rolston notes, but in determining ethical guidelines for nature, environmentalists must remember that this is a "cultural metaphor and cannot be interpreted in censure of natural history."[32] From Rolston's perspective, the wildest, most untouched places on earth are less, rather than more, tainted by sin than cultivated areas. Yet it is precisely in wild nature—nature reddest in tooth and claw—that some theologians discern the clearest symptoms of nature's fall. Perhaps, Rolston suggests, "theologians will not be able to figure out what they believe until they have studied biology."[33]

Rolston has serious reservations regarding the proposal that we minister to nature with the ethics of Jesus in mind. The feeding of hungry persons advocated in certain biblical passages such as Matthew 25 does not require feeding wild animals, Rolston argues. "Pointless suffering in culture is a bad thing and ought to be removed where possible." But a similar ethic of compassion toward the suffering individual is not appropriate in nature because "pain in wild nature is not entirely analogous to pain in an industrial, agricultural, and medically skilled culture."[34] Humans are no longer buffeted by the forces of natural selection as our wild "neighbors" are; for better or for worse, we have largely succeeded in removing ourselves—and our domesticated animals—from these refining fires. Natural suffering, however, "remains in the context of natural selection; it is pain instrumental to the survival of the species," Rolston notes, "even after it becomes no longer in the interest of the pained individual."[35] Redemption, then, cannot mean saving individual organisms from conditions of natural suffering, which would only undermine evolution's way of adapting species to changing environments, and possibly create more suffering.

Yet there is a sense in which redemption of nature is meaningful in environmental ethics. Environmental conservation might be seen as a kind of redemption, Rolston argues, ensuring that natural goods are saved and passed on. Regenerative and restorative powers of life, he suggests, might also be interpreted as a form of redemption, both at systemic levels such as an ecosystem, and at cellular levels within each organism. Redemption understood in this way is a form of "biotic self-renewal," in Aldo Leopold's phrase, an indication that the system is functioning well. Redemption as regeneration may thus be a sign of health. Rolston also acknowledges, however, that redemption qua regeneration may be linked to suffering, but not in the sense that redemption is a remedy for sin and suffering. Rather, it is more accurate to say that pain is sometimes a prelude to, an inauguration of, redemptive, regenerative processes, as with the birth process.[36] Environmental destruction resulting from human carelessness and short-sightedness might be termed "sinful." But nature itself is not sinful in the sense that its present structure—trophic relations between predator and prey, for example—emerged as a postlapsarian phenomenon. On the contrary, those trophic interactions are the very means by which nature continually regenerates and transforms itself. "Life overleaps any particular parts and is reincarnated through passing material sequences." In this process, "corporate and conformational biological identity persists where individual and physical identity is transient."[37] Along with this form of biological persistence there is also a persistence of value—a capturing and smearing out of the individual's value in the larger system. Beyond this, however, there is no further redemption or reward for the individual's sacrifice, no eschatological resurrection of its individual identity.

From a theological standpoint, then, the question we ought to pose about the existence of "disvalues" such as suffering and death is "not whether Earth is a well-designed paradise for all its inhabitants, nor whether it was a former paradise from which humans were anciently expelled." Rather, we should ask whether nature exhibits "significant suffering through to something higher."[38] But what does Rolston mean by "something higher"? Clearly, he is not guaranteeing that a more perfect, compensatory order awaits each sufferer, human and animal alike. "I am not sure that I know what that means," Rolston confesses.[39] Nor is he suggesting that suffering organisms strengthen their moral fiber through experiences of hardship, though certainly they learn from painful experiences. He accepts that life is a "passion play" of sorts, one in which lifeforms are perpetually perishing. The death of organisms brings with it a passing over whereby the death itself contributes to, and is part of, other processes. And "what," Rolston asks, "is so disvaluable about that?"[40]

However, a good deal hinges on what Rolston means by suffering through to something higher and many of his critics have found this phrase, and others like it, problematic on both theological and biological grounds. I am in partial agreement with these critics, though perhaps for different reasons. What makes such phrases problematic, in my view, is their close resemblance to concepts central to process thought which tends (in its ecotheological manifestations) to view the value of non-human life through an anthropocentric, pan-experiential and pan-psychic lens, á la Alfred North Whitehead. Though Rolston objects to the extent to which process thought decouples God from natural processes, he sometimes uses similar language to describe God's action in "luring" nature upslope; likewise, as I will argue, his discussion of disvalues partially resonates with process views of life as an inseparable mix of pleasure and pain or, in process language, "enjoyment and discord." Rolston largely affirms the open-endedness of evolutionary development described in process thought—the possibilities and novelties, and the actualizing of potentials achieved through new evolutionary pathways. In fact, while Rolston prefers to describe nature as "storied history" rather than process, he sounds very much like process-oriented ecotheologians such as Charles Birch and John Cobb, in maintaining that a degree of unpredictability and randomness permits the ongoing production of projective and emergent values—values implying possibilities of discovery inherent in the evolutionary process.[41]

Crucial differences between Rolston's theology and process theology become apparent in the way in which each grounds and defends moral obligations in environmental ethics. As I discuss below, Rolston accepts that evolution has produced charismatic species with rich subjective lives, and that processes such as predation have, over the course of natural history, propelled organisms to "higher" trophic levels. But he critiques subjectivist (and, at root, anthropic) ethics that privilege organisms with high-grade, rich experiences. This is apparent in cases of value conflict in environmental ethics that Rolston discusses in *Conserving Natural Value* and elsewhere. "Process theists want to interpret evolutionary nature from the conclusions reached in ourselves," Rolston argues, "to attenuate the intensity of the psychical experiences as one moves rearward and downward on the phylogenetic scale."[42] This is an accurate characterization of the hierarchy of value endorsed by Birch and Cobb: applying a "richness of experience" criterion to nature stipulates that our obligations are greatest toward (individual) lifeforms that exhibit the richest subjective life and possess the greatest capacity to suffer. "In proportion to [organisms'] capacity for rich experience," Birch and Cobb argue, "we should respect them and give consideration to making this experience possible."[43] Such ethics remain psychologically rather than biologically informed, Rolston argues, and thus fail to break out of anthropocentric—and more specifically, "anthropopathic"—paradigms. Once again we see Rolston refusing to align himself with ecotheologians that give priority to the suffering, sentient individual at the expense of processes that produce those (admirable) capacities in organisms.

SYSTEMIC VALUE, HOLISM, AND CRUCIFORM NATURE

By now the broad contours of Rolston's account of systemic value may be apparent, but his argument deserves closer scrutiny. At times Rolston presents his theory of systemic value in a theological light and at times from a more secular, philosophical perspective, but we can be sure that his theological commitments are integral to this interpretation of natural values. I present here only a brief summary of a much more complex and nuanced argument in Rolston's work.

The values that emerge from processes such as natural selection and ecosystemic interactions are "systemic" and "emergent"—that is, values projected out from the whole system rather than residing individually in each part of it. This is not to say that subordinate parts, such as individuals, have no value or have only instrumental value. On the contrary, Rolston argues that systems, as well as their subordinate parts, have intrinsic value and these values are objectively real and not merely anthropogenic constructions. Our duties toward nature (Rolston's ethic is largely deontological) are derived from nature's intrinsic value (he denies the "is-ought fallacy").

However, the value of the constitutive parts, including individual organisms, is less than the value of the whole, even when those individual parts are "tallied up." This is so because much of what is valuable about nature goes on in the interstices, in the processes that propel the system forward, not merely in the products that are relatively fleeting instantiations of the process. There can be properties belonging to the whole that are not manifest in the individual parts. A species, for example, has a collective "gene pool" which confers some flexibility in adapting to changing environments; yet any given representative of that species possesses only a subset of those genes. Thus an organism may or may not be a good evolutionary fit, but the species line has "options." It possesses a degree of creativity, a prolific (pro-life) tendency that representatives of the species lack individually. In evolution, "organisms are edited so that from many options, the well-adapted survive, and this results, among other things, in advancing ecosystemic and evolutionary creativity."[44] Species and ecosystems have emergent properties and therefore emergent values. Our duties are not just to species, but to processes such as speciation.

Interventions in such processes are warranted primarily when humans' past interventions have disrupted natural function: damage to nature (and attendant suffering) caused by humans should be addressed, but always with the goal of managing nature so that it can manage itself. Genuine conflicts of intrinsic value occur infrequently in nature because natural values do not so much clash as blend or merge, becoming "transmuted" into something else. It is important to understand that in Rolston's view, many of the conflicts that generate "unjust" suffering in nature, i.e., suffering that ought to be redressed, are the result of past human actions that imperiled parts of nature. Seeking to resolve those conflicts engages us in a complicated effort to mimic holistic, natural processes and this means, at times, treating the parts as relatively expendable, vis-à-vis the whole, just as nature appears to. In *Conserving Natural Value*, Rolston affirms the controversial actions of wildlife managers who sometimes resolve human-created conflicts by culling overrepresented (but complex and sentient) organisms for the sake of protecting a far less sophisticated but endangered species. It is here that he clearly parts company with process-oriented environmental ethics.[45]

Rolston's account of value, and his vision of wildlands management, draws some of its inspiration from the holism and biological realism of Leopold, who was himself a devotee of Darwin. Leopold understood ecosystemic values such as stability, integrity, and beauty, to be inseparable from the "trophic pyramid"—the structure of eating and being eaten—that is the engine of nature. Clearly, then, emergent values of the system are bound up with and generated by the "disvalues" such as suffering. Yet if we take a long view, a systemic perspective, it is clear that local disvalues are often (not always, Rolston concedes) transformed. Rolston examines several "candidate disvalues"—brutal facts of nature—in order to suggest some ways in which they are muted and transmuted systemically. At least two disvalues are particularly relevant to the "problem" of the expendable, suffering individual in nature. One, of course, is predation. Another is the surplus reproduction of organisms that dooms so many to premature death.

Predation is obviously a disvalue for the organism that is killed but clearly of value to the predator, since the "violent death of the hunted means life to the hunter." The value of that life is not so much lost through predation as captured, Rolston emphasizes. Moreover, the individual prey that is consumed obviously "loses all; but the species may gain as the population is regulated, as selection for better skills at avoiding predation takes place."[46] Darwin hints at all this in his discussion of the pleasures and pains that drive evolution. But Rolston argues further that the existence of predation, and heterotrophy in general, drives evolutionary values onward and upward, toward greater complexity, intelligence, even beauty. "Heterotrophs must be built on autotrophs [organisms that synthesize their own food abiotically], and no autotrophs are sentient or cerebral."[47] In this sense, not only are disvalues muted systemically, but evolution appears to be value-enhancing. Evolutionary "upstrokes," as Rolston calls them, are instances of value capture: lower intrinsic values are caught by the system and, in a sense, locked in, even though evolution does not anticipate these upstrokes. The predatory process allows species to "rise higher on the trophic pyramid, funded by capturing resources from below for greater achievements in sentience, cognition, and mobility."[48] The upstrokes do not signal an end to an evolutionary process (either in the sense of completing it or in the sense of a teleological propulsion toward those developments) but once they are in place, further enhancements become possible. The human species is not the goal of evolution but could not have come to exist without predation; however distasteful predation seems to us, a world without bloodshed would be a world without humans as well, Rolston notes. Without predation in our past, we would not be here to "deliberate" about nature's disvalues.

Rolston also addresses the disvalue of nature's apparent "wastefulness." Seen from within a human moral framework, it seems unnecessarily cruel that for many species, far more offspring are produced than can survive and flourish. Nature is "ridiculously prolific." Oak trees, for example, may generate "200 million acorns" to replace themselves once or twice; a pair of birds may produce 30 eggs to replace themselves. But, Rolston points out, the largest surpluses of offspring occur among the least sentient species: acorns, for example, do not suffer when dined upon by squirrels. Moreover, it is not quite accurate to say that these disconcertingly prolific displays constitute a "waste" or "surplus" in the system. "One organism's waste is another organism's treasure" and the food chain—generally, but again not always—absorbs this excess. As with dysfunctional pain, there are anomalies, massive die-offs, surges in populations—"an overkill of muskoxen, a horde of locusts."[49] But "systemically, on average" wherever energy and biomass are available, life tends to use them. "Nature's exuberance is also nature's economy."[50] In order to see these values transmuted, we must "transpose the question from one of human preferences into one about systemic processes."[51]

Perhaps Darwin would be somewhat comforted by Rolston's argument for the capture and transformation, not mere loss, of values in evolution. But he might well have asked whether God could not have designed a better system, such that these disvalues did not need to be transformed. Rolston's general response is that a world without processes that produce suffering might be possible—a world of all flora and no fauna, autotrophs and no heterotrophs—but it would be impoverished. Of course, we could have a world in which complex, sentient creatures sprang fully formed from the earth, as in the Genesis stories of creation, and remained unchanged throughout time. But as Robin Attfield has argued, such a world (while not logically impossible) would require constant intervention from God to maintain nature and all its lifeforms according to divine specifications.[52] A world subject to supernatural intervention and maintenance at any given moment, Attfield further notes, would be fundamentally irregular and unpredictable. Paradoxically, a perfect world might be one humans could never comprehend—a point that Darwin, with his fondness for "fixed laws," might well have appreciated. The system we have "is probably the only kind of non-interventionist world-system which could give us the capacities found in nature that we value," Attfield argues.[53] Attfield's claim that a perfect world devoid of disvalues requires constant intervention is particularly interesting in light of ecotheologies that long for a peaceable kingdom in nature and likewise endorse interventionist imperatives (feeding, healing) as a step toward realizing that eco-vision.

Assuming that Darwin would still side with atheism after hearing the Rolston-Attfield proposal, he might yet be brought around to Rolston's conviction that natural selection is essentially cruciform and deiform. While Rolston will not offer assurances that every case of individual suffering is transvaluated, much less redeemed in an eschatological sense, he can at least suggest that natural selection demonstrates a type of redemptive transformation through suffering and that the existence of this kind of system is not incompatible with Christianity.[54] Whether or not Darwin would perceive Rolston's God as sufficiently benevolent is another issue. Rolston's proposal is not perfect, nor is he the only philosopher to recognize that suffering lies at the heart of both Darwinism and Christianity.[55] But his approach is perhaps the best one available to those wishing to reconcile God and evolution in ways that generate practical environmental ethics.[56]

Rolston examines the issue Darwin struggled with—the adaptiveness or maladaptiveness of pain in nature. He acknowledges the position that suffering "torments the possibility of divine design, reducing natural history to a desolate evil scene."[57] But he follows Darwin's insights a few steps further and in doing so he begins to reconnect pain in evolution to Christian meaning. Extrapolating over generations, or even millennia, pain and the drive to avoid it, can alter evolutionary history, as when prey species evolve structures or behaviors enhancing their ability to elude predators. Of course, predators too must hone their hunting skills if they are to survive, and thus pain is never eliminated from the system. Like Darwin, Rolston maintains that counterproductive pain is relatively rare in evolutionary biology. He echoes Darwin in stressing that "any population whose members are constantly in counterproductive pain will be selected against and go extinct or develop some capacities to minimize pain."[58] In biological terms, we can say that such pain tends to be maladaptive while functional pain tends to be favored by selection. In more theological terms, we might say that life as a whole suffers through to something higher. The evil of suffering in evolution is largely "self-eliminating, except as it is productive of a consequent good."[59]

> We cannot show this in the detail of every case, and there are troublesome anomalies. Nevertheless, this is a general tendency in the evolutionary system. But this position is not inconsistent with a theistic belief about God's providence; rather, it is in many respects remarkably like it.[60]

This is so because the "Judeo-Christian faith never teaches that God eschews suffering in the achievement of the divine purposes," he reminds us.[61] Christians have misunderstood the tradition if they perceive religion as a shield against suffering; rather at the core of Christian teaching is "a call to suffer and to be delivered as one passes through it."[62] In the very concept of a Messiah, for Christians as well as Jews, we see life produced out of death, crucifixion and resurrection intertwined, new life rising out of old, "life suffering through toward something higher."[63] By the same token, to paraphrase Kingsolver's Baptist minister, God does not deliver his creatures out of suffering but through it. "God creates by tolerating mistakes and slowly reconciling them over time, moving through the pains of growth, but not taking mistakes and pain away."[64] The struggle for existence, which Darwin deemed to be "universal," underscores life's goodness. Nature is cruciform and deiform, Rolston insists, because of struggle, not merely in its absence. However, seeing the transformed value, the deiform essence of nature, demands expanding our usual, anthropocentric perspective, to gain an appreciation of a broader, projective kind of value. Most of all, it demands that we consider the possibility that the way nature works is the way God works, even if it is not our way: "*Via naturae est via crucis*." In nature we see that "God writes straight with crooked lines."[65]

Critics may object that Christianity loses its meaning if it fails to offer a promise of compensation for all cases of struggle and sacrifice and Darwin might still side with such critics. I will examine some further objections to Rolston's proposal shortly. But first, consider one last argument for deiform nature and systemic value. Ecotheologians believe that extending the promise of redemption to the rest of nature illustrates their nonanthropocentric orientation. But Rolston might direct their attention—and Darwin's—to passages in the Bible, such as the book of Job, that suggest a divine engagement with the larger processes of life, particularly those that involve struggle and predation, and those that have little to do with human affairs and concerns. It is significant that in the story of Job, which dwells in such vivid and moving detail on natural processes known only to God, we also discern the outlines of a theodicy. Job's demand that God justify himself is met with an extended, slightly sarcastic critique of the human misperception that the world revolves around them.

> Then the Lord answered Job out of the whirlwind: "Who is this that darkens counsel by words without knowledge? Gird up your loins like a man, I will question you, and you shall declare to me. Where were you when I laid the foundation of the earth? Tell me, if you have understanding. Who determined its measurements—surely you know!
> Can you hunt the prey for the lion, or satisfy the appetite of the young lions, when they crouch in their dens, or lie in wait in their covert? Who provides for the raven its prey, when its young ones cry to God, and wander about for lack of food?"

As Bill McKibben argues in *The Comforting Whirlwind*, the bottom line of the book of Job is the decentralization of humans. God's response to Job's rebuke contains a long argument for the nonanthropocentric value—indeed, the theocentric value—of nature. It also implies that God enjoys wildness, something Rolston has suggested.[66] For all of these reasons, McKibben argues, God's answer to Job remains one of the oldest and best pieces of nature writing we have. I would argue that it ought to play a more central role in ecotheology and evolutionary theodicy than it has.

CRITIQUES OF ROLSTON FROM EVOLUTIONARY THEODICY

There are, of course, many who find fault with Rolston's proposal for reconciling God and nature, and we must consider some of these critiques and the amendments and alternatives proposed by those who issue them. It is important to keep in mind that we are seeking here a form of reconciliation of nature and God that also generates environmental imperatives that are workable in the natural world as we currently understand it. Thus, I am not concerned with evaluating all proposals (and there are many admirable ones) for bringing evolution and religion together, but only those that are developed in the direction of environmental ethics.

The ecotheologians I have alluded to thus far do not, for the most part, explicitly critique Rolston's work, not because they agree with it but because many are apparently unfamiliar with it (Northcott and—especially—McDaniel are exceptions). However a substantial amount of commentary has emerged from a particular subset of scholars who might more accurately be described as evolutionary theodicists than ecological theologians, though, clearly, these areas overlap. These theodicists, many of whom are more closely connected to the discipline of philosophy than theology, engage Rolston's work directly. Within this camp, some see his overarching theodicy, his particular delineation of terms such as redemption, and his account of suffering through to something higher, as entirely too human-centered. When we examine these charges more closely, we find that they are, once again, tied to concerns about the status of individual, suffering organisms. Jill Le Blanc, for example, claims that Rolston "ignores" and "evades" this issue; Christopher Southgate follows Rolston up to a point, but then charges his theodicy with "casting aside" individual animal lives. Even Attfield, who finds favor with much of Rolston's account of disvalues transmuted into values would be more convinced if Rolston would emphasize individual, and not just systemic, flourishing in natural processes.

Let us begin by putting a very basic challenge to Rolston: How do we know that the system is on the whole more valuable than disvaluable? In order to appreciate this question, consider the following. A predator that eats its prey benefits not only its species line (by perpetuating its successful predator genes through survival and reproduction); it also clearly benefits itself and, according to Rolston's account, receives individual pleasure. The prey organism that is eaten, on the other hand, suffers a personal loss (pain and death), though inadvertently and unwittingly it confers benefits to its species line by having its unsuccessful prey genes terminated. Note that the prey organism thus benefits its own species line, plus the species line of the predator that ate it, as well as the individual predator. There seems to be a strange asymmetry built into this system, and that asymmetry is reproduced in any evolutionary theodicy constructed upon this interpretation. The predator species accrues benefits both individually and collectively. Is God primarily on the side of predators? This sort of asymmetry of values might lead us to conclude that, on balance, the system of nature is not clearly good, as Rolston maintains, but at best only "valuationally neutral."[67]

Robin Attfield argues that while Rolston has shown, perhaps, that the evils (or disvalues) in the system are muted, even cancelled out, he has not shown that they are outweighed by the goods (or values). This is particularly so if we add to the prevalence of "local"[68] disvalues the existence, however anomalous and rare, of cases of intense, non-functional pain. This brings us back to Darwin's worry (and, perhaps, to a way of alleviating it). Though from a broad perspective nature appears generally beneficent, there are certain angles from which it looks distinctly cruel and horrific, with more disvalue than value.[69] Rolston might emphasize that evolution is not merely value-transforming but (as I have noted above) value-enhancing: in other words, we should not simply consider the quantity of values versus disvalues but the greater quality of values, the "higher values," generated in the transformative process. In fact, it is hard to know how discrete values and disvalues might be "tallied up" and compared in a theory of systemic value (but this could as easily be a point against Rolston's argument as one in favor of it). In any event, Rolston's frequent description of disvalues as "muted" and "transmuted" may have the unfortunate effect of suggesting that disvalues are diluted or toned down, rather than stressing more positively (as I think he means to) that values are actually enriched and increased.

Still, Attfield's point is well taken. Instead of pulling back to take in the broadest possible view of evolution's straight lines, Attfield wants to zoom in on the crooked lines with which God writes, and ask whether there isn't perhaps more happiness (and value) in those individual lives than Rolston has noticed.[70] Attfield wishes to supplement Rolston's case with some "additional considerations" that give more weight to the lives of individuals (prey included) without disputing the "impressive" case for systemic value. He concurs with Rolston's defense of values that are inseparable from disvalues of predation, but points out that "what should be added to the balance of the argument here is the intrinsic value of the flourishing of each flourishing creature that has ever lived."[71] If I follow him, Attfield's locution here, "the flourishing of each flourishing creature," is not simply repetitious. Attfield is not saying (as some ecotheologians do) that the system is a just one if, and only if, both the individual and the larger community can flourish simultaneously. He accepts that pain and predation are indispensable parts of the system and does not find fault with a natural world that "does not observe social justice or compensate individual sufferers."[72] Rather, he is emphasizing that each organism that has in fact flourished has flourished owing to the existence of this sort of system; the overall system is a precondition for the flourishing of individuals (including the individual prey), just as much as the individual organisms provide the conditions for the ongoing, systemic value of the whole.

Moreover, if I have further understood Attfield correctly, he sees a distinction between flourishing on the one hand, and simple fitness (in the Darwinian sense) on the other. An organism can flourish—have a rich, pleasurable life—up to the moment when it is brought down by a predator, for example, or until the winter comes that finally proves too harsh. "The failure of most individuals to live long enough to reproduce," Attfield points out, "does not betoken that there are no elements of flourishing in their lives."[73] Attfield stresses that these considerations do not "undermine" Rolston's natural theology but, rather, provide more evidence for some of his claims.

I think Attfield is right, both in his general point about flourishing and in his belief that this point is not inconsistent with Rolston's account. While such pre-mortem flourishing may not count for much in evolutionary theory where differential survival rates and relative fitness are objectively calculated and compared, it ought to count for something in an evolutionary theodicy, where creation's (and the Creator's) goodness is at stake, regardless of the organism's inability, ultimately, to elude predators, survive an extremely harsh winter, and pass on its genes. Assuming that Darwin would still remain largely skeptical of Rolston's theodicy as I have presented it, Attfield's amendment here, combined with all foregoing arguments, might help tip the scales back toward some form of theism.[74] Darwin himself insists (or at least hopes) that while widespread suffering is a fact of life in the wild, there are "other considerations" that "lead to the belief that all sentient beings have been formed so as to enjoy, as a general rule, happiness."[75] Perhaps he just needed a little convincing.

However, Attfield raises a different concern here as well, one that has been developed by other critics. Rolston's form of theodicy might be predicated too much upon the evolutionary emergence of higher lifeforms—particularly humans—in order to complete the transformation of disvalues into values. In light of this objection, it might be better (contrary to what I suggested above) if Rolston did not stress the higher quality, and not merely quantity, of certain values as proof of nature's overall goodness, for this potentially lands him in a different sort of problem. One could argue that since Rolston's theodicy demonstrates that human evolution required predation, then disvalues associated with predation are redeemed only in light of "*human* capacities which it indirectly facilitates through evolutionary processes, and on the value of personal development thus facilitated."[76]

Saying that predation produced higher forms, including humans, is not the same as saying that these developments in evolution are valuable only because humans arose. Nevertheless, highlighting the values inherent in individual flourishing places less pressure on Rolston to redeem disvalues by making them preconditions for human evolution. For his own part, Attfield does not seem to think that Rolston's theodicy necessarily depends in this way on human evolution. But Rolston can easily be misread on this point. He readily admits to endorsing a loose form of teleology, but I do not think he intends that humans are the goal of value-producing natural processes. He argues, for example, that nonhuman animals and their capacities are "not valuable as poor imitations of what is later achieved" but have their own virtuosity and excellences that are alien to us.[77] Moreover, he claims that evolution is not orthogenetic but opportunistic: "Just as chance favors the prepared opportunist, chance favors the selective system that gropes for life and more life and locks in the upstrokes."[78] Yet, it is in instances such as these that his use of process language, however nuanced, becomes a liability, as when he claims that through evolution God has "lured the ascent" of life, turning "protozoans into persons."[79] I think Rolston would do well to avoid these sorts of claims, particularly given his attempts elsewhere to distance his views from process thought's pan-experiential, anthropic framework.

A critique of Rolston's overarching theodicy as essentially human-centered (or perhaps human-culminating) has been more fully voiced by Jill Le Blanc who offers a slightly different take on the issue.[80] There are two related parts to her critique. The first is that "lower" forms of life and their suffering have only a vicarious role to play in "suffering through" to higher forms such as humans and, again, those individual lives (or deaths) remain uncompensated. Second, the value of the grand, evolutionary narrative Rolston describes is value of the sort that only humans could appreciate. Let us take the second point first. Le Blanc argues that values such as "creativity" and "narrative" in evolution—the dramatic dialectic of pleasure and pain, conflict and resolution, adversity and achievement—seem suspiciously anthropocentric (one might even say protestant, though Le Blanc doesn't say it). The wolverine cannot appreciate this compelling storyline as it struggles to stay alive; the "value that wolverines find in the world is not the opportunity for character development."[81] The "narrative" we discern is a presumptuous imposition. Rolston's presumption is even more evident, she notes, in the way that he offers to "forgive" the injustice done to these individual suffering organisms. Here Le Blanc alludes to Dostoevsky's critique of forgiveness-by-proxy (as voiced by Ivan Karamazov); she insists that one "cannot forgive on another's behalf . . . it belittles the suffering of anyone but oneself. It is as though one says 'He suffered, but I'm better for it.' "[82] In much the same way, Rolston ignores the suffering of the individual. Even more worrisome to Le Blanc is the possibility that "pragmatically" Rolston's type of forgiveness on others' behalf might encourage us to give up fighting on behalf of individuals who are victims of injustice. Forgiving may lead all too easily to forgetting; Rolston's form of theodicy could justify responses to disvalue that are convenient but erroneous, even immoral.

I am not sure that Rolston can be charged with forgiving and ignoring injustices simultaneously. More important, Le Blanc's account seems to confuse an issue that Rolston's generally keeps clear. As examples of responses to disvalues, she moves from a case of humans opting for widespread spraying of DDT (bad idea, as it turned out), to one of humans responding to polio with widespread dissemination of a new vaccine (good idea). These examples tend to conflate disvalues in nature and in culture (humans in culture responding to disvalues in nature; humans responding to disvalues within their own culture, etc.). Rolston is quite clear that we have different ethical obligations in human cultural contexts and in the context of wild nature. Compassion is always "morally required" in human culture, and often misapplied in (to) nature.[83] The suffering of a wolverine, if it is natural, is not an "injustice" at all and, thus, there is nothing to "forgive." Values and disvalues in nature, on Rolston's account, are objective but they are nonmoral values.[84] It is precisely for this reason that we should not censure nature or demand human forms of justice—social justice—on behalf of nature. Attfield appreciates this point, even while he would redirect Rolston's attention to the value of all the flourishing lives in nature. Le Blanc does not appreciate it, and this leads to what strikes me as a false accusation, namely that Rolston presumes to forgive natural "injustices" where in fact he would not discern an injustice at all.

Interestingly, in attributing this presumption to Rolston, she is implying that he usurps the role of God: forgiving sins done against someone other than oneself is "God's métier."[85] But the alternative she proposes, namely "fighting" these alleged injustices in nature, just as we ought to fight homelessness in society, might be presumptuous as well. Le Blanc suggests that our responses to "evil" ought to be roughly analogous, whether in nature or culture. "In the case of human evil, if I think the world ought to be different, I should do what I can to change it—feed the hungry, for example, or shelter the homeless . . . In the case of disvalue in nature, I must likewise do what I can in particular cases." What, particularly, ought we to do? It depends on the situation, she argues, but in general, we can "create good or value" by actions that include chaining ourselves to threatened old growth trees, or providing "food or drugs to starving or sick wild animals."[86] This brings us back to the well-intentioned but misguided imperatives of ecotheologians who wish to correct nature by recreating it as a community where the hungry are fed and the sick healed. For Le Blanc, as for many ecotheologians, these actions flow from a "motivation of love based upon an appreciation of our felt, lived commonality with other members of the world."[87] But this sort of environmental ethic fails to consider that nature's value, its wildness and otherness, may be compromised by "loving" actions; again, it also overlooks the possibility that nature's ways, while not necessarily our ways, might well be God's ways—a possibility Rolston takes very seriously, as we have seen.

Like Le Blanc, Southgate sees Rolston's definition of redemption for individual sufferers as incomplete and insufficient. Southgate has recently proposed a system of biological and theological redemption in the here-and-now that harmonizes in some respects with Rolston's work.[88] However, he argues, Rolston's form of redemption qua regeneration "does not of itself 'redeem' the suffering experienced by individuals . . . regeneration does not comprehend all that is connoted by the word redemption, and the suffering of individual organisms, even if it promotes the flourishing of others, must still remain a challenge for theodicy."[89] Southgate proposes to meet this challenge head-on, by constructing a genuinely Darwinian form of Christian theodicy that still ensures redemption for the individual victims of natural selection.

This proposal sounds extremely ambitious, if not impossible—and, in fact, it is not very successful in my view. I have argued that ecotheologies and evolutionary theodicies often critique either God or nature. Southgate's solution engages in a bit of both. He proposes that we can adhere to a thoroughgoing evolutionary perspective without having to submerge entirely the soteriological claims needed to sustain belief in a benevolent deity.[90] Like Darwin, Southgate cannot reconcile the unredeemed suffering of millions of individual organisms with the existence of a God who remains worthy of our worship. He attempts to restore God's worthiness by means of two postulates: (a) God does not abandon victims of evolution and (b) Humans are called to become healers and co-redeemers of the world. Specifically, our calling engages us in nothing less than "the redemption of evolution" in "real time."[91] This involves a healing "that allows humans to take up their place not merely as 'created co-creators' with their God . . . but as co-redeemers with Christ of the whole evolutionary process."[92] The role of humans as co-redeemers, he argues, is a more "positive" role than most offered by current environmental ethicists who generally advise us to take a hands-off approach. Southgate's characterization of contemporary environmentalism—which is predominantly and inordinately interventionist in its call for a restored and liberated creation—illustrates that ecotheologians and evolutionary theodicists may need to spend more time talking to each other.

To his credit, Southgate acknowledges that his proposal of redeeming evolution sounds "very problematic"—after all, we would not similarly expect to redeem "the law of gravity," he notes. But the idea "warrants further exploration," albeit with deep humility.[93] For example, we might adopt a vegetarian lifestyle as a form of real-time redemption; at the very least, we should seek more humane methods of animal husbandry. Note that neither of these suggestions deals directly with evolutionary processes but rather with cultural practices regarding domesticated food animals. However, Southgate also has in mind "wise conservation and repair of the environment" as marks of "redeemed human wisdom."[94] It is difficult to imagine how we might "conserve" nature, or anything valuable therein, while at the same time shutting down the whole evolutionary process.

As if redemption of evolution were not a tall enough order, Southgate further endorses retroactive redemption, so that we might "wipe the tears from the eyes of evolution's myriad past victims."[95] In other words, even if we could halt the violence of natural selection today, we would still need somehow to make it up to all evolution's casualties, stretching back to the beginning of life on earth. Clearly this lies beyond human powers, so Southgate espouses as "necessary" the existence (or creation) of an animal heaven in order to ensure that all of evolution's victims "are able to fulfil their being."[96] As co-redeemers of evolution, we do our part on earth and hope that God will follow suit and do his part in heaven.

Rather than humility, this proposal unfortunately resembles blasphemy. The crux (literally in this case) of Southgate's disagreement with Rolston lies in his conviction that the Cross of Jesus has its meaning in "the transformation of creation" understood as the healing of the evolutionary process.[97] Echoing Jay McDaniel's claims, he argues that "a God of loving relationship could never regard any creature as a mere evolutionary expedient" (McDaniel, it should be noted, does not hold that animal heaven is necessary but only desirable to complete redemption and restore God's worthiness).[98] But Rolston's interpretation of systemic value already entails a form of real-time transformation. The difference is that it actually lies within natural processes, processes whose time scale may not be the same as ours and whose form of redemption may not satisfy us. Rolston would argue that we have to shed an exclusive preoccupation with individual suffering and redemption in order to see that value is not lost but transformed. Yet, of course, as humans we never do see the whole narrative of evolution but only glimpses of it. To be sure, Rolston's claim that life, as a whole, suffers through to something higher is partly an article of faith. But nothing that we know about the natural world contradicts this claim, as far as I can see, and there is much that seems to support it, both in nature and in the Christian tradition.

In seeking to "redeem" evolutionary processes and render God more worthy of worship, Southgate would put an end to the very system that creates and maintains value, beauty, sentience, and even, perhaps, intelligibility in the world we inhabit. Such proposals for reconciling evolution with religion, while holding tenaciously to human perspectives on suffering and justice, require so many additional theological postulates and refinements of biological data that they become unwieldy and unconvincing. Like the Ptolemaic model of astronomy requiring numerous epicycles, gears upon gears, in order to maintain the conviction that everything revolves around Earth, these solutions become cumbersome in their attempts to "save the phenomena."[99] The simpler, and more elegant, solution would be to admit that humans, and their particular concerns and judgments and experiences, are not the center. This, as I see it, is the answer Rolston offers environmental ethicists. To my mind, no one has yet offered a better one—with the possible exception of God who said essentially the same thing to Job.

* * *

In the scene from *The Poisonwood Bible* with which I began, the minister and his daughter eventually solve the mystery of their garden's fruitlessness: though the mission is virtually overrun with a fearsome array of insect species, not one of them is the sort that pollinates a Kentucky Wonder bean. All of their labor has been for naught. In a fit of anger over his recent string of failures, the minister suddenly turns on the pet parrot Methuselah, a native of the Congo passed on to them from a previous missionary. He pulls the terrified creature from its cage and hurls it skyward. With some instinctive memory of flight still lingering in clipped wings, Methuselah manages a soft landing atop their Wondrous vines. For months afterward, he hovers pathetically near their home, no longer able to survive in the wild. One day, a pile of bloodied feathers confirms that Methuselah's life has been cut short by an anonymous predator.

In Kingsolver's narrative, Methuselah's ill-fated "freedom" is symbolic of the newly gained, dubious independence of the Republic of Congo. But at another level, the story contains an important message about wildness and domestication, nature and culture, and the way in which these overlapping categories complicate our human sense of fairness. It is a story about humans raging against, and sometimes blaming God for, the wrong things. Kingsolver, I think, is suggesting that if sin has any consistent meaning, it is found in certain patterns of human behavior: a general tendency to see injustice where there is none, to create it where it did not exist, to fail to detect it where it most needs redressing. If anyone is wronged in this scenario it is surely Methuselah—not so much because he was devoured by a predator, which is a common enough fate of jungle birds, but because humans stripped him of his natural powers and cast him off. The minister's misdirected anger is really a rebuke of God for not providing a natural world that conforms to his own expectations, his sense of logic and justice.

Environmental ethicists ought to be more circumspect about censuring nature—or God—for processes that fail to match human preferences. Perhaps as Rolston suggests, God is also interested in the larger currents of life, the overall sorting rather than the local shuffling, the supervival of the system as a whole, rather than the survival of each individual part. "One cannot detect the historical God full-scale in each local event, any more than one can detect natural selection in each individual life."[100] Perhaps the "conflict" we perceive between individual flourishing and systemic flourishing is a problem only for human culture—not for nature or for God either.[101] Rolston acknowledges that it is not easy to construe the world from nonanthropocentric perspectives, to accept nature as it is. We have to struggle to see God in nature's struggle, to see how God writes straight with crooked lines. But it is not our place to erase or re-write those lines, even if we could. By holding fast to this conviction, Rolston manages to bring evolution and theism closer together while moving environmental ethics away from human-centered concerns.

ACKNOWLEDGEMENT

Thanks to Robert Crouch, William Howarth, Greg Mikkelson, Richard Miller, Wayne Ouderkirk and Holmes Rolston for helpful comments on an earlier draft. I am also grateful to Rolston for bringing to my attention some recent arguments on the subject of evolutionary theodicy.

This essay is dedicated to my brother Philip Sideris.

NOTES

[1] Mark Wynn, "Natural Theology in an Ecological Mode." *Faith and Philosophy* 16.I (1999): 27–42, p. 39.

[2] Darwin vacillated somewhat even after publishing the *Origin of Species*, though not so much between atheism and theism as between atheism and agnosticism. He attended church services all his life and

some of his letters to friends and family show that he continued to think about the possibilities of reconciling belief with evolutionary theory. See James Rachels, *Created From Animals: The Moral Implications of Darwinism* (Oxford: Oxford University Press, 1990) for a discussion of Darwin's theistic, atheistic, and agnostic tendencies.

[3] Charles Darwin, *Autobiography of Charles Darwin*, Nora Barlow, ed. (New York: Norton, 1958), p. 87.

[4] Donald Fleming, "Charles Darwin, the Anaesthetic Man." *Victorian Studies* 4 (March 1961): 219–36, pp. 230–31.

[5] Darwin, *Autobiography*, p. 87.

[6] It is questionable whether Darwin was a devotee of Paley's design arguments even in those early years when Paley was becoming something of a straw man for naturalists proposing new mechanisms for adaptation in nature.

[7] Darwin, *Autobiography*, p. 87.

[8] Ibid., p. 90.

[9] Ibid., p. 90.

[10] Ibid., p. 90.

[11] James Rachels observes that Darwin's theory highlights what remains *the* problem for theodicy, the suffering of nonhuman animals. Rachels largely agrees with Darwin that no robust form of Christian theism can survive these and other challenges . See Rachels, *Created from Animals*.

[12] Darwin, *Autobiography*, p. 89.

[13] Charles Darwin, *The Origin of Species* (London: Penguin, 1987), p. 129.

[14] Darwin, *Autobiography*, p. 88.

[15] Jay McDaniel, *Of God and Pelicans: A Theology of Reverence for Life* (Louisville: Westminster/John Knox, 1989), p. 41.

[16] Darwin, *Autobiography*, pp. 87–8.

[17] Darwin himself rejected any notion of nature's sinful fallenness, in part because the late appearance of humans on the evolutionary scene makes this explanation untenable, but also because he wanted to dissociate his own theory from theories of "devolution" that claimed certain lifeforms (including "savage" human tribes) had degenerated from a more perfect, former condition.

[18] Lisa Sideris, *Environmental Ethics, Ecological Theology and Natural Selection* (New York: Columbia University Press, 2003).

[19] All of the ecotheologians cited here claim that their ecological views are scientifically informed. In fact some explicitly make claims to a Darwinian worldview.

[20] Rosemary Radford Ruether, *Gaia and God: An Ecofeminist Theology of Earth Healing* (San Francisco: HarperCollins, 1994), p. 213.

[21] Sallie McFague, *Super, Natural Christians* (Minneapolis: Fortress, 1997) p. 158

[22] Jürgen Moltmann, *God in Creation* (Minneapolis: Fortress, 1993), p. 102.

[23] Michael Northcott, *The Environment and Christian Ethics* (Cambridge: Cambridge University Press, 1998), p. 194.

[24] Charles Birch, "Christian Obligation for the Liberation of Nature." In Charles Birch and William Eakin, eds., *Liberating Life: Contemporary Approaches to Ecological Theology* (Maryknoll: Orbis, 1990), p. 67.

[25] Tom Regan, "Christianity and Animal Rights." In Birch and Eakin, eds., p. 87.

[26] Larry Rasmussen, *Earth Community, Earth Ethics* (Orbis: Maryknoll, 1996), p. 347.

[27] Ibid., p. 347.

[28] McFague, *Super, Natural Christians*, p. 16.

[29] Ibid., p. 169.

[30] Holmes Rolston, "Does Nature Need to be Redeemed?" *Zygon* 29.2 (1994) 205–29, p. 205.

[31] Holmes Rolston, "Wildlife and Wildlands." In Dieter Hessel, ed. *After Nature's Revolt* (Minneapolis: Fortress, 1992), p. 131.

[32] Ibid., p. 131.

[33] Rolston, "Does Nature Need to be Redeemed?" p. 208.

[34] Rolston, "Wildlife and Wildlands," p. 135.

[35] Holmes Rolston, "Challenges in Environmental Ethics," In Michael Zimmerman, ed., *Environmental Philosophy: Animal Rights to Radical Ecology* (Englewood Cliffs: Prentice Hall, 1994), p. 140.
[36] This is a suggestive parallel between the birth process and other processes in nature that cause pain, in light of the fact that the Genesis account explains pain in childbirth as a result of, and punishment for, sin.
[37] Holmes Rolston, *Science and Religion: A Critical Survey* (Philadelphia: Temple University Press, 1987), p. 86.
[38] Ibid., p. 142.
[39] Rolston, "Does Nature Need to be Redeemed?" p. 227.
[40] Holmes Rolston, "Disvalues in Nature." *The Monist* 75 (1992): 250–78, p. 259.
[41] Rolston does not agree with the more extreme claim that God *is* process, and he disagrees that "persuasion" (without, say, compulsion) sufficiently describes the parameters of God's action in creation (see *Science and Religion*)
[42] Rolston, *Science and Religion*, 321.
[43] Charles Birch and John Cobb, *The Liberation of Life: From the Cell to the Community* (Cambridge: Cambridge University Press, 1981), p. 153.
[44] Rolston, *Science and Religion*, p. 110.
[45] Process-oriented ethics grant primarily instrumental value to plants, because of the lack of experiential richness that (intrinsically valuable) vertebrates and other higher organisms possess. Despite the emphasis on "process," such approaches give greater attention to individual experiencing organisms than to the ecological and evolutionary processes that carry life as whole forward.
[46] Rolston, "Disvalues in Nature," pp. 253–54.
[47] Ibid., p. 253.
[48] Ibid., p. 254.
[49] Ibid., p. 268–69.
[50] Ibid., p. 269. Wayne Ouderkirk notes the similarity between Rolston's arguments regarding waste and predation in nature and the principle of plenitude and aesthetic themes proposed by theologians such as Augustine and Irenaeus, particularly the latter. "Can Nature be Evil? Rolston, Disvalue, and Theodicy." *Environmental Ethics*, 21 (1999), 135–150.
[51] Rolston, "Disvalues in Nature," p. 255.
[52] Robin Attfield, "Evolution, Theodicy, and Value." *Heythrop Journal* 41 (2000): 281–96.
[53] Ibid., p. 288.
[54] Rolston's proposal for reconciling Christianity and evolution is in stark contrast to "creationist" views that acknowledge the unproblematic existence of "microevolution" (e.g. antibiotic resistance) but dispute the validity of macroevolution, which is seen as incompatible with Christianity.
[55] For all their disagreements, Rolston and Michael Ruse make a similar move in bringing Darwinism and Christianity together with the issue of suffering as the crucial vehicle. In *Can a Darwinian Be a Christian?*, Ruse argues that (in contrast to Darwin's own misgivings about pain) "if you are a Darwinian looking for religious meaning, then Christianity is a religion which speaks to you. Right at its centre there is a suffering god, Jesus on the Cross. This is not some contingent part of the faith, but the very core of everything . . . Darwinism, a science which so stresses physical suffering, looks to Christianity, a religion which so stresses physical suffering and the divine urge to master it" (p. 134). Ruse is not writing specifically on environmental issues, of course. It is this urge to master suffering that becomes problematic when humans attempt to discern ethical obligations toward nonhuman life.
[56] James Rachels, for example, points out that mere *compatibility* or *consistency* of Christianity and Darwinism is not particularly significant. "There are innumerable absurd beliefs that might be held without self-contradiction; that is too weak a test to be of much interest" (*Created from Animals*, p. 124). Rachels contends that *deism*—and deism alone—remains most compatible with evolution, but its notion of God is too "abstract" and devoid of content to allow any sort of personal relationship with God. Rolston would dispute this point: in *Science and Religion* he argues for the possibility of a "personal" and "present" God who is nevertheless invested in larger processes of life, including those that entail suffering of individuals. Indeed, he is dissatisfied with the process account of God partly because

of its impersonal nature. See Rolston's account of "transscientific theism" (*Science and Religion*, pp. 322–35).

Rachels seems convinced that the core of Christianity consists in what he calls the "image of God thesis"—divinely bestowed human uniqueness—and that this thesis crumbles under the weight of evolutionary theory. But it should be noted that while Rachels rejects the anthropocentrism of the image of God thesis, he advocates a form of *moral individualism* that may well be anthropocentric. Rachels argues that "it is the richness and complexity of the *individual* life that is morally significant" (p. 189). The connection between Darwinism and moral individualism is much stronger, he claims, than the mere consistency of Darwinism and some forms of theism. Even if Rachels is right about this, moral individualism is of little use in dealing with *environmental* issues (distinct from animal welfare questions about the use of animals for food, experimentation, or products; Rachels cites, for example, trapping of civet cats for perfume manufacture). Environmental ethics involves adjudicating between collective entities or between collective entities and individual organisms. Surely we cannot make sound environmental decisions based on the richness and complexity of individuals. This proposal resembles the "richness of experience" criterion posed by process thinkers who remain preoccupied with individual mental capacities. Moral individualism's "principle of equality," as Rachels calls it, fails as a *biodiversity* principle. Strangely, Rachels believes that Darwinism effectively shifts focus *away* from the species level.

57 Rolston, *Science and Religion*, p. 133.

58 Holmes Rolston, *Genes, Genesis and God* (Cambridge: Cambridge University Press, 1999), p. 305.

59 Rolston, *Science and Religion*, 143.

60 Ibid., p. 143.

61 Ibid., p. 145.

62 Ibid., p. 221.

63 Ibid., p. 135.

64 Ibid., p. 334.

65 Ibid., p. 46, p. 129.

66 Bill McKibben, *The Comforting Whirlwind: God, Job, and the Scale of Creation* (Grand Rapids: Eerdman's Publishing, 1994). For a similar view in Rolston, see *Science and Religion*, p. 334 ff.

67 Attfield, "Evolution, Theodicy, and Value," p. 292.

68 Wayne Ouderkirk argues that Rolston discounts disvalues too much after acknowledging them. To describe disvalues as "local" plays down their ubiquity, their "endemic and significant" existence "beyond the merely local." "Can Nature be Evil?" p. 144.

69 Rolston himself observes this ambivalence in Darwin, who admires the grandeur of nature while pronouncing it wasteful, clumsy, and cruel (see Rolston "Disvalues in Nature")

70 Attfield's argument seems to be the opposite of Ouderkirk's in this respect—i.e., Rolston has made too much rather than too little of these local disvalues. Both agree that given Rolston's own account of this system, the verdict on nature's goodness *in toto* could go either way, but Attfield's amendment tries to coax it in the direction of goodness.

71 Attfield, "Evolution, Theodicy, and Value," pp. 292–3.

72 Ibid., p. 294.

73 Ibid., p. 294.

74 This suggestion seems similar to Jay McDaniel's claim that even individual organisms that suffer and die prematurely nevertheless experience richness while they live and their richness may contribute to the divine life, to God understood as an "ecological Whole."(*Of God and Pelicans*, 48). Though McDaniel *hopes* for an eschatological resurrection of the individual, he does not think it necessary in order to show the goodness of God. I suspect Rolston might be wary of McDaniel's process-influenced, close identification of God with the "ecological Whole," however.

75 Darwin, *Autobiography*, p. 88.

76 Attfield, "Evolution, Theodicy, and Value," p. 293

77 Holmes Rolston, *Conserving Natural Value* (New York: Columbia University Press, 1994), p. 104.

78 Rolston, *Science and Religion*, p. 127.

79 Ibid., p. 119. See also "Does Nature Need to be Redeemed?" p. 228.

[80] Jill Le Blanc, "A Mystical Response to Disvalue in Nature." *Philosophy Today* 45.3 (Fall 2001): 254–65.

[81] Ibid., p. 259. Le Blanc here cites Allen Carlson and Jim Cheney who also note the presence in Rolston's writing of a version of character-building and "free-will-defense" rationales for evil.

[82] Ibid., p. 256.

[83] Rolston, "Wildlife and Wildlands," p. 135. See also Rolston's "Natural and Unnatural; Wild and Cultural." *Western North American Naturalist* 61.3 (2001): 267–76. Rolston holds that humans are for the most part post-evolutionary (no longer directly impacted by forces of natural selection as our animal kin are) but not post-ecological. In other words, we have the ability to impact nature dramatically, even while the direct impact on us by nature is buffered by our cultural developments, medicine, technology and general lifestyle. This strikes me as common sense, though other environmentalists, most notably J. Baird Callicott, dispute Rolston's nature/culture distinction.

[84] Darwin himself might question this argument, since he discerned, at the very least, incipient moral instincts in animals. I am not entirely convinced that it is accurate to regard nonhuman animals as nonmoral. It might be more accurate to say that they (or at least some) have a *different* form of morality that we ought to respect. Call it a nonhuman morality rather than nonmoral animality, but in either case we ought not to censure it from human cultural perspectives.

[85] Le Blanc, "A Mystical Response to Disvalue in Nature," p. 256.

[86] Ibid., p. 264. Protecting old growth forests from human destruction would be quite different from protecting wild animals from *naturally* occurring disease or starvation; perhaps she has in mind human-introduced conditions, in which case these two situations are similar and might both be considered "injustices." These kinds of issues need to be clarified in Le Blanc's argument, as they do in the arguments of most ecotheologians.

[87] Ibid., p. 264.

[88] Christopher Southgate, "God and Evolutionary Evil: Theodicy in the Light of Darwinism. *Zygon* 37.4 (2002): 803–24.

[89] Ibid., p. 805.

[90] It should be noted that Southgate's solution balances ontological, teleological, and soteriological claims and his proposal consists of four main points, which I will not recapitulate here. The fourth of these—his two-part soteriological claim—is of particular relevance.

[91] Ibid., pp. 815–19.

[92] Ibid., p. 818.

[93] Ibid., p. 818.

[94] Ibid., p. 819.

[95] Ibid., p. 820.

[96] Ibid., p. 820. Southgate makes specific mention of "pelican heaven" following Jay McDaniel who discusses the problem of the "back-up chick," the neglected second-born chick who usually fails to survive but serves as insurance for the parents when the first born dies (see McDaniel, *Of God and Pelicans*. Rolston also discusses the back-up chick as a "discordant" case in evolution in *Science and Religion*.)

[97] Southgate, "God and Evolutionary Evil," pp. 820–21.

[98] Ibid., p. 821.

[99] Intellectual honesty compels me to note that historians of science routinely point out that the Copernican model eventually became quite cumbersome as well, perhaps as cumbersome as the Ptolemaic model it replaced—at least until Kepler clarified things.

[100] Rolston, *Science and Religion*, p. 325.

[101] It may not even be accurate to say that this "conflict" has posed a problem for *all* human cultures. I am grateful to Greg Mikkelson for pointing this out.

ALLEN CARLSON

7. "WE SEE BEAUTY NOW WHERE WE COULD NOT SEE IT BEFORE": ROLSTON'S AESTHETICS OF NATURE

INTRODUCTION

Holmes Rolston's substantial and significant research in environmental philosophy is most commonly associated with environmental ethics. In this chapter, I consider his contribution to environmental aesthetics. I argue that his reflections on our aesthetic experiences of the natural environment constitute a deep and important dimension of his thought. I investigate the ways in which this aesthetic dimension is implicitly present in a number of his discussions of environmental values and comes to fruition in an explicit and comprehensive theory of the aesthetics of nature. I thereby make clear how Rolston, in addition to being one of the founding fathers of environmental ethics, is also a pivotal figure in the development of environmental aesthetics.

I follow a more or less chronological order in my presentation and consideration of Rolston's aesthetics of nature. In the first main section, I briefly demonstrate how the aesthetic dimension in Rolston's thought manifests itself in some of his very earliest work in environmental philosophy. In this section, I not only introduce Rolston's deep appreciation of the beauties of nature, but also show that Rolston's approach stands in line with the naturalist tradition of philosophizing about nature, a theme that resurfaces at various points throughout my discussion. In the next section, I turn to the account of natural value that Rolston advanced in a series of articles published throughout the 1980s and comprehensively presented in his 1988 volume *Environmental Ethics: Duties and Values in the Natural World*.[1] Since aesthetic values are not the main focus of this account, I bring out the fact that aesthetics does nonetheless play a significant, even if frequently overlooked, role in Rolston's thought. I do this by considering the ways in which aesthetic values function in Rolston's treatment of three problems, the is/ought problem and what I call the new problem of evil and the after-Darwin problem. In light of this backdrop, I next consider Rolston's main contribution to environmental aesthetics, the comprehensive theory of the aesthetics of nature that he introduces in *Environmental Ethics* and develops in subsequent writings. I elaborate his theory as involving versions of the positions that have become known in the aesthetics literature as scientific cognitivism and positive aesthetics. In this section, I also consider some philosophical objections to Rolston's theory of the aesthetics of nature, suggesting ways in which these objections may be addressed. In the last main section of the chapter, I follow up some of Rolston's concerns about the possible subjectivity of aesthetic value, arguing that his account lends itself to a more objectivist treatment than might be initially thought. I conclude with some general observations about Rolston's contributions to the aesthetics of nature.

C.J. Preston and W. Ouderkirk (eds.), Nature, Value, Duty, 103–124.

NATURE APPRECIATION AND THE NATURALIST TRADITION

In the late 1960s and throughout the 1970s, Rolston published a set of essays that might be characterized as explorations into both the appreciation of natural environments and the nature of those who appreciate them. For example, in "Mystery and Majesty in Washington County" and in "Hewn and Cleft from this Rock: Meditation at the Precambrian Contact" reflections on the geology, the flora, and the fauna of, respectively, Washington County, Virginia, and Carter County, Tennessee, give rise to meditations on the nature of humanity and humanity's place within nature.[2] Likewise, in "Lake Solitude: The Individual in Wildness" a solitary traveler's thoughts about an "environmental encounter" at Lake Solitude in Rocky Mountain National Park develop into a philosophical reflection on the human presence in nature.[3] And in "The Pasqueflower" this blossom's rebellious, yet nurturing relationship with winter are contemplated as a metaphor for the relationship between human culture and wild nature.[4] The text for this last essay is Aldo Leopold's observation that "the chance to find a pasqueflower is a right as inalienable as free speech."[5] The reference is appropriate, not simply because of environmental philosophy's debt to Leopold, but also because in these writings Rolston blends knowledge and appreciation to achieve essays in natural history that are comparable in many respects to some of Leopold's classic pieces.

In addition to their intrinsic interest and wonderful ambience, these early essays are important in that they make clear the extent to which Rolston's research in environmental philosophy, perhaps more so than that of most modern contributions, stands in a direct line with an older tradition of philosophizing about nature. This tradition had its high point in the nineteenth century in the writing of individuals such as John Muir and may be called the naturalist tradition. In its prime, it laid the groundwork for an essentially aesthetic appreciation of nature by bringing together science, ethics, and a touch of religion. Much of Rolston's environmental philosophy utilizes a similar combination. The above-mentioned essay, "The Pasqueflower," is a good example. Here scientific description, moral metaphor, and religious symbolism are interwoven to engender a profound aesthetic appreciation not only of the flower, but also of the vast ecological and seasonal systems that nurtures it. It is appropriate that the piece originally appeared in the "Naturalist at Large" section of *Natural History* magazine.

The way in which these early essays place Rolston within the naturalist tradition is especially noteworthy here, since it indicates the extent to which the aesthetic experience of nature is a significant and deeply embedded dimension of Rolston's thought right from the outset. Moreover, these essays frequently demonstrate this point in an even more direct way. Consider, for example, the following passage from the earliest of those mentioned above, "Mystery and Majesty in Washington County":

> Once upon a July twilight I was gifted all undeserved with the loveliest of sunsets. The sun plunged behind the old snag that stood as a sentinel halfway up the western knob on the skyline near home, terminating the day with a blaze of glory that fired an orange red across half the spacious firmament. It was as though the sinking sun had resolved to exhaust itself in activity and color. Dusk was spent in hues of crimson and violet

> that lined the deep purple stratocumulus, then to yield to encroaching darkness, but not before a bit of July became part of me forever.[6]

VALUE THEORY AND THE GENESIS OF ROLSTON'S AESTHETICS OF NATURE

Although the aesthetic dimension in Rolston's thought can be detected in his earlier writings, the dimension itself only appears as a distinctive philosophical theme in his research on values in nature. Initially in his groundbreaking "Is there an Ecological Ethic?" and in other major essays, such as "Values in Nature," "Are Values in Nature Subjective or Objective?" and "Values Gone Wild," Rolston develops a comprehensive account of values in nature, which receives its most systematic treatment in his book, *Environmental Ethics: Duties to and Values in the Natural World*.[7] The account is first and foremost a substantive attempt to provide a foundation for environmental ethics. As such, the details of the theory are not essential here. Rather what is important is the way in which, frequently only implicitly, it makes reference to and relies upon the aesthetic values of nature. Since aesthetic values are not the explicit nor the primary focus of the account, their place and importance are not always obvious and seemingly become most evident only when Rolston's attempts to deal with certain problems that either traditionally challenge accounts such as his or seem to be generated by the account itself. Consequently, in order to bring out this frequently overlooked role of aesthetic values in Rolston's general theory of values in nature, I examine three of such problems, the traditional is/ought problem as well as two others, which I label the new problem of evil and the after-Darwin problem. I argue that concerning each of these problems, at least one of Rolston's lines of reply appeals substantially if not exclusively to aesthetic values. In this way Rolston's treatment of these problems importantly foreshadows and thereby deepens our appreciation and understanding of his comprehensive theory of the aesthetics of nature, which I elaborate later in this chapter.

A traditional theme of the naturalist tradition, the flavor of which, as noted above, permeates Rolston's work, is the welding of knowledge and appreciation—of facts and values. This theme is developed in Rolston's research on values in nature such that it, as much other work in environmental ethics, seemingly runs directly afoul of the venerable dogma of the logical gap between is and ought. Rolston addresses this problem in a number of places, describing it in "Can and Ought We to Follow Nature?" as "the basic cleavage that runs through the middle of the modern mind dividing every study into the realm of the *is* and the realm of the *ought*."[8] What I take to be Rolston's more standard approach to the problem is developed in "Values Gone Wild" and in *Environmental Ethics*, where he appears to embrace a form of naturalism. He proposes the idea of nature itself as a "source of values," characterizing "wild value" as "storied achievement" in an evolutionary ecosystem.[9]

However, apart from the idea of "value as storied achievement," Rolston seems to have another, somewhat distinct approach to the is/ought problem that puts less emphasis on nature itself as a "source of values" and more on our scientific descriptions of nature, in particular ecological descriptions, as such a source of value. In "Is there an Ecological Ethic?" he claims that "values seem to be there as soon as the facts are fully in" because:

> Ecological description finds unity, harmony, interdependence, stability, etc., and these are valuationally endorsed, yet they are found, to some extent, because we search with a disposition to value order, harmony, stability, unity. Still the ecological description does not merely confirm these values, it informs them; and we find that the character, the empirical content, of order, harmony, stability is drawn from, no less than brought to, nature. . . . the earlier data are not denied, only redescribed or set in a larger ecological context, and somewhere enroute our notions of harmony, stability, etc., have shifted too and we see beauty now where we could not see it before.[10]

If the fact/value gap is bridged in this fashion, it is not because soft, whimsical values are simply defined in terms of hard, scientific facts, but rather because both facts and values are equally given by and within the framework of ecological science. On this view, science gives one "redescription" of nature that has as inseparable components both facts and values. For example, science does not simply inspect the relationship between two organisms, describe it as symbiotic, and then evaluate it as stable and harmonious. Rather both the relationship's symbiosis and its stability and harmony are parts of the same redescription—if the relationship were not stable and harmonious, it would not be symbiotic. Thus, the redescription is a single scientific account of the natural world, which produces both facts and values that are therefore equal in status to one another. Whether or not this line of thought is promising concerning the is/ought problem or, if distinct from it, a more or a less satisfactory approach to the problem than Rolston's conception of value as "storied achievement" is not important here. However, from the point of view of Rolston's aesthetics, this line of thought is especially significant, for the specific values involved, such as order, harmony, stability, and unity, are essentially *aesthetic* values, and the redescription of nature given by science yields the result, as Rolston himself puts it, that "we see *beauty* now where we could not see it before." Thus, in this reply to the is/ought problem, aesthetic factors are in fact playing the dominant role, a role that they also play in response to the other two problems considered in this section and, more importantly, a role that lies at the very heart of Rolston's comprehensive theory of the aesthetics of nature, as becomes clear later in this chapter.

In addition to the is/ought problem, the naturalist tradition also worried about another issue that is relevant to Rolston's account of values in nature. That tradition, as noted, brought together science, ethics, aesthetics, and religion, and thus it occasionally confronted the problem of evil: the question of how to deal with the presence of evil in a world created by an all-knowing, all-powerful, and all-moral God. A similar problem, which might be called the new problem of evil, arises in light of an approach to natural value such as Rolston's. This new problem is the question of why there should be any evil in a purely natural world if the correct account of positive value in nature ties that value tightly to its naturalness. Rolston suggests the similarity of this problem to the traditional problem of evil in "Can and Ought We to Follow Nature?" where he remarks: "It is no coincidence that the ecological perspective often approaches a religious dimension in trying to help us see the beauty, integrity, and stability of nature within and behind its seeming indifference, ferocity, and evils."[11] The problem is that Rolston, in spite of his account of natural value, finds himself forced "to concede that there are gaps in this account" in that he does not "find nature meaningful everywhere, or beautiful, or valuable" but rather is, for example, "moved to horror by malaria, intestinal parasites, and genetic deformities."[12] He summarizes: "My concept of the good is not coextensive with the natural, but it does greatly overlap it; and I find my estimates steadily enlarging the overlap."[13] Nonetheless, given his account of natural value, this perceived lack of "coextensiveness" requires an explanation.

As in the case of the is/ought problem, Rolston seems to offer at least two different solutions to the new problem of evil. However, only one is of direct interest here in that it relates to his aesthetics of nature.[14] In "Values Gone Wild" and in *Environmental Ethics*, Rolston offers a solution to the problem that seemingly involves "enlarging the overlap" of the concepts of the good and the natural until they are in fact completely "coextensive." Reminiscent of the nineteenth-century's theologian-naturalists' obsession with the "ichneumon fly" and its relationship with caterpillars and spiders, Rolston's example is, as suggested above, endoparasitism: "Once as a college youth I killed an opossum that seemed sluggish and then did an autopsy. He was infested with a hundred worms! Grisly and pitiful, he seemed a sign of the whole wilderness, . . . too alien to value."[15] This kind of example is easy to amplify; consider this passage from George Gaylord Simpson's Patagonian journal, *Attending Marvels*:

> Like almost all wild creatures, ostriches are always diseased. In every one of the many we opened, the intestines were choked with worms, and external parasites are equally abundant. Since this condition is the rule, perhaps it should not be considered as a sort of disease but as a normal ostrich condition, and the animals may be thought of as small migratory worlds, densely populated.[16]

Interestingly enough, Rolston's solution to the new problem of evil in "Values Gone Wild" and in *Environmental Ethics* is similar to that which is implicit in this remark by the great evolutionary theorist. It is also similar to a somewhat standard reply to the traditional problem of evil: the reply that what appears to be evil from our view is in fact not evil from a God's-eye view. In Rolston's reply, however, the God's-eye view is the view of ecology and evolutionary theory, which broadens our perspective across time and space such that what we initially see only as one creature's evil is now properly seen as someone or something else's good. On his account, as in the reply to the is/ought problem, scientific redescription is again at work—but in this case it works such that we "begin to get a new picture painted over the old, although some of the old picture still shows through."[17] The result of the new picture is an "enlarged" notion of value as that which "makes a favorable difference to an ecosystem, enriching it, making it more beautiful, diverse, harmonious, intricate," such that "a disvalue to an individual may be a value in the system and will result in values carried to other individuals."[18] Rolston concludes that: "Excellence is not a matter of encapsulated being, but of fittedness into a pervasive whole," for the "recommendation that one ought to value these events follows from a discovery of their goodness in place."[19] The upshot is: "The giddy, wild experience of valuing nature shed of culture has led us through alien values and threatened chaos, but we have transformed a negative evaluation into a positive one."[20]

As noted, this reply to the problem is somewhat similar to Rolston's treatment of the is/ought problem and, although it is addressed to what I call the new problem of *evil*, it is most plausible when it is viewed in a manner parallel to that way of dealing with the is/ought problem. This is by strongly emphasizing not moral values but rather the kind of values that are given by scientific redescription, that is, order, unity, harmony, stability, and so forth. The reply requires only that apparent evils be "transvalued," that they be eliminated by being transformed into positive values, but not that these positive values be moral values. An emphasis on moral values makes the reply implausible in that it seemingly requires us to see, for example, the relationship between Rolston's opossum and its worms as morally good. But, in other connections Rolston rightly suggests that such a requirement would be inappropriate, saying that our "moral conduct exceeds nature" and that "nature is nonmoral."[21] In this sense, our moral values and predicates "exceed" Rolston's opossum and its worms: their relationship is simply not a moral matter. The worms do not have to become—in fact, cannot become—morally good in order to overcome the fact that, as Rolston puts it, "the systemic source which they of necessity inhabit seems ugly, evil, wild."[22] Their seeming evil can be overcome simply by recognizing their relationship with the opossum as nonmoral and then seeing it as, for example, stable, balanced, harmonious—that is, as having some of the positive values yielded by scientific redescription of nature. When the emphasis is put on such values, this reply to the new problem of evil is similar to Simpson's suggestion that the infested ostriches of Patagonia be considered not diseased, but "as small migratory worlds, densely populated." It also stands directly in the line of the naturalist tradition and its stress on the aesthetic appreciation of nature, for, as in the case of the response to the is/ought problem, the positive values in question are by and large *aesthetic* values. Again, as in the treatment of the is/ought problem, in this solution to the new problem of evil, aesthetic factors are playing the central role, which foreshadows Rolston's comprehensive theory of the aesthetics of nature discussed later in this chapter.

In addition to the is/ought problem and the new problem of evil, there is a third, very closely related problem that arises in light of Rolston's views about natural value and, perhaps even more so than those two problems, aids in the appreciation of his aesthetics of nature. However, it does not have as long a philosophical lineage as do the other two problems, although it has obvious connections with the naturalist tradition. The problem came into being in the heyday of that tradition; as noted, I call it the after-Darwin problem. Rolston's most concise statement of it is in "Is There an Ecological Ethic?": "After Darwin (through misunderstanding him, perhaps), the world of design collapsed, and nature, for all its law, seemed random, accidental, chaotic, blind, crude, an 'odious scene of violence.' "[23] And in "Values Gone Wild" he develops it in such a way that it takes on the dimensions of the new problem of evil gone completely wild:

> The Darwinian revolution has revealed that the governing principle is survival in a world thrown forward in chaotic contest, with much randomness and waste besides. The wilderness teems with its kinds but is a vast graveyard with a hundred species laid waste for one or two that survive. . . . Wildness is a gigantic food pyramid, and this sets value in a grim, deathbound jungle. All is a slaughterhouse, with life a miasma rising over the stench. Nothing of the compassion of which we value in culture is found there.[24]

Thus, Rolston asks "can there be value in the wild holocaust, any reason for society to preserve or admire it?"[25] It seems as if after Darwin the problem is not simply how to account for some evil in nature, but rather how to find any value in it whatsoever!

The solution to the after-Darwin problem begins with Rolston's aside "through misunderstanding him, perhaps." We misunderstand, or at least mis-emphasize, if, for example, we go from the Darwinian observation that "everything born is born to eat and be eaten" to "All is a slaughterhouse, with life a miasma rising over the stench." Anyone who bothers to look, either before or after Darwin, can see that there is a lot of eating and being eaten going on. Darwin's evolutionary theory does not in any important sense *reveal* this; rather it demonstrates that all this eating and being eaten can be viewed in a way that reveals values such as, once again, balance, harmony, order, and stability. The way in which Darwin's theory makes nature seem a "random," "chaotic," and "wild holocaust" is the way in which the correct solution to a problem, especially if the problem is only dimly perceived, can make the problem itself seem much worse. It is precisely in being shown how nature is not after all random and chaotic that some individuals come to mistakenly see it as random and chaotic. This solution to the after-Darwin problem is similar to the above-discussed reply to the new problem of evil, for here again scientific redescription of nature helps us find positive values in nature. Rolston suggests this kind of solution when he notes that "science has been resurveying the post-Darwinian natural jungle and has increasingly set its conflicts within a dynamic web of life . . . Its harmony is often strange, and it is not surprising that in our immaturity we mistook it, yet it is an intricate and delicate harmony nevertheless."[26]

This response to the after-Darwin problem brings out again what is noted above concerning both the is/ought problem and the new problem of evil: that the positive values that scientific redescription helps us find in nature, those that provide a possible solution to each of these problems, are primarily aesthetic rather than moral values. Scientific redescription may turn the whole natural "slaughterhouse," as the worms in Rolston's opossum, into "an intricate and delicate harmony," but it cannot turn it into something moral—even though it may help to show that it does not involve moral issues at all. This way of interpreting Rolston's treatment of these problems may seem to overemphasize the role of aesthetic value rather than moral value. However, in many instances even Rolston's own arguments point primarily to aesthetic rather than moral values. For instance, consider this claim:

> As we progress from descriptions of fauna and flora, of cycles and pyramids, of stability and dynamism, on to intricacy, planetary opulence and interdependence, to unity and harmony with oppositions in counterpoint and synthesis, arriving at length at beauty and goodness, it is difficult to say where the natural facts leave off and where the natural values appear.[27]

Here it seems plausible to say that we arrive at "beauty," but much less so at "goodness," if "goodness" is understood as moral goodness. Indeed, in such passages, the addition of goodness to beauty seems somewhat strained, and in others, Rolston does not bother to add it. For example: "Natural value is further resident in the vitality of things, in their struggle and zest, and it is in this sense that we often speak of a reverence for life, lovely or not. Or should we say that we find all life beautiful?"[28]

SCIENTIFIC COGNITIVISM AND POSITIVE AESTHETICS

The foregoing investigation of the sources and the significance of aesthetic values in Rolston's general theory of natural value demonstrates the central, even if not always explicit, role that aesthetic factors play in Rolston's thought. Moreover, in light of this investigation, it is no surprise to find that in his major discussions focusing specifically on nature's aesthetic values, Rolston endorses versions of the positions that have become known in the aesthetics literature as scientific cognitivism and positive aesthetics. Both scientific cognitivism and positive aesthetics follow almost directly from his theory of values in nature and from the ways in which he attempts to meet the problems discussed in the preceding section. The first, scientific cognitivism, is the view that, roughly put, scientific knowledge is necessary for appropriate aesthetic appreciation of nature.[29] And, as indicated above, it is by reference to scientific knowledge, especially ecological knowledge, that Rolston attempts to ground the value of nature, and, moreover, the values in question are more comfortably thought of as aesthetic rather than as moral. Positive aesthetics is the view, again roughly put, that all virgin nature has only positive aesthetic values—as Muir expressed it: "None of Nature's landscapes are ugly so long as they are wild"—and, as noted above, Rolston's concept of positive value "greatly overlaps" that of the natural and he finds himself "steadily enlarging the overlap" to the point of "coextensiveness"—the point where he asks, "should we say that we find all life beautiful?"[30]

Although they are, as noted above, foreshadowed in his earlier essays, Rolston's versions of both scientific cognitivism and positive aesthetics make their explicit debut in Chapter 6 of *Environmental Ethics*, in which he directly addresses the topic of the aesthetic value of nature. However, the strongest and clearest statement and defense of the former position, scientific cognitivism, is in his "Does Aesthetic Appreciation of Landscapes Need to be Science-based?"[31] Here Rolston contrasts aesthetic appreciation of nature that is informed by science with that which is, on the one hand, influenced by mythological beliefs about nature and, on the other, shaped by dwelling within the natural environment. Insofar as both of the latter help to "constitute" the landscape of the appreciator, Rolston applauds them, for in doing so they contribute to participatory aesthetic appreciation of nature.[32] The participatory theme in his aesthetics of nature is developed more fully in Rolston's "Aesthetic Experience in Forests," where he argues that: "Aesthetic appreciation of nature, at the level of forests and landscapes, requires embodied participation, immersion, and struggle."[33] Nonetheless, although participation is deemed necessary for appropriate aesthetic experience of nature, Rolston's central theme remains that "landscape perception needs to be science-based, as well as participatory," for science is the "primary avenue for perceiving landscapes, better than any other."[34] He concludes that without scientific understanding of "what is really taking place" in the natural environment, an appreciator's experience of nature, even though it may have "on-the-ground immediacy, lacks depth, and this deeper beauty is what science can unfold."[35] He sums up concisely in "Beauty and the Beast: Aesthetic Experience of Wildlife," by claiming that "the more we know the more there is to see, and the more we see, the more there is to be admired."[36]

The idea of scientific knowledge "unfolding" the "deeper beauty" of nature and giving us "more" to appreciate points directly to Rolston's version of positive aesthetics. In *Environmental Ethics* he follows up the suggestions of Muir and William Morris that nothing in wild nature is ugly, weighing considerations both pro and con and concluding "not that virgin nature is invariably aesthetically positive in immediate detail, but that it is *essentially* so."[37] He holds that, as he states in "Aesthetic Experience of Forests": "Like clouds, seashores, and mountains, forests are never ugly, they are only more or less beautiful: the scale runs from zero upward with no negative domain."[38] Moreover, although he seemingly draws the line at "destroyed" and "ruined" nature, his positive aesthetics yet stretches far beyond "easy cases."[39] In both *Environmental Ethics* and "Does Aesthetic Appreciation of Landscapes Need to be Science-Based?" he considers, like his opossum infested with worms, "the rotting carcass of an elk, full of maggots," holding that, although many would find it "revolting," he can "see through" the "environment of the moment into the Environment quintessential" and find it "aesthetically exciting."[40] Likewise, in "Aesthetics in the Swamps," Rolston takes on the "damp, marshy, overgrown, rank, dismal, gloomy" swamp, bog, and mire, the images of which, as he puts it "have 'ugliness' built into them."[41] Nonetheless, he skillfully shows, in a mere 14 pages, how scientific understanding can enhance aesthetic experience such that even these most aesthetically challenging of landscapes are revealed as things of great beauty. Indeed, the essay is a model demonstration of how science can nurture positive aesthetic appreciation.

Turning now to Rolston's explicit line of argument for science-based positive aesthetic appreciation of nature, we find its most systematic development in *Environmental Ethics*. In general, the argument is similar to the line of thought that I outlined above concerning the is/ought problem and especially the new problem of evil and the after-Darwin problem. As with these problems, Rolston appeals to science, especially ecology and evolutionary theory. However, now scientific knowledge is required to directly underwrite positive aesthetic appreciation of the whole natural environment. As in the case of the solutions to those problems, science "enlarges our scope" and puts the apparently ugly parts of nature in a larger context: "The rotting elk returns to the humus, its nutrients recycle; the maggots become flies, which become food for birds," and so on.[42] The general position is that "these processes, abhorrent from the perspective of my individuality, may not be ugly at all in the system, where they are the recycling of resources."[43] The upshot: "ugliness is contained, overcome, and integrated into positive, complex beauty."[44] Rolston further notes the obvious ramification of this view, which is that such appreciation is "not so much a matter of *sight* as of *insight*."[45] In focusing aesthetic experience on the ecological recycling process, the appreciator "has to appreciate what is not evident. There are lots of marvelous things going on in dead wood, or underground, or in the dark."[46] And to achieve this "all-seeing" level of appreciation, here Rolston, again following in the naturalist tradition exemplified by Muir, explicitly makes the move to the God's-eye view: In line with his "aesthetic redemption" of the "dismal swamp," Rolston quotes Muir concerning its most infamous inhabitant, the alligator: "doubtless these creatures are happy and fill the place assigned to them by the great Creator of us all. Fierce and cruel they appear to us, but beautiful in the eyes of God."[47]

This general line of argument, when utilized in the ways noted above, that is, to address the is/ought problem and especially the new problem of evil and the after-Darwin problem, has generated little criticism from individuals interested in the aesthetics of nature. In part, this is because in addressing these problems the *aesthetic* implications of Rolston's ideas, as noted, are mainly implicit. Perhaps it is also because as a solution to these kinds of problems the line of thought is addressed mainly to local nastiness, such as the worm-infested opossum, or to more general ideas, such as that of "nature, red in tooth and claw," which I suspect few environmentally acute individuals any longer find *morally* worrisome. Indeed, perhaps the reason for such "moral evolution" has much to do with the work of individuals such as Rolston. Be that as it may, my primary reason for introducing and discussing these problems, apart from their intrinsic philosophical interest, is not only to bring out the very significant role of aesthetic values in Rolston's general theory of natural value, but also to indicate the genesis of his science-based positive aesthetics in that theory of value and in the solutions to these kinds of problems, especially when these solutions are given an "aesthetic twist" by arguing, as I do above, that they work better with aesthetic values than with moral values. However, when the line of argument is explicitly and fully developed, as Rolston does in *Environmental Ethics*, as a defense of versions of scientific cognitivism and positive aesthetics, then the argument becomes the justification for a general theory of the aesthetic value of the natural world. As such, it becomes the focus of a number of aesthetic criticisms that are worth considering.

Some of the main lines of criticism of Rolston's justification of science-based positive aesthetics are developed by Yuriko Saito and Malcolm Budd. Noting that Rolston appears to hold that "the presumed negative aesthetic value of the dead elk with maggots stems from isolating these objects from a larger context," Saito points to three problematic aspects of this view.[48] First, the view makes unclear exactly what the objects of aesthetic appreciation are meant to be, although seemingly they must be the objects that are claimed to be the bearers of beauty, that is, ecosystems and similar "larger units." However, this leads directly to the second and third worries, the first of which is that "even if we agree that the whole is aesthetically positive, it does not follow that the beauty of the whole implies the beauty of the parts."[49] And third, this change in the objects of aesthetic appreciation seeming places the beauty in question "beyond our ordinary perceptual experience."[50] These last two concerns are echoed by Budd, who points out that the "idea that each ecosystem . . . has a positive overall aesthetic value implies nothing about the aesthetic values of the natural items it contains" and, moreover, that attributing the aesthetic value to the ecosystem precludes "the realistic possibility of one's appreciating that value, no matter how much one's perception of things or events in it might be informed by relevant ecological knowledge."[51]

What these criticisms make clear is that the root difficulty with Rolston's justification for science-based positive aesthetics is that it apparently shifts the focus of aesthetic appreciation from the objects of our perceptual experience, the elk with maggots, to something else, seemingly the whole ecosystem, that is normally beyond perceptual experience. But this is problematic for, as Budd rightly notes: "Perhaps the only requirement imposed by the idea of the aesthetic is that events integral to a system's aesthetic value should be perceptible."[52] The upshot is that given Rolston's way of developing this justification, the positive values in question come dangerously close to being outside of the realm of the aesthetic altogether.[53] Thus, Saito concludes her critique by stressing "that the *aesthetic* value of the elk with maggots is not simply our conceptual understanding of its role in the ecosystem."[54] In order for the experience to be an aesthetic experience, such conceptual understanding "has to be brought back to these individual objects."[55] In some way or other, the understanding has to be "*presented* in the visual composition, as well as in the smell and texture of the decaying animal carcass and the movements of the maggots."[56] It is clear that the issue here is similar to that discussed above concerning the new problem of evil and the after-Darwin problem, although now the stakes are somewhat higher, for the concept of the aesthetic itself requires not simply that the wormy opossum, the rotting carcass, the maggots, and all the rest simply be made out to be acceptable or "good" in some vague sense or other, but rather that *they themselves* must become objects of positive aesthetic experience: they must become "things of beauty."

This is a formidable requirement. However, I think that at least a partial answer to this challenge lies in the above-outlined solutions to the new problem of evil and the after-Darwin problem and especially in the solution to the is/ought problem. The issues raised by Saito and Budd concerning Rolston's justification for science-based positive aesthetics arise from the fact that in developing his particular line of justification Rolston does not fully work out the more radical implications of these earlier ideas. Consider again the following quote from his early essay, "Is there an

Ecological Ethic?":

> Ecological description finds unity, harmony, interdependence, stability, etc., and these are valuationally endorsed, yet they are found, to some extent, because we search with a disposition to value order, harmony, stability, unity. Still the ecological description does not merely confirm these values, it informs them; . . . the earlier data are not denied, only redescribed or set in a larger ecological context, and somewhere enroute our notions of harmony, stability, etc., have shifted too and we see beauty now where we could not see it before.

There are two distinct ideas in this passage. One is that ecological description sets earlier data "in a larger ecological context" and the other is that it "redescribes" such data. The first idea is that which Rolston uses in reply to the new problem of evil and the after-Darwin problem and emphasizes in his explicit justification of science-based positive aesthetics. It is this idea that opens his justification to the objections developed by Saito and Budd. However, the second idea concerning scientific redescription, which is the one primarily involved in the reply to the is/ought problem, is not only more radical, but is also, I suggest, more fruitful in the development and defense of science-based positive aesthetics. It is noteworthy that in summarizing the general line of thought in "Is there an Ecological Ethic?" Rolston emphasizes the second idea: "The ecological revolution . . . is undeniably at work reilluminating the world."[57]

The significance of the idea of science redescribing and thus reilluminating the natural world is that such redescription of things can make them appear to have—reilluminate them as appearing to have—new and different aesthetic values. Consider again the example of science examining the relationship between two organisms and redescribing it as symbiotic. As noted in connection with the is/ought problem, science does not first describe the relationship as symbiotic and then evaluate it as stable and harmonious. Rather in redescribing it as symbiotic, it is automatically, as it were, reilluminated as stable and harmonious. The relationship's symbiosis and its stability and harmony are parts of the same redescription—if the relationship were not stable and harmonious, it would not be symbiotic. The upshot is, to paraphrase Rolston: "we see stability and harmony now where we could not see it before." Moreover and more important, it is not just that we *see* stability and harmony. If the redescription of the organisms' relationship is correct, or at least more correct than previous descriptions, then the aesthetic values of stability and harmony are not only ones that the organisms' relationship appears to have, but also the aesthetic values that it does in fact have, or at least closer to the ones it in fact has than those it appeared to have under previous less correct descriptions.[58] And, contra the worries of Saito and Budd, it is not the larger ecosystem that has these aesthetic values, but the organisms' relationship itself.

If this is a plausible way of understanding scientific cognitivism—of how science plays an essential role in the appreciation of nature's aesthetic values—what then is its relevance to positive aesthetics? How does the fact that when science "redescribes" nature it also necessarily "reilluminates" it as having new aesthetic values provide a justification for the claim that nature's aesthetic values are essentially positive aesthetic values? The justification lies in the fact that the values involved, such as order, unity, harmony, stability, are to a large extent *science's own values*, the values because of which it is sometimes said to be ultimately an aesthetic enterprise. If science could not describe the natural world such that it had at least some of such values, it would not yet have achieved a correct or at least a more correct *scientific* description of nature. As noted above concerning the example of organisms in a symbiotic relationship, if the relationship were not stable and harmonious, it would not be symbiotic. Thus, when such a description is achieved, nature necessarily has some of these values. In this sense, science reads its values into nature; in describing the facts, it does so in such a way that positive aesthetic values are necessarily present. The basic idea is that as science more completely and fully redescribes the natural world it makes it appear more aesthetically good: "we see beauty now where we could not see it before."[59] Moreover, as science more *correctly* redescribes the natural world, it reveals it to be in fact increasingly aesthetically good.[60]

Is this general line of thought adequate to justify positive aesthetics? I remain undecided. Moreover, although the justification seems to sidestep the above-noted problems developed by Saito and Budd, positive aesthetics is yet open to other objections.[61] Much depends on how the idea of positive aesthetics is understood. For example, Budd argues that this kind of position is "unconvincing" if positive aesthetics is interpreted as the doctrine that "all natural things have *equal* positive aesthetic value."[62] However, Rolston seems to clearly avoid this problem, holding that positive aesthetics "does not find all places equally or perfectly beautiful; it maps them on a scale that runs from zero upward but has no negative numbers."[63] Saito also raises another different kind of objection, holding that positive aesthetic appreciation of natural events that cause harm to humans, such as earthquakes and hurricanes, is not "morally appropriate."[64] Whether or not this idea is plausible in itself, as an objection to Rolston's version of positive aesthetics, it, as Budd's, seems to miss the point. As noted above, Rolston's line of thought has its genesis in his account of *natural* value and thus seemingly does not address the issue of nature's effects on humans. Moreover, as also noted in connection with the new problem of evil, on Rolston's view moral concerns "exceed nature," for "nature is nonmoral." A purely natural thing, such as the relationship between Rolston's opossum and its worms, is simply not a moral matter. The same seems to be even more clearly the case concerning Saito's earthquakes and hurricanes, even if humans happen to get in their way.[65] Be that as it may, the question of the opossum and its worms as well as of the elk and its maggots still remains: Can they, even with the most powerful, persuasive, and correct ecological redescriptions, become objects of positive aesthetic experience: things of beauty?[66] Perhaps the best we can do here is simply to agree with Rolston when he claims that: "We do not live in Eden, yet the trend is there, as ecological advance increasingly finds in the natural given stability, beauty, and integrity."[67]

SUBJECTIVISM CONCERNING AESTHETIC VALUE

In the two preceding sections, I have suggested that Rolston's insights about how scientific, and especially ecological, redescriptions help us to see the aesthetic value of the natural world constitute resources not only for defending positive aesthetics, but also for addressing problems such as the is/ought problem, the new problem of evil, and the after-Darwin problem. However, as noted, Rolston himself does not always fully develop these insights in treating these issues. Perhaps this is because of the subjective reputation of aesthetic value. In *Environmental Ethics*, Rolston notes the "need to split aesthetic value off from many other values carried by nature," in part because "some interpreters" find "aesthetic experience to be inevitably subjective."[68] He adds that aesthetic value does not constitute a model of other natural value, the theory of which must have "a more foundational, biologically based account."[69] And in his earlier "Are Values in Nature Subjective or Objective?" he expresses concern about Samuel Alexander's proposal that aesthetic values are tertiary properties, saying "most judges become convinced that these tertiary qualities are overlays, not really there in the natural world. Rainbow-like, only more so, they are gifts of the spectator's mind."[70]

Given Rolston's overriding interest in providing a substantive foundation for environmental ethics, such concerns about the subjective reputation of aesthetic value are understandable. However, if we take seriously Rolston's insights about how scientific, and especially evolutionary and ecological, redescriptions help us to see the aesthetic values of the natural world, then aesthetic value also has what can be characterized as a "foundational, biologically based account." When the aesthetic values of nature are understood in this way, as "imposed by science," concerns about subjectivity are less germane, for the aesthetic values given in the scientific "overlay" are no more "gifts of the spectator's mind" than are the facts. To return to the example of two organisms in symbiotic relationship, the framework of science provides a description of the relationship as symbiotic and as stable and harmonious and such description is ultimately the means by which we both understand and appreciate the natural world. The aesthetic values that are given, like the facts, are not, therefore, the whimsical gifts of some spectator's mind, but rather the long-sought and well-considered "findings" of the scientific worldview. As Rolston himself puts it at the conclusion of "Disvalues in Nature," "a better description of what is objectively taking place makes . . . better evaluation possible, an evaluation that is as objective as is the description."[71]

To fully appreciate this point, it is informative to compare, as Rolston does in "Does Aesthetic Appreciation of Landscapes Need to be Science-Based?," the descriptions of science with those that are offered by other worldviews, such as various religions, mythologies, and ideologies.[72] For example, compare a religious framework that describes the relationship between two organisms as the embodiment of the intertwined souls of two lost lovers with a scientific one that describes it as symbiotic, stable, and harmonious. There is a sense in which the two descriptions are on an even footing, for both are in some sense cultural descriptions. As Rolston rightly notes: "Natural science is our latest and perhaps most sophisticated cultural achievement."[73] Nonetheless, even though both descriptions are in some sense or other cultural, the latter scientific one, being the product of our "most sophisticated cultural achievement," is the superior description in a number of ways. As noted, in "Does Aesthetic Appreciation of Landscapes Need to be Science-Based?" Rolston holds that science is our "primary avenue for perceiving landscapes, better than any other." Scientific description is, as it were, both thinner and better fitting than any other kinds of description; we might think of it as being like a leather glove rather than a woolen mitten. As the glove allows a more objective and perhaps an objective enough view of the hand beneath, so too we might claim that the thin and close-fitting descriptions of science allows an objective enough view of the natural world. And, given the position that scientific description helps us to see the aesthetic values of nature, it follows that those aesthetic values are likewise objective enough.[74]

Nonetheless, even though they are in this way "underwritten" by science, the aesthetic values of nature still might not seem quite "objective enough" to have an important place in a philosophy of the environment. Many might still have qualms about giving aesthetic values rather than moral values the place of prominence that I have suggested throughout this chapter or have misgivings about allowing them to play a role in grounding environmental ethics. In his most recent discussion of the aesthetics of nature, "From Beauty to Duty: Aesthetics of Nature and Environmental Ethics," Rolston notes that: "Aesthetic imperatives are usually thought less urgent than moral imperatives."[75] This is certainly true and it must be admitted that there is, of course, an important and relevant sense in which moral and aesthetic values differ. While it is a moral fault not to recognize moral value, it does not seem to be a similar kind of fault not to recognize aesthetic value; the latter is at most some kind of personal fault.[76] Consequently, there is an imperative not only to do the morally valuable, but also to recognize it, while there is seemingly not a corresponding imperative to recognize the aesthetically valuable. However, be that as it may, once the aesthetically valuable is recognized, there is yet a straightforward imperative concerning it. "This is beautiful, so eliminate or ignore it" is almost as close to "contradiction" as is "This is morally good, so do not do it." Similarly, "This is ugly, so prevent it" is almost as close to "tautology" as is "This is morally bad, so do not do it." The point is that concerning the aesthetically valuable there is an imperative similar to the imperative to *do* the morally valuable. According to this aesthetic imperative, once recognized, ugliness is to be prevented and beauty is to be appreciated and preserved.

The significance of the imperative to appreciate and preserve aesthetic value can be seen by noting its role in our treatment of and regard for works of art and other cultural artifacts of aesthetic value. Once the aesthetic value of such things is recognized, there is seldom any question of whether or not they should be appreciated and preserved—they are immediately secured in museums, galleries, and private collections. In fact, the enthusiasm with which cultural objects of aesthetic value are appreciated and preserved indicates not only that there is an aesthetic imperative, but also that we find it easy to act upon this imperative, in most instances easier than acting upon standard moral imperatives. Indeed, the appreciation and preservation of aesthetic value is quite natural for humans; at one point Rolston rightly speaks of "the human genius for appreciation."[77] Typically, it is not the appreciation and preservation, but rather the discovery and recognition of aesthetic value that is the more difficult. However, it seems that, as with works of art, once individuals truly recognize—truly discover for themselves—the aesthetic value of nature, they find it easy to act upon the aesthetic imperative concerning it. Other than the ever-present problems of competing interests and desires, the difficulty is typically not in convincing individuals to prevent ugliness and to appreciate and preserve natural beauty, but rather in helping them initially to discover the full extent and the great diversity of nature's aesthetic value. This is certainly part of what Leopold had in mind when he concluded the "Conservation Esthetic" by saying that the task is "not of building roads into lovely country, but of building receptivity into the still unlovely human mind."[78]

CONCLUSION

In this chapter I have focused on the genesis and the development of, as well as some of the problems facing, Rolston's aesthetics of nature. I have attempted to bring together the ideas that, although we "do not live in Eden," "the trend is there, as ecological advance increasingly finds in the natural given stability, beauty, and integrity"—and thus that we can "see beauty now where we could not see it before" and perhaps should simply say "we find all life beautiful." These are powerful and important ideas. However, in conclusion, it is also appropriate to return to the somewhat less philosophical themes mentioned at the outset of the chapter and to thereby re-emphasize what is nonetheless an equally important dimension of Rolston's work: the extent to which his writings, like Leopold's, "build receptivity into the still unlovely human mind." As exemplified so well by his early essay "The Pasqueflower" and by his very recent "Aesthetics in the Swamps," his writings nurture, teach, and promote the appreciation of nature—they helps us to discover nature's aesthetic value, its beauty, as much as they philosophically analyze it. In this, as in other ways noted above, Rolston's work stands in a direct line with the great naturalist tradition. In the introduction to *Environmental Ethics*, Rolston offers himself as "a wilderness guide" to lead us into what he calls in "Values gone Wild" "the deep wilderness of values."[79] Along the way, he notes that the discovery of values here "is not so much by argument as by adventure."[80] The image is appropriate and can be extended to his work as a whole—it constitutes both a philosophical and a practical guide for the adventure of discovering and appreciating values in nature.[81]

NOTES

[1] Holmes Rolston, III, *Environmental Ethics: Duties to and Values in the Natural World* (Philadelphia: Temple University Press, 1988).

[2] Holmes Rolston, III, "Mystery and Majesty in Washington County," *Virginia Wildlife* 29 (November 1968): 6–7, 22–23; "Hewn and Cleft from this Rock," *Main Currents in Modern Thought*, 27 (January–February, 1971): 79–83. These two essays along with others, including all those cited in this chapter that were published prior to 1986, are reprinted in Rolston's *Philosophy Gone Wild* (Buffalo:

Prometheus Books, 1986). Some of the ideas presented in this chapter were initially developed in my review of *Philosophy Gone Wild* in *Environmental Ethics* 8 (1986):163–177.

3 Holmes Rolston, III, "Lake Solitude: The Individual in Wildness," *Main Currents in Modern Thought*, 31 (March– April, 1975): 121–126.

4 Holmes Rolston, III, "The Pasqueflower," *Natural History* 88 (April, 1979): 6–7, 13–16.

5 Aldo Leopold, quoted in ibid., 6.

6 Rolston, "Mystery and Majesty," 22.

7 Holmes Rolston, III, "Is there an Ecological Ethic?" *Ethics* 85 (1975): 93–109; "Values in Nature," *Environmental Ethics* 3 (1981): 113–128; "Are Values in Nature Subjective or Objective?" *Environmental Ethics* 4 (1982): 125–151; "Values Gone Wild," *Inquiry* 26 (1983): 181–207; *Environmental Ethics: Duties to and Values in the Natural World.*

8 Holmes Rolston, III, "Can and Ought We to Follow Nature?" *Environmental Ethics* 1 (1979): 7–30; the quote is at 16.

9 Rolston, "Values Gone Wild," 197; *Environmental Ethics*, 222. Although this particular approach to the problem is not my main focus here, since it has little bearing on Rolston's aesthetics, it is worth noting that it involves an account of value that an advocate of the is/ought gap might see as little more than a persuasive definition: "*Value is storied achievement.* With this definition we reach a fundamental motif, which could be widely woven through culture and might be deployed even into inorganic realms, though we are tracing it here only organically." (See "Values Gone Wild," 192.)

10 Rolston, "Ecological Ethic," 100–101. It is suggested to me by Christopher Preston that the line of thought indicated in this quote does not constitute another distinct approach to the is/ought problem, but is rather simply another way of putting the same point, in that the "storied achievement" in an evolutionary ecosystem is "identical" to the "stability, integrity, and beauty" of the system. This may well be correct. However, it does not diminish the significant role of aesthetic factors in what is in that case an alternative way of stating the reply. Indeed, it may indicate that aesthetic factors play an even much larger role in Rolston's general account of value than I suggest in this section.

11 Rolston, "Follow Nature," 29.

12 Ibid., 28.

13 Ibid.

14 Rolston's other solution to the new problem of evil is initially developed in "Can and Ought We to Follow Nature?" and parallels a standard reply to the traditional problem of evil. In this response he recognizes the existence of natural evil and finds it necessary for the existence of good. He claims that the natural world "is nursed in struggle" requiring "both an assisting and a resisting reality"— "a nature where the evils are tributary to the goods." (See 29.) Likewise, this solution is evident in "The Pasqueflower": "This pasqueflower springs forth in its particular form of early beauty as much because of the winter as to spite it: it buds and blossoms because it is blasted. Without the wind, there would be no windflowers, and without the advancing of death, there is no advancing of life." (See 16.) This same kind of solution is also the main line of thought in Rolston's most extensive treatment of this issue, which is in his "Disvalues in Nature," *The Monist* 75 (1992): 250–278. Here Rolston discusses in turn predation, parasitism, selfishness, randomness, blindness, disaster, indifference, waste, struggle, suffering, and death. In most cases in which he recognized the "disvalue" as a natural evil, he suggests that it is necessary for value. For example, of randomness he opines: "It does not seem possible for the world to be otherwise if there is to be autonomy, freedom, adventure, success, achievement, emergents, openness, surprise, and idiographic particularity." (See 259.) The solution to the new problem of evil I discuss in the text of this chapter is also evident in "Disvalues in Nature," but it has a smaller role and is not explicitly related to aesthetic values in that essay. For an extended discussion of Rolston's position in "Disvalues in Nature" and its similarity to traditional attempts to deal with the problem of evil, see Wayne Ouderkirk, "Can Nature be Evil? Rolston, Disvalue, and Theodicy," *Environmental Ethics* 21 (1999): 135–150. As I, Ouderkirk finds two distinct, if interrelated, lines of thought in Rolston's treatment of nature's "evils."

15 Rolston, "Values Gone Wild," 192. It is interesting to note that "parasitism" is second only to "predation" of the "disvalues" Rolston considers in "Disvalues in Nature."

16 George Gaylord Simpson, *Attending Marvels: A Patagonian Journal* [1934] (Chicago: University of Chicago Press, 1982), 199–200.

17 Rolston, "Values Gone Wild," 195.

18 Ibid., 197.

19 Ibid.

20 Ibid., 198; see also *Environmental Ethics*, 220–223.

21 Rolston, "Follow Nature," 28; "Disvalues in Nature," 255. In the latter essay, he cautions concerning nature's value that we must: "Notice carefully that the appropriate evaluative category is not nature's *moral* goodness." (See. 252.) For Rolston's more recent thoughts on the general relationship between human culture and nature, see his "Natural and Unnatural; Wild and Cultural," *Western North American Naturalist* 61 (2001): 267–276.

22 Rolston, "Values Gone Wild," 193.

23 Rolston, "Ecological Ethic," 107. The phrase "odious scene of violence" is from John Stuart Mill's essay "Nature" [1874] and, as Rolston puts it, "characterizes Mill's estimate of nature."

24 Rolston, "Values Gone Wild," 193.

25 Ibid.

26 Rolston, "Ecological Ethic," 107.

27 Ibid., 101.

28 Rolston, "Follow Nature," 23–24. Throughout this section, I have argued that the values that Rolston appeals to concerning the is/ought problem, the new problem of evil, and the after-Darwin problem, that is, values such as harmony, stability, and unity, are by and large aesthetic values rather than moral values. This does not mean, however, that the values involved must be exclusively aesthetic values. Rolston's conception of "natural value" as "storied achievement in an evolutionary ecosystem," that is, as "systemic value," seems to involve an evaluative category broader than simply the conjunction of moral and aesthetic values. But, we might ask, if "systemic value" is not exclusively aesthetic value and not appropriately considered to be moral value, what kind of value is it? In "Disvalues in Nature," Rolston states that the "appropriate category is one or more kinds of nonmoral goodness, better called nature's *value*," adding that: "One appropriate category . . . will be its biological goodness, the extent to which the natural system is pro-life, prolific." (See 252.)

29 I initially sketch out scientific cognitivism, under the name "the natural environmental model of nature appreciation" in Allen Carlson, "Appreciation and the Natural Environment," *Journal of Aesthetics and Art Criticism* 37 (1979): 267–276, reprinted as Chapter 4 of Allen Carlson, *Aesthetics and the Environment: The Appreciation of Nature, Art and Architecture* (London: Routledge, 2000). Other presentations of similar views include: Yuriko Saito, "Appreciating Nature on its Own Terms," *Environmental Ethics* 20 (1998): 135–149; Marcia Eaton, "Fact and Fiction in the Aesthetic Appreciation of Nature," *Journal of Aesthetics and Art Criticism Special Issue: Environmental Aesthetics*, eds. A. Berleant and A. Carlson, 56 (1998): 149–156; Patricia Matthews, "Scientific Knowledge and the Aesthetic Appreciation of Nature," *Journal of Aesthetics and Art Criticism* 60 (2002): 37–48; Glenn Parsons, "Nature Appreciation, Science, and Positive Aesthetics," *British Journal of Aesthetics* 42 (2002): 279–295. The essays by Carlson, Saito, and Eaton, along with a number of others discussing scientific cognitivism, are reprinted in Allen Carlson and Arnold Berleant, eds., *The Aesthetics of Natural Environments* (Peterborough: Broadview Press, 2004).

30 The quote from Muir is from "The Wild Parks and Forest Reservations of the West," in John Muir, *Our National Parks* (Boston: Houghton Mifflin, 1916), 6–7. I develop the positive aesthetics position and consider a number of justifications for it in Allen Carlson, "Nature and Positive Aesthetics," *Environmental Ethics* 6 (1984): 5–34, reprinted as Chapter 6 of *Aesthetics and the Environment*.

31 Holmes Rolston, III, "Does Aesthetic Appreciation of Landscapes Need to be Science-Based?" *British Journal of Aesthetics* 35 (1995): 374–386.

32 The participatory theme in aesthetics of nature is most fully developed by Arnold Berleant; see his *The Aesthetics of Environment* (Philadelphia: Temple University Press, 1992) and *Living in the Landscape: Toward an Aesthetics of Environment* (Lawrence: University Press of Kansas, 1997).

[33] Holmes Rolston, III, "Aesthetic Experience in Forests," *Journal of Aesthetics and Art Criticism Special Issue: Environmental Aesthetics*, eds. A. Berleant and A. Carlson, 56 (1998): 157–166; the quote is at 162. This essay is reprinted in Carlson and Berleant, *The Aesthetics of Natural Environments*.

[34] Rolston, "Science-Based," 380.

[35] Ibid., 383.

[36] Holmes Rolston, III, "Beauty and the Beast: Aesthetic Experience of Wildlife," in *Valuing Wildlife Resources: Economic and Social Perspectives*, eds. D. J. Decker and G. Goff, (Boulder: Westview Press, 1987), 201.

[37] Rolston, *Environmental Ethics*, 244–245. This quote continues as follows: "when even the ugliness is embraced by the sublime." This brings out the fact that in *Environmental Ethics*, 243–245, when dealing with certain kinds of "hard cases," Rolston makes use of, in addition to beauty, the traditional aesthetic category of the sublime. I do not have space to pursue this dimension of his aesthetics in this chapter, but I employ the sublime in a similar manner in Allen Carlson, "Soiden Ihaileminen: Kosteikkojen Vaikea Kauneus" (Admiring Mirelands: The Difficult Beauty of Wetlands), in *Suo on kaunis*, ed. L. Heikkila-Palo (Helsinki: Maakenki Oy, 1999), 173–181.

[38] Rolston, "Forests," 164.

[39] However in "Disvalues in Nature" Rolston explicitly discusses the forces that destroy and ruin nature and here even seems to bring such "disasters" under the umbrella of positive aesthetics. He claims that "violent forces of nature are as much to be celebrated for their creativity as for their destruction" and approvingly quotes Muir's response to the Inyo earthquake that struck Yosemite Valley in 1872: "A noble earthquake! . . . a terrible sublime and beautiful spectacle." (See 265–266.) The quote from Muir is from *The Wilderness World of John Muir*, ed. E. W. Teale (Boston: Houghton Mifflin, 1954), 166–167.

[40] Rolston, *Environmental Ethics*, 238; "Science-Based," 384–385.

[41] Holmes Rolston, III, "Aesthetics in the Swamps," *Perspectives in Biology and Medicine* 43 (2000): 584–597; the quotes are at 584–585. For comparable treatments of the aesthetics of wetlands, see Carlson, "Soiden Ihaileminen: Kosteikkojen Vaikea Kauneus" (Admiring Mirelands: The Difficult Beauty of Wetlands), and J. Baird Callicott, "Wetland Gloom and Wetland Glory," *Philosophy and Geography* 6 (2003): 33–45.

[42] Rolston, *Environmental Ethics*, 239.

[43] Ibid., 241.

[44] Ibid.

[45] Ibid.

[46] Ibid., 239.

[47] Ibid., 240. The quote from Muir is from John Muir, *A Thousand-Mile Walk to the Gulf* (Boston: Houghton Mifflin, 1916), 98.

[48] Yuriko Saito, "The Aesthetics of Unscenic Nature," *Journal of Aesthetics and Art Criticism Special Issue: Environmental Aesthetics*, eds. A. Berleant and A. Carlson, 56 (1998): 101–111; the quote is at 103.

[49] Ibid., 104.

[50] Ibid.

[51] Malcolm Budd, "The Aesthetics of Nature," *Proceedings of the Aristotelian Society*, 100 (2000): 137–157; the quote is at 152–153.

[52] Ibid., 152.

[53] It may come as no surprise if given Rolston's way of developing the justification for science-based positive aesthetics, the positive values in question come dangerously close to being outside of the realm of the aesthetic altogether—especially since, as outlined above, the line of thought has its genesis outside the "realm of the aesthetic" or, at the very least, in a theory of value that considers value in much broader terms than simply aesthetic value, and thus Rolston seemingly does not necessarily conceptualize aesthetic values as "tied to the perceptible" in the way in which the concept of the aesthetic demands.

[54] Saito, "Unscenic Nature," 104.

[55] Ibid.

[56] Ibid.

[57] Rolston, "Ecological Ethic," 108.

[58] I elaborate this version of scientific cognitivism in Allen Carlson, "Nature, Aesthetic Judgement, and Objectivity," *Journal of Aesthetics and Art Criticism* 40 (1981): 15–27 and in "Nature and Positive Aesthetics," reprinted as Chapters 5 and 6 of *Aesthetics and the Environment*. That a somewhat parallel view as applicable to art appreciation is implicit or explicit in much literature in aesthetics. A concise presentation is Kendall Walton's seminal article "Categories of Art," *Philosophical Review*, 79 (1970): 334–367. In the above-noted essays, I elaborate scientific cognitivism utilizing Walton's general framework. For an excellent account of this view in terms of its Waltonian underpinnings, see Parsons, "Nature Appreciation, Science, and Positive Aesthetics."

[59] It might be thought that this whole line of thought, and even the key Rolstonian idea that in light of ecological redescription "we see beauty now where we could not see it before," is undercut by the recent ideas about ecosystem stability developed by ecologists and biologists such as Daniel Botkin. See his *Discordant Harmonies: A New Ecology for the 21st Century* (New York: Oxford University Press, 1990). However, I think we will increasingly find the "discordant harmonies" of nature just as aesthetically pleasing as we have come to find the comparable seemingly discordant, chaotic, and contingent works of 20th Century art and music. Rolston addresses the way in which the new ecology seems to challenge Leopold's land ethic in "The Land Ethic at the Turn of the Millennium," *Biodiversity and Conservation* 9 (2000): 1045–1058, contending that although the new ecology "revises Leopold," it "retains relatively ordered ecosystems," which "order has enough beauty to make environmental ethics a responsibility." (See 1053.) I suspect that the "after the new ecology problem" will turn out to be similar in structure and solution to the after-Darwin problem, and, in the last analysis, no more of a worry. However, if it does turn out to be more of a worry, it will be a worry not just for Rolston's position and for positive aesthetics, but also for, as Christopher Preston points out, the very integrity of science itself.

[60] I develop the kind of relationship between scientific cognitivism and positive aesthetics sketched in this paragraph in "Nature and Positive Aesthetics" and in Allen Carlson, "Appreciating Art and Appreciating Nature," in *Landscape, Natural Beauty and the Arts*, eds. S. Kemal and I. Gaskell (Cambridge: Cambridge University Press, 1993), reprinted as Chapters 6 and 7 in *Aesthetics and the Environment*. For a concise statement of the relationship, see Allen Carlson, "Hargrove, Positive Aesthetics, and Indifferent Creativity," *Philosophy and Geography* 5 (2002): 224–234. However, how sharp a distinction between the two positions should be maintained is open to question. Parsons persuasively argues that "positive aesthetics is not a thesis that stands independent of our account of appropriate aesthetic appreciation of nature, something that we have to show *follows* from that account." Rather he proposes that "positive aesthetics be made 'internal' to the theory of appropriate aesthetic appreciation." (See Parsons, 294.)

[61] In addition to the difficulties posed by Saito and Budd, other issues concerning positive aesthetics are raised by Eugene Hargrove and Stan Godlovitch; see Hargrove's *Foundations of Environmental Ethics* (Englewood Cliffs: Prentice Hall, 1989), Chapter 6, and "Carlson and the Aesthetic Appreciation of Nature," *Philosophy and Geography* 5 (2002): 213–223; and Godlovitch's "Evaluating Nature Aesthetically," *Journal of Aesthetics and Art Criticism* 56 (1998): 113–125 and "Valuing Nature and the Autonomy of Natural Aesthetics," *British Journal of Aesthetics* 38 (1998): 180–197.

[62] Budd, 147. This objection, as well as some others raised by Budd that I do not discuss here, is in fact aimed more at some of my remarks about positive aesthetics than at Rolston's. In "Appreciating Art" I follow up Hans Arp's claim that "in nature a broken twig is equal in beauty and importance to the clouds and the stars" and consequently misleadingly imply Budd's interpretation of positive aesthetics by saying that all natural objects "seem equally aesthetically appealing" or are "more or less equally appreciable." However, I suggest that these remarks be read as claiming only that all natural objects, since all have positive aesthetic value, are equally worthy of appreciation, and not that they all are actually equal in positive aesthetic value. I think the latter claim would be difficult to justify with any argument. Given his view that aesthetic appreciation of nature is "endowed with a freedom denied to artistic appreciation," Budd finds more plausible a version of positive aesthetics maintaining "that each natural thing, at some level of observation, has a positive aesthetic value." This version seems plausible enough, but is, I think, too weak. It does not accommodate what seems to me undeniably true: that, even granting

Budd's promiscuity concerning not only "levels of observation" but also "conditions of observation and time," each natural thing, at many, if not almost all, levels and conditions of observation, has substantial positive aesthetic value and little, if any, negative aesthetic value. (See Budd, 155–156.)

63 Rolston, *Environmental Ethics*, 237.

64 Saito, "Unscenic Nature," 109.

65 This is not to say that there are not other moral issues concerning these kinds of cases. Ned Hettinger rightly points out to me that: "There is the moral question about whether or not we ought to alleviate the suffering of the opossum and in that sense it is a moral matter. There is also the moral question of whether or not we ought to let ourselves positively aesthetically appreciate this situation. Is it morally appropriate to find a hurricane aesthetically positive while it destroys people's homes and lives? While the hurricane is not a moral agent that is appropriately held responsible, we are moral agents and can be held morally responsible for our aesthetic tastes/responses. So I don't think the fact that nature is amoral is sufficient to respond to the objection that we morally ought not appreciate natural phenomenon that cause great suffering" (personal correspondence). Hettinger's comments are in line with Saito's remark to the effect that aesthetic appreciation in such cases may not be "morally appropriate," but nonetheless they do not counter the contention that Saito's examples miss the point as objections to Rolston's version of positive aesthetics. At most they only show that it may not always be morally acceptable for humans to aesthetically appreciate every case of positive aesthetic value. For an elaboration of Hettinger's views on positive aesthetics, see Ned Hettinger, "Allen Carlson's Environmental Aesthetics and the Protection of the Environment," *Environmental Ethics* 27 (2005): 57–76.

66 A number of other kinds of putative exceptions to the positive aesthetics position are discussed in the literature. In addition to "natural disasters of massive scale and power, such as a hurricane, earthquake, tornado, avalanche, tidal wave, volcanic eruption, flood, and the like," Saito also considers "fleas, flies, cockroaches, and mosquitoes." She argues that many of such cases are not, in the final analysis, true exceptions, but that concerning the former positive aesthetic appreciation may be neither psychologically possible nor, as noted, morally appropriate. (See Saito, 106–107, 109.) For a follow-up discussion, see Robert S. Fudge, "Imagination and the Science-based Aesthetic Appreciation of Unscenic Nature," *Journal of Aesthetics and Art Criticism*, 59 (2001): 275–285. Budd considers cases of living things with a condition that is "diseased or malformed or indicative of approaching death." I think Budd's cases are closer to true exceptions, or at least the second of the three conditions is, in that "grossly malformed living things," more so than either diseased or dying ones, might be viewed as "nature's mistakes." (See Budd, 147, 149.)

67 Rolston, "Ecological Ethic," 108.

68 Rolston, *Environmental Ethics*, 233.

69 Ibid.

70 Rolston, "Subjective or Objective," 128; see also *Environmental Ethics*, 208–210.

71 "Disvalues in Nature," 275.

72 I discuss this issue in "Landscape and Literature," in *Aesthetics and the Environment*.

73 Rolston, *Environmental Ethics*, 9. There are those who would argue that the recognition that science is a "cultural achievement" undercuts its claim to objectivity. Rolston addresses this concern in "Natural and Unnatural; Wild and Cultural" and especially in "Nature for Real: Is Nature a Social Construct?" in *The Philosophy of the Environment*, ed. T. D. J. Chappell (Edinburgh: University of Edinburgh Press. 1997). For example, in the latter he considers Richard Rorty's view that "we must resist the temptation to think that the redescriptions of reality offered by contemporary physical or biological science are somehow closer to 'the things themselves.' " See Richard Rorty, *Contingency, Irony and Solidarity* (New York, Cambridge University Press, 1989), 16. But, we might ask, do philosophical worries about getting closer to "the things themselves" undercut the view that scientific description is objective enough? (See note 74.)

74 How objective is objective enough? I tend to agree with Jenna Thompson, who argues that what is required is that "we can and do give reasons for our aesthetic judgments," which constitute "grounds that rational, sensitive people can accept." She adds: "To insist that value judgments must be objective in this sense does not require us to suppose that beauty and other aesthetic values are real properties of

objects," since "it is not necessary to settle metaphysical questions about beauty in order to defend the idea that our aesthetic judgments are, or can be, objective." See Jenna Thompson, "Aesthetics and the Value of Nature," *Environmental Ethics* 17 (1995): 291–305; the quote is at 292–293. As Thompson's remarks suggest, any stronger sense of objectivity may require realism concerning aesthetic properties. For a recent defense of such a realist view, see Eddy Zemach, *Real Beauty* (University Park: Pennsylvania State University Press, 1997). For a critique of some of Zemach's arguments, see Glenn Parsons and Allen Carlson, "Critical Notice of Zemach, *Real Beauty,*" *Canadian Journal of Philosophy*, 29 (1999): 635–654.

75 Holmes Rolston, III, "From Beauty to Duty: Aesthetics of Nature and Environmental Ethics," in *Environment and the Arts: Perspectives on Environmental Aesthetics*, ed. A. Berleant (Aldershot: Ashgate, 2002), 127. Remarks such as this, as well as others quoted above from *Environmental Ethics*, suggest that Rolston's account of aesthetic value may not lend itself to the objective treatment I suggest in this section. Indeed Eugene Hargrove explicitly criticizes Rolston's aesthetics, in contrast with his general theory of natural value, mainly on the grounds that it is not adequately objective. See Eugene Hargrove, "Rolston on Objective and Subjective Beauty in Nature," *This Volume*, pp. 125–142. However, in both *Environmental Ethics*, 235–237, and in "From Beauty to Duty," 132–134, Rolston distinguishes between what he calls "aesthetic properties" and "aesthetic experience" or "aesthetic capacities." He holds that the former, as opposed to the latter, "lie objectively in natural things," and, since he includes among them the kinds of aesthetic properties discussed throughout this chapter, such as order, symmetry, and unity, his position would seemingly justify the claim that aesthetic values are "objective enough" to play a role in grounding environmental ethics. Indeed, in "From Beauty to Duty," he reaches this conclusion. After considering both sides of the question "Can aesthetics be an adequate foundation for an environmental ethic?," he concludes, in line with what I have argued here: "Yes, increasingly, where aesthetics itself comes to find and to be founded on natural history." (See 140.)

76 For example, in "From Beauty to Duty," although Rolston contends that it is "not an option" to say that a beautiful and graceful running impala is awkward and ugly, he yet adds that to do so would "simply reveal my ignorance, my insensitivity." (See 132.)

77 Rolston, "Values Gone Wild," 190.

78 Aldo Leopold, "Conservation Aesthetic" [1948], *A Sand County Almanac, with Essays on Conservation from Round River* (New York: Ballantine, 1966), 295.

79 Rolston, *Environmental Ethics*, xiii; "Values Gone Wild," 181.

80 Ibid., 198.

81 I greatly appreciate and have benefited from observations by Ned Hettinger, Wayne Ouderkirk, Glen Parsons, and Christopher Preston.

EUGENE HARGROVE

8. ROLSTON ON OBJECTIVE AND SUBJECTIVE BEAUTY IN NATURE

Beauty appears to be very important to Rolston's environmental ethics, and he has written extensively on it.[1] In some important respects, he has followed the emphasis of Aldo Leopold in *A Sand County Almanac*, where Leopold writes: "A thing is right when it tends to preserve the integrity, stability, and beauty of the biotic community. It is wrong when it tends otherwise."[2] In his book *Environmental Ethics*, Rolston writes, "When humans gain a description of how ecosystems work, Leopold believes that a prescription arises to respect the beauty, integrity, and stability of such systems."[3] "The beauty, integrity, stability of an ecosystem can put constraints on appropriate human conduct in both small and larger ways."[4] Moreover, he has repeatedly listed beauty or aesthetic value as a key value in his quest to demonstrate the presence of values in nature.[5]

I argue here that Rolston's treatment of beauty in his book is inconsistent and incompatible with the general direction of his views on value in nature. Generally, Rolston holds that value in nature is objective. He provides hints early in the book that beauty will be objective too. However, in the third section of chapter six, he separates beauty from the rest of natural value, treating it as subjective, rather than objective, as something arising out of human perception. Although Rolston generally argues for a notion of autonomous intrinsic value in nature against a truncated subjectivist conception held by J. Baird Callicott, ultimately he adopts a view of natural beauty that in significant ways mirrors Callicott's conception.

In an article published in 1982, Rolston announced an agenda that can be viewed as a major direction of his writings and research to the present:

> My strategy in what follows is to fight a way through how we know what we know (what philosophers call epistemological issues surrounding the terms subjective and objective) in order to reach the state of affairs in the real world and to be able to defend the existence of value there (what philosophers call ontological issues surrounding subjectivity and objectivity). In doing this, I keep the whole discussion as close to science as I can, while demanding a full-blooded, no nonsense account of the phenomenon of value in, and valuing of, the natural world. Earlier on, I admit to some inescapable blending of the subjective and objective, but later on, after this admission, I defend all the objectivity I can for natural value.[6]

Rolston has undertaken this agenda to show that values exist objectively in nature because he fears that as a result of the relativism of quantum physics, "the question of values being objectively 'out there' seems hardly discussible. The subjectivists

C.J. Preston and W. Ouderkirk (eds.), Nature, Value, Duty, 125–142.

have won all the chips. . . . What hope is there for value theory to do anything more than record what appears and seems? . . . Subjectivity has eaten up everything, even the fact/value distinction."[7]

In his book *Environmental Ethics*, Rolston continues his defense of objective value in an extended criticism of Callicott's theory of truncated intrinsic value. He quotes Callicott as follows:

> The source of all value is human consciousness, but it by no means follows that the locus of all value is consciousness itself. . . . An intrinsically valuable thing on this reading is valuable for its own sake, for itself, but is not valuable in itself, i.e. completely independently of any consciousness, since no value can in principle . . . be altogether independent of a valuing consciousness. . . . Value is, as it were, projected onto natural objects or events by the subjective feelings of observers. If all consciousness were annihilated at a stroke, there would be no good and evil, no beauty and ugliness, no right and wrong; only impassible phenomena would remain.[8]

Rolston objects to Callicott's view, which he characterizes as merely "projecting intrinsic value on nature," on the grounds that this account leaves no room for any conception of value in nature. He writes: "All the attributes under consideration are objectively there before humans come, but the attribution of value is subjective." He adds, "Despite the language that humans are the source of value which they locate in the natural object, no value is really located there at all. The only new event is that these properties are registered in—translated into felt values by—the perceptual apparatus of the beholder."[9]

Early passages in *Environmental Ethics* suggest that Rolston is committed to an objective sense of natural beauty, as, for example, when he writes: "When I delight in the wild hawk in the windswept sky, that is not a value that I invent but one that I discover—and follow."[10] Perhaps the clearest example comes in a passage in a section titled "Value Ownership":

> That values are actualized in real things, often natural things, seems to warrant the view that valuing is sometimes in part a form of knowing where we register properties—aesthetic properties of the Grand Canyon, for instance, in the appreciating mind—notwithstanding what we may add in the appreciating process. Otherwise, we commit the fallacy of misplaced location and ascribe to the viewer what is really in the scene, or at least what comes relationally. But if we fail to recognize, on the other hand, that the canyon has no aesthetic capacities, we tumble into the pathetic fallacy, projecting onto natural things sentiments that exist only in the human mind.[11]

This passage appears to be very strong support for the idea that Rolston believes there is objective aesthetic value in nature. He speaks of the aesthetic properties of a nonliving natural object, the Grand Canyon, as if they exist in the natural object. He warns against committing the "fallacy of misplaced location," in which properties existing in one place (the world) are falsely considered to exist in another (the human mind). Further, his remarks about "projecting onto natural things sentiments that exist only in the human mind" seems to be a tie into his coming criticism of Callicott's truncated intrinsic value theory as one that projects value on nature.[12]

This view is reiterated again in the chapter, "Endangered Species," where he warns against the fallacy of misplaced wonder with regard to the aesthetic appreciation of species:

> We might say that humans ought to preserve an environment adequate to match their capacity to wonder. All species might be resources for wonder. But this is to value the experience of wonder, rather than the objects of wonder. Valuing merely the experience seems to commit a fallacy of misplaced wonder, for speciation is itself among the wonderful things on Earth.[13]

This fallacy appears to be a special case of the fallacy of misplaced location, a misplacement of wonder (or the sublime), attributing wonder to the human mind rather than to objects in the natural world.

Given that these passages seem strongly to be in support of the idea that natural beauty is objective, "out there," and not just in the mind, it was jolting and disheartening to me, and I believe to many other readers, to find that Rolston ultimately rejects objective natural beauty in the "Valuing Aesthetic Beauty" section of chapter six, "The Concept of Natural Value":

> Phenomena such as life and life support, nutrition, resource capture and recycling, photosynthesis, oxygen transport by hemoglobin, warblers regulating insect populations, genetic mutations producing varying phenotypes in novel niches, speciation—all of these and many of the values they carry we may concede to be objectively present in nature before humans come. But beauty? With beauty we cross a threshold into a realm of higher value: the experience of beauty is something humans bring into the world. Just as there is no creature with a world view and an ethic before humans arrive, nothing has any sense of beauty. Humans ignite beauty, rather as they ignite ethics in the world.[14]

Clearly, Rolston denies any objectivity to natural beauty. It is not "objectively present in nature before humans come." It exists entirely within human experience and this experience "is something that humans bring into the world." As a result, beauty becomes subjective in his view—dependent on the subjective experiences of human beings.

In connection with this denial of objectivity for beauty, Rolston draws a parallel between ethics and aesthetics. He claims that ethics comes into the world with human beings and that aesthetics follows the same pattern. Just as no nonhuman animals have a sense of right and wrong, no nonhuman animal has a sense of beauty. This analogy between ethics and aesthetics does not really rescue Rolston, however, because ethics and aesthetics seem to work differently in Rolston's text. In Rolston's view, ethics does arise with the appearance of humans because nature is amoral or immoral.[15] Thus, although Rolston discovers or finds the aesthetic value of the windswept hawk in nature, there is no parallel ethical value to be found in the hawk or anything else. Likewise, while there are aesthetic properties actualized in the Grand Canyon waiting to be registered in the appreciating human mind, there are no comparable ethical properties to be found in it waiting for the ethical human mind to discover them. Thus, the alleged parallel between ethics and aesthetics fails. While there is something to be discovered in nature aesthetically, there is nothing to find ethically.

Although I am not a dedicated supporter of Rolston's general position on objective value in nature, believing that he overemphasizes objectivity, I am very interested in the role of natural beauty in environmental ethics or environmental philosophy; and I consider his position on nature aesthetics, which denies any significant objectivity to natural beauty, to be not only inconsistent with his general position in environmental philosophy, but also damaging to the quest for an environmental ethic that can be generally accepted by ordinary people. The problem can be most easily seen in reference to his attempts to tie his environmental ethic to Leopold's land ethic. Leopold spoke of "integrity, stability, and beauty."[16] Rolston reordered these values listing them as "beauty, integrity, and stability," seemingly placing a stronger emphasis on beauty.[17] However, if beauty is totally subjective, something that humans bring to the world, it has no place in determining "what is right" in Leopold's sense with regard to the natural world. Rolston seems to be fully aware that he is undercutting Leopoldian beauty for he writes just before the fateful passage: "Stability and even integrity (wholeness) are objectively present in biotic communities, but not beauty."[18]

The most obvious objection to Rolston's position can be expressed in the context of his discussion of the fallacy of misplaced wonder, quoted above. In that passage, the error that invokes the fallacy is a focus on the experience of wonder without proper recognition of the objects of wonder. In the passage tying beauty to the appearance of human consciousness, Rolston asks questions in terms of beauty and answers those questions in terms of the experience of beauty: "But beauty? With beauty we cross a threshold into a realm of higher value: the experience of beauty is something humans bring into the world." One might, for example, paraphrase the passage on the fallacy of misplaced wonder as a criticism of Rolston on beauty as follows:

> We might say that humans ought to preserve an environment adequate to match their capacity for beauty. All natural objects might be resources for beauty. But this is to value the experience of beauty, rather than the objects of beauty. Valuing merely the experience seems to commit a fallacy of misplaced beauty, for natural objects, living and nonliving, are themselves among the beautiful things on Earth.

To this paraphrased objection, if his original objection means anything, I submit, Rolston has no reply, since the wonderful (the awesome, the sublime) is a specific kind of beauty (or aesthetic value). If there is a fallacy of misplaced wonder, then there is no good reason not to have a fallacy of misplaced beauty as well. Rolston has simply replaced objective natural beauty with the experience of natural beauty in humans, making a parallel mistake. One might even be tempted to call his view the truncated theory of beauty. It is the experience of something that he says does not actually exist in nature, something that is brought to the world by human beings and projected on it.

The pronouncement that beauty does not exist in nature, however, does help explain the earlier discussion of the aesthetic properties of the Grand Canyon and the fallacy of misplaced location. At the end of that paragraph, Rolston adds: "But if we fail to recognize, on the other hand, that the canyon has no aesthetic capacities, we tumble into the pathetic fallacy, projecting onto natural things sentiments that exist only in the human mind."[19] Apparently, in light of the denial that beauty exists in nature, the distinction between aesthetic properties and aesthetic capacities places the aesthetic properties in nature and the aesthetic capacities in the human mind. Beauty is created by the human mind receiving the aesthetic properties through sensation, processing them, and then projecting them back on the natural object. This interpretation, however, completely changes the significance of the passage. Although the reader thinks on first reading that Rolston is arguing that natural beauty exists objectively in the world, in retrospect he apparently holds the opposite view, that it is projected onto the world. Seemingly, aesthetic properties in the world are not enough to establish the existence of natural beauty. Aesthetic capacities in a human mind are required and the beauty ignited through those capacities remains outside of the world in the mind alone.

One might think that Rolston's conclusion comes without any clear preparation. A clue to such preparation, however, can be found in the paragraph immediately before Rolston's declaration that beauty exists only in the human mind, where he writes: "Only by accident (epiphenomenon) does nature excite the subjective, human experience of beauty. . . . Nature is not an artist: it only sometimes chances to echo our aesthetic tastes."[20] Earlier in the chapter "Natural Value," Rolston discusses five kinds of natural value: natural value as epiphenomenon, as echo, as emergent, as an entrance, and as an education.[21] Rolston refers to the first two of the five E's in the section titled "From Epiphenomenal to Educational Value." Rolston's reference to beauty as epiphenomenon is not flattering, for natural value as epiphenomenon is not "enlightening," "a nonexplanation," "an epigenetic anomaly."[22] As initially presented, natural value as echo is little better. As an echo, natural value happens "to mirror the sweep and line of [our] subjective preference." It is "unecological."[23] Basically, Rolston's echo position is an acceptance of Callicott's more general position, but in light of the section "Valuing Natural Beauty," applied to beauty alone. Although natural value in general does not arise with human consciousness, and natural value in general is not projected on the world by humans, such is not the case with regard to natural beauty. With regard to beauty, Rolston adopts Callicott's view. Humans project beauty on the world.

Having established that Rolston elects to apply a rival theory, of which he generally disapproves, to account for natural beauty, the appropriate question to ask is whether the application of the rival theory is appropriate and justifiable. I would argue that Rolston's application of Callicott's view to natural beauty is weak from the outset, for in the final paragraph of his discussion of natural value as echo, Rolston writes: "A valuing system would be an odd benefit indeed unless it better fitted us to our home niche. The echoing is most often working the other way around; the human valuer is reflecting what is actually there."[24] These remarks obviously undercut the conclusions that Rolston reaches later in the chapter, that beauty is projected onto the world. If "the human valuer is projecting what is actually there," then he or she is not projecting something that is not there on the world. If the echo model reflects what is "actually there," and beauty is an echo, then beauty does not come into the world with humans, but is already there. The issue thus becomes whether echoing is "working the other way around" in the case of beauty. Rolston says no, but how does he know? How can he tell that he is projecting value on nature rather than reflecting what is already there?

Rolston tries to resolve this question, as far as I can tell, in part by trying to determine whether any other living being also has a sense of beauty. Curiously, Rolston makes some interesting concessions on the page immediately following his claim that beauty is something that humans bring into the world. He writes: "Unless we think that birds and beasts have no experience at all, it is difficult to deny them the precursors of aesthetic experience." Then he adds: "So we might conclude that beauty is not objectively present in nature; it is in the eye of the beholder—mostly in human eyes, perhaps nascent in the eyes of birds and beasts."[25] Although this uncertainty about whether beauty comes into the world with other organisms should undermine Rolston's claim that beauty comes into the world with the emergence of humans alone, it does not. Rolston invokes the distinction between aesthetic capacities and aesthetic properties and goes on to say that "when we value [the projects of projective nature] aesthetically, our experience is being *superposed on* natural properties."[26] The next paragraph then produces, without any special justification, the claim that Rolston is supposedly trying to prove: "In a sense, systemic nature even has an aesthetic power, since it is able to produce aesthetic properties, even though nature does not have—until it produces humans—the capacity for aesthetic experience."[27] Perhaps, Rolston feels that work is being done by the word *superposed*. This word means "to lay or place on, over, or above something else," or geometrically, "to make (one figure) coincide with another in all parts, by or as if by placing one on top of the other." Rolston presumably intends less than *superimpose*, which means "to put, lay, or stack on top of something else," or more strongly, "to add as a dominant or unassimilated feature."[28] Clearly, *superposed* and *superimposed* are not identical terms, but possibly the addition of *on* to produce *superposed* on bridges the gap to superimposed, to justify the otherwise unexplained "until [nature] produces humans." Whatever is going on, however, Rolston actually gains little on a close reading, for the very next sentence includes the same ambiguity found at the end of the section on the echo model: "When humans arrive and value wild nature, we are sometimes valuing a projective nature that we are discovering, more than we are projecting our values onto nature. Nature carries aesthetic properties objectively, and these are ignited in the subjective experience of the arriving beholder."[29] In the first of these sentences, Rolston states that natural beauty is either discovered in nature (it is already there) or projected on nature (and is therefore not there at all). In the next sentence, natural beauty is not out there in nature (only aesthetic properties are there) and the natural beauty is ignited in the subjective beholder when it arrives in sensation. Because Rolston's own account acknowledges that natural beauty is sometimes discovered in nature (for example, the windswept hawk) and sometimes produced out of the ignition of aesthetic properties arriving in sensation, he has not made his case that natural beauty comes into the world with the arrival of human beings. His actual position, considering his contradictory claims, is that natural beauty sometimes comes into the world with the arrival of human beings and that sometimes it is discovered in nature by human beings (implying that it is already objectively there).

I frankly do not understand Rolston's reluctance to recognize a sense of beauty in some nonhuman organisms. The most likely reason seems to be that he is displaying an unwarranted prejudice in favor of the superiority of human beings over other animals. For example, such a prejudice may be reflected by his remark that "With beauty we cross a threshold into a realm of higher value: the experience of beauty is something humans bring into the world."[30] Perhaps higher value is a code word for "beyond the capabilities of nonhuman animals."

The idea that some nonhuman animals do not have a sense of natural beauty seems to me to be undercut by Rolston's discussion of ants in the section on natural value as an emergent. Here Rolston claims, "Sugar is not sweet all by itself, but it is sweet both to humans and to ants."[31] I confess that I have no idea whether ants find sugar to be sweet or not, and I can think of no evidence that could support the claim. Granted, ants may prefer to carry sugar back to the ant hill in preference to other white substances, but whether they do so because they find it or judge it to be sweet is for me an open question. Rolston's defense of his claim seems to be that ants have experiences and that things that have experiences make value judgments: "Valuing dilutes across the simplifying of the central nervous system, but if we rely entirely on the emergent account, value is never extraneural. Where there are no centers of experience, valuing ceases and value vanishes."[32] It seems to me that if ants can judge sugar to be sweet, then it is no more problematic to conclude that other insects judge flowers to be beautiful and that birds make aesthetic judgments about the opposite sex in their species during mating. Indeed, these two cases seem to me to be easier than the ants and sugar case because the preferences of insects and birds, conscious or not, produced the beauty that we find in flowers and birds. Although this beauty can properly be called emergent, since it has evolved within nature, it is also the product of choice—indeed, subjective choice if Rolston is correct that ants and other creatures have experiences.

As noted above, Rolston's claim that no nonhuman animal has a sense of beauty is in part based on an analogy between aesthetics and ethics, in which he claims that no nonhuman has an ethic: "Just as there is no creature with a world view and an ethic before humans arrive, nothing has any sense of beauty. Humans ignite beauty, rather as they ignite ethics in the world." While it may be true that before humans arrived, there was no nonhuman animal that knew it had an ethic, it is not true that there was no ethical-like behavior. Obviously, nonhuman social animals do engage in social behavior and this behavior can be viewed as a forerunner of human ethics thought of as a code of social behavior for humans. This behavior can be attributed to evolutionary factors, and apparently appeared quite early, since social behavior is common among such insects as ants and termites. While human ethics can be regarded as very sophisticated in comparison with the social behavior of other animals, there is ethics among other social animals to the degree that there are recognizable standards of right and wrong behavior, recognizable not only to human ethologists, but also to the individual animals within each community. The claim that there is no ethics in nature is true in the sense that other animals do not have theories of ethics as we do. However, the claim is a matter of degree only, since most humans operate in terms of social standards, without an ethical theory, much as other social animals do. Viewed this way, the analogy with ethics does not rule out the evolution of nature aesthetics in much the same way that standards of social behavior have evolved among some animals.

In the early pages of *Environmental Ethics*, Rolston toys with two views of the beauty of wildflowers. In one view, the beauty is in the subjective experience of humans. Thus, to protect the flowers, it is appropriate to put up signs asking people to leave the flowers for others to enjoy. In the other view, the focus is on the welfare of the flowers themselves. The wording for the sign representing this view is "Let the flowers live!" Reflecting on an occasion in which the first signs were replaced by the second, Rolston writes:

> Will the new signs be more effective? Do they represent a shifting environmental ethic? Is this only an aesthetic appeal? "Let lovely things be?" Is it a psychological appeal? "Don't vandalize!" Are the new signs subtly trying to recommend an experience? "Appreciate beautiful things?" Or is there a respect for life, replacing what on the earlier signs was only a respect for persons? (Would you recommend replacing signs that read, "Don't crosscut switchbacks," with new signs: "Give Earth a chance!"?) Perhaps the signs mean, "Let the flowers have their own standing!"

Rolston answers his own question at the beginning of the next paragraph as follows: "There seems no reason why such own-standing normative organisms are not morally significant. That is, a moral agent in deciding his or her behavior ought to take account of the consequences for other evaluative systems."[33] Seemingly, the wildflower counts in terms of its autonomous objective intrinsic value. This conclusion is further confirmed a few pages later where he writes: "The old signs, 'Leave the flowers for others to enjoy,' were application signs using a humanistic ethic. The new ones invite a change of reference frame."[34]

However, at the end of the chapter, in a section titled "Human Interests and Organismic Values," Rolston takes a different turn with the introduction of a new principle, "the principle of the nonloss of goods," intended as a supplement to his earlier "principle of the nonaddition of suffering."[35] After affirming that "We eat plants, as we eat animals," Rolston writes:

> In making judgments at this level, however, the principle of the nonaddition of suffering will not work, since there is no suffering. We can substitute a principle of the nonloss of goods. The goods preserved by the

> human destruction of plants must outweigh the goods of the organism destroyed; thus, to be justified in picking flowers for a bouquet one would have to judge correctly that the aesthetic appreciation of the bouquet outweighed the goods of the flowers destroyed. One might pluck flowers for a bouquet but refuse to uproot the whole plant, or pick common flowers (daisies) and refuse to pick rare ones (trailing arbutus) or those that reproduce slowly (wild orchids).[36]

The introduction of this principle, it seems to me, is a 180-degree turn away from the theme of the "Let the flowers live!" signs. Here the subjective feelings of the individual human moral agent rule. Here the pleasure of the individual human outweighs the survival welfare of the flower if that individual concludes that picking the flower will be adequately pleasurable. Here intrinsic value as "psychological interest satisfactions desired without further contributory reference, pleasures good in themselves,"[37] trump the autonomous objective intrinsic value that Rolston more generally defends. The wildflower is thus not like the warbler, whose "life is defended intrinsically, without further contributory reference."[38]

The difference between the warbler and the wildflower seems to be that the warbler is an "object-with-will" whereas the wildflower is not. Rolston writes: "There is an object-with-will, even though there is no subject-with-will. The organism is genetically programmed to argue, to probe, to fight, to run, to grow, to reproduce, to resist death."[39] Apparently, because the wildflower cannot resist death, it lacks autonomous objective intrinsic value, or at least enough of it to be valued on the basis of its autonomy.

Curiously, Rolston seems to provide some justification for the implications of his principle of the nonloss of goods at the beginning of the section in which he criticizes Callicott's theory of truncated intrinsic value. There he writes that "humans may value sequoias as timber but may also value them as natural classics for their age, strength, size, beauty, resilience, majesty." He then adds:

> Let-the-flowers-live valuing is of this kind; humans make no instrumental, consumptive use of the flowers. They do not pick them. But they do view them. This viewing constitutes the flowers' value, a value not previously present in the flowers independent of the human presence. Still it is a value that, when it appears as a product of subjective awareness, is attached objectively to the flowers flourishing in the meadow, not attached instrumentally in relation to some resource use humans may make of them—for instance, as a bouquet. Value thus requires subjectivity, since only subjectivity can coagulate it in the world. But the value so coagulated, we will claim, is objectively intrinsic to the nonsentient life and not merely instrumental.[40]

Rolston seems to be saying here that the flower has no objective intrinsic value of its own (or at least very little) until the human comes to observe it and finds it pleasing. If the human observer merely looks at and admires the flower, the human engages in a "nonconsumptive use" of the flower. If, however, the human picks the flower for a bouquet or for any other reason, the flower is now being consumptively used and presumably loses its intrinsic value, becoming "merely instrumental" to the intrinsic value exemplified as human pleasure (a "psychological value satisfaction"). The flower acquires or loses its objective intrinsic value depending on the use that the human selects, either consumptive or nonconsumptive.

It might be objected that the passage above is Callicott's view, not Rolston's own. However, the passage concludes with a sentence containing the words, "we will claim," indicating that, however much the view in the paragraph might be similar to Callicott's, it is really adjusted or fixed by the fact that the value discussed is "objectively intrinsic" and "not merely instrumental." Nevertheless, the objective intrinsic value Rolston refers to here cannot be his autonomous intrinsic value because the value as "a product of subjective awareness" that is "attached objectively" to the flower is not so attached because of whatever autonomous intrinsic value it may have. Rolston merely states that it is objective because it is not instrumental. I cite this passage because it is a passage that could easily be taken to be a presentation of Callicott's view, which nevertheless helps explain Rolston's rationale for the principle of nonloss of goods, suggesting that there is little difference in the views of Callicott and Rolston as far as aesthetics is concerned.

Because the principle of nonloss of goods provides a justification for the sacrifice of wildflowers for the subjective pleasure of a single human, aesthetics seems to offer little protection of individual wildflowers. The greater the beauty of a wildflower, the greater the pleasure for the human, and the greater the chance that the flower will be picked and prevented from reproducing. Rolston does cite one case at length in which plants, specifically wildflowers, have some status that elevates them above such aesthetic judgment of humans, but it is not in the context of their aesthetic value, but rather in reference to their possible extinction. Rolston holds that endangered plants deserve special treatment such that animals can justifiably be killed to protect the plants from being eaten. The specific case is three species of endangered wildflowers (San Clemente bush mallow, San Clemente Island Indian paintbrush, and San Clemente Island larkspur) on San Clemente Island, where large numbers of feral goats were shot to protect the plants. Rolston explains as follows:

> Following the theory worked out in Chapter 3, a goat does have more intrinsic value than a plant, although plants have more instrumental value in ecosystems than goats. So if the tradeoff were merely 1,000 goats for 100 plants, regardless of instrumental, ecosytemic, and species considerations, the goats would override the plants. But the picture is more complex. Out of place from their original ecosystems, goats are degrading the ecosystems in which they currently exist, producing the extinctions of species that are otherwise well fitted and of instrumental value in those ecosystems. At this point the well-being of plants outweighs the welfare of the goats.[41]

Here we find a hierarchy in terms of amounts of objective intrinsic value in which plants have the least intrinsic value, and are therefore the most expendable, animals have more, and are therefore less expendable, and humans with the most are the least expendable. However, endangerment to the species, rather than endangerment to the individual within the species, trumps the better treatment that organisms with more intrinsic value normally get, permitting their sacrifice for the good of the lower-level species. (In a videotaped debate with Bernie Rollins some years ago, Rolston hinted that these conclusions applied not just to animals, but humans as well. When Rollins stated that Rolston would be willing to kill humans to protect wildflowers, Rolston with a grin replied that he would do everything that he could to persuade the humans to stop picking the flower, but if they didn't, come nightfall, they would have to look out for themselves.)

This example, while it appears to provide some moral status for the wildflowers, at least collectively, it does so in a way that is not primarily concerned with natural objective intrinsic value. The individual plant has no status at all. Thus, the individual objective intrinsic value is not a factor in the decision to kill the goats. Collectively, however, the plants are "well fitted and of instrumental value in those ecosystems." The reference to instrumental value is a reference to function in the system. Here the object of concern is not the species but the system of which it is a part. The claim that the species are well fitted could be related to a subjective intrinsic valuing of the system, that, for instance, its historical value to humans would be decreased if the endangered species disappeared from the system.

Concerning Rolston's claim that the valuing of the human produces "a value that, when it appears as a product of subjective awareness, is attached objectively to the flowers flourishing in the meadow,"[42] the claim is not very convincing especially in light of Rolston's own remarks a few pages later. First, there is his remark, in seemingly the same context, that "But nothing is really added intrinsically to the object at all; everything in the object remains what it was before. Despite the language that humans are the source of value which they locate in the natural object, no value is really located there at all."[43] Then on the next page, he adds:

> But by now we begin to suspect that the anthropogenic account of intrinsic value is a strained saving of what is really an inadequate paradigm, that of the subjectivity conferral. For all the kindly language about intrinsic value in nature, the cash value is that, "Let the flowers live!" really means, "Leave the flowers for humans to enjoy," because the flowers are valuable—able to be valued—only by humans even though when properly sensitive humans come along they do value these flowers for what they are in themselves.[44]

This criticism, which at this point in the book, appears to be a criticism of Callicott's theory of value projection on nature, can also be taken as a criticism of Rolston's own view about wildflowers. The goats were killed off not simply because the plants were endangered but because the plants were endangered wildflowers. If the plants did not flower, and flower in a manner to call special attention to them among humans, it is unlikely in real life that humans would have given them another thought. If so, then the wildflowers are primarily instrumentally valuable in making possible the aesthetic pleasure viewing them induces in humans, and that pleasure is a central focus of the intrinsic value related to the aesthetic value of the wildflowers. Rolston might object that the wildflowers are not being instrumentally valued as a trigger for intrinsically valued pleasure. Nevertheless, most likely the goats were killed to protect the flowers—because the flowers were in this aesthetic value relationship with humans—and the goats would not have been exterminated otherwise.

The aesthetic nature of the San Clemente case is not highlighted in Rolston's text because he simply refers to the wildflowers as endangered plants. Likewise, he introduces a second wildflower example in which Santa Barbara live-forever, also on an island, Santa Barbara Island, were endangered by rabbits, again without stressing that the plants were wildflowers. If he had made these connections, the extremely important role of aesthetics in his discussion of endangered species would have become more obvious, for the two wildflower examples are sandwiched in between discussions of two other endangered species, the Florida panther and the California golden trout, both of which are presented as species worthy of preservation, largely for aesthetic reasons. Rolston writes that the panther is "thought by many to be the most aesthetically exciting animal on the North American continent." With regard to the golden trout, he says that "The justification for [poisoning the California rivers infested by brown trout, and thereby killing everything in the rivers] was partly aesthetic, partly for quality fishing, partly out of respect for the state fish."[45]

The chapter on endangered species, which follows the chapter on organisms, seems to be presenting more complex cases involving the principle of nonloss of goods, which in terms of the example given (deciding whether to pick the flower), appears to be focused on showing how the subjective aesthetic feelings of humans can override the autonomous intrinsic value of nonhuman entities more generally. Sacrifice of organisms with more autonomous intrinsic value and sacrifice of the nonbasic interests of some humans can occur on behalf of species with less autonomous intrinsic value when those species are endangered and are a source of aesthetic pleasure for humans—the panthers in Florida, the wildflowers of San Clemente and Santa Barbara Islands, and the golden trout in California. However, the members of the species can still be sacrificed individually to achieve aesthetic pleasure in humans (especially the wildflowers and trout, with the panthers being more problematic, perhaps because of their greater intrinsic value, given their greater complexity).

In addition to the argument given in the book to justify the picking of wildflowers, Rolston has provided me with two additional ones through personal communication. Once on a visit to my home, upon arrival, Rolston set his luggage down in my driveway and bend over to pick a bluebonnet from my yard. Astonished, I asked for an explanation. Rolston replied that he had picked the flower for scientific reasons and that it was permissible in addition because I had so many bluebonnets blooming in my yard. Apparently, Rolston wished to compare a specimen of the species of bluebonnet in my yard *(Lupinus texensis)* with the bluebonnet that grows in Colorado *(Lupinus sericeus)* and he considered this "scientific inquiry" adequate to justify the killing of one plant in my yard. The second justification, that there were more than enough bluebonnets in my yard is implicit in the goat/plant example. The goats are free to eat the plants as long as doing so does not endanger the species. Likewise, Rolston is free to pick the flowers in my yard as long as his doing so does not endanger the (wildflower) species, which was the case. My anthropocentric objection that I owned the wildflowers and did not want them to be picked fell on deaf ears. He was also not particularly moved by my request that he let the flowers live. I did not think to suggest that he should have considered whether the aesthetic loss that I suffered subjectively might be greater than the gain to science achieved by his picking of my flower. (Happily, Rolston has since refrained from picking wildflowers in my presence, although I suspect he still advances the cause of science when I am not with him.)

In evaluating Rolston's treatment of natural beauty, it is important to realize that he deviates significantly from his general position that autonomous, objective, natural intrinsic value exists in nature. In the general position, there is intrinsic value in the world because it is autonomous, which is then discovered by the human moral agent and valued intrinsically. It is objectively in nature, and it is objective value because it is autonomous. Natural beauty, however, does not work this way. Humans bring beauty into the world subjectively by igniting it. At the beginning of the criticism of Callicott's truncated intrinsic value theory, however, Rolston promises, as noted above, that "the value so coagulated, we will claim, is objectively intrinsic to the nonsentient life and not merely instrumental."[46] This promise never seems to have been fulfilled as far as natural beauty is concerned. Rolston's best opportunity, it seems to me, came when he noted that "it is difficult to deny [birds and beasts] the precursors of aesthetic experience" and that beauty might be subjectively "nascent in the eyes and experiences of birds and beasts," a position which he nevertheless then rejects.[47] If he had not rejected this position, he would at least have had subjectivity with regard to beauty in the world and could perhaps have then said that that subjectivity was objectively in the world (meaning not simply trapped in human subjectivity). Such a view could then have been construed as a way to differentiate himself from Callicott's position with regard to aesthetics and fulfill the promise that aesthetic value is objectively intrinsic to nonsentient life through the subjectivity of other nonhumans.

In my own book, *Foundations of Environmental Ethics*, I follow Harvard geologist Nathaniel Shaler in arguing that flowers reflect the aesthetic preferences of insects and that birds through their aesthetic preferences in mating have turned themselves into living works of art.[48] Shaler's view is that a sense of beauty has been carried forward through biological evolution through insects and birds and other creatures, ultimately, manifesting itself in human beings. The difference between this view and Rolston's is only a minor one. Rolston says "perhaps" and then declares that humans somehow invented natural beauty in a nonevolutionary way. The conclusion that aesthetics has been passed forward through evolution, however, does not seem to be any more contentious than Rolston's claim that the genetic set is a propositional set that represents a logical set, no less than a biological set, that ultimately is a normative set.[49] If a genetic set can specify what ought to be, that "what ought to be" could include some contribution to what ought to be beautiful, which could then reveal itself in the aesthetic preferences of both humans and prehumans. Taking this route to nature aesthetics would be more consistent, in my view, with Rolston's general position on value (and valuing) in nature. In accordance with this approach, aesthetic properties in nature would count more than they do now in Rolston's account and the valuing of humans would be part of nature (since it is a product of evolution), rather than something distinctly human and separate from nature.

Rolston may have been reluctant to follow this path because it is still not possible to talk of the evolution of beauty in nonliving entities as such. In other words, while it may be true that the beauty of flowers has evolved over time in accordance with insect preferences, there can be no similar evolutionary development with regard to the beauty of mountains and other nonliving objects. Mountains are not and cannot be beautiful because of the shaping influence of the aesthetic preferences of living organisms. One could argue, nonetheless, that the natural world is the context within which the aesthetic preferences of organisms are shaped. This view is a Humean one, that our idea of beauty comes from impressions of the world.[50] In accordance with this view, natural beauty is contingent on the actual natural history of this Earth. If the natural history had been different, then the context in which natural beauty comes to be would be different and natural beauty (in terms of the preferences of organisms) would be different. In part, too, the natural beauty of nonliving things could depend on a sense of beauty as it evolved in living creatures on this planet. In terms of Rolston's dualism between subject and object, one could argue that the subjects and the objects came to be as a result of the biological and geological evolution that constitutes the natural history of the Earth. While the aesthetic preferences of organisms would not be set by evolution, the ability to generate aesthetic preferences in the context of the world as it has come to be could be a feature of evolution, not just something that happened when humans arrived on the scene.

It is possible that Rolston developed his actual position, that beauty is in the eye of the beholder and not objectively in the world, in part as a response to Allen Carlson's claim that all of nature is aesthetically positive or good. To the contrary, Rolston insists that nature can be naturally ugly:

> If hikers come upon the rotting carcass of an elk, full of maggots, they find it revolting. Here is a bad example of its kind, disharmony, a putrid elk. Any landscape looked at in detail is as filled with dying as with flourishing things. Everything is in some degree marred and ragged—a tree with broken limbs, a crushed wildflower, an insect-eaten leaf. An eagle chick plagued with ticks is not a pretty thing. Sometimes there are disfigured, even monstrous animals. So why is this not ugliness in the landscape? It is! We do not enjoy such experiences. Tourists take no photographs of these eyesores.[51]

This paragraph seems to me to be ill-conceived, since, if beauty came into the world with the arrival of humans, then ugliness must have arisen in the same way and at the same time. Should we then say that ugliness emerges out of the human sense of ugliness? If there are properties igniting ugliness, are they also aesthetic properties, or non-aesthetic properties, or anti-aesthetic properties? The idea behind positive aesthetics is not that everything in nature is straightforwardly beautiful, but rather that everything can be viewed as beautiful if viewed in an appropriate manner. For Carlson, for example, positive aesthetics means the creation of aesthetic standards by which natural things can be evaluated.[52]

In the next section of the chapter, "Ugliness Transformed in Ecosystemic Perspective," Rolston seems to be moving in precisely the direction that Carlson indicates for positive aesthetics. He writes:

> We can expect that humans, like other animals, will have been naturally selected to find certain things repulsive, those things (rotting carcasses, excrement) that they as individuals need to avoid in order to survive. But these processes, abhorrent from the perspective of my individuality, may not be ugly at all in the system, where they are the recycling of resources. Environmental ethics stretches us out from our individualistic, self-centered perspectives into a consideration of systemic beauty.[53]

He continues in the next paragraph with "There is ugliness, but, even more, there are transformative forces that sweep toward beauty in perpetual perishing."[54] These statements are a good example of the positive aesthetics he rejects in the previous section, suggesting that his objections were merely rhetorical.

Rolston concludes his discussion of nature aesthetics with a final section, "Beyond Beauty to the Sublime." Here, still on the theme of positive aesthetics, Rolston tries to deal with the relationship of ugliness, the picturesque or the scenic, and the sublime. He begins:

> Aesthetic properties in nature push the beholder toward the experience of the sublime, something larger than beauty. At the beginning, we search for something pretty or colorful, for scenic beauty, for the picturesque.[55]

He concludes:

> The upshot is not that virgin nature is invariably aesthetically positive in immediate detail but that it is essentially so when even the ugliness is embraced by the sublime. . . . Beauty does not require permanence. Within landscapes there is ugliness in the detail, but at the systemic level, at the dynamic scene, softened by perspective from a distance, there is sublime beauty.[56]

While this section continues his rhetorical acceptance of positive aesthetics, it does so by blurring the distinctions between aesthetic categories—the sublime and the picturesque. Nature aesthetics historically began with a focus on the sublime as the only aesthetic property in nature. Beauty was then conceived as the opposite of the sublime and attributed to art. In light of developments in landscape painting, especially in terms of the paintings of Claude Lorraine, a conception of picturesque beauty was developed and also attributed to nature. In this historical sequence, the movement was from the sublime to the beautiful to the picturesquely beautiful. Perhaps people do come to an aesthetic appreciation from the pretty to the sublime and then embrace the ugly as sublime, but if so, it would be good to see a study demonstrating the transitions. *This issue*, it seems to me, is confused because most of what was formerly considered *sublime* is now considered beautiful and the term *sublime* and its related term *awesome* have fallen out of common use.

Because of Rolston's unusually ambiguous writing style, in which he frequently writes as if he is holding a position that he later rejects, it is difficult to evaluate and critique his positions, given that seemingly conflicting statements may actually be statements explicating a view Rolston does not actually hold together with statements criticizing that view. Such may be the case in his discussion of ugliness and positive aesthetics, since his insistent claim that parts of nature can be ugly is followed by two sections softening that claim.

My criticism of Rolston's general position, that natural beauty comes into the world with the appearance of human consciousness, is plagued by similar problems. For this reason, much of my explication of Rolston's views has been modified by the addition of "seemingly" or "apparently." Nevertheless, it is clear (from personal communication) that Rolston holds the view that natural beauty is subjective and that it did not exist before the arrival of humans on the evolutionary scene. Further, although he is ambivalent about whether other organisms have some nascent sense of beauty, his uncertainty is not great enough to have any effect on this claim. After expressing the possibility, he simply reiterates his view that natural beauty arrived with human beings. Finally, although he acknowledges aesthetic properties in nature, their presence does not indicate for Rolston the presence of natural beauty objectively in nature in accordance with his general theory of autonomous intrinsic value. Natural beauty does not arise until the viewing of the aesthetic properties in nature ignite natural beauty in the human mind subjectively.

This subjective account of natural beauty is, as I have indicated above, a significant anomaly in Rolston's environmental ethics because more generally he insists that the presence in nature of something being valued intrinsically is enough to justify the claim that intrinsic value exists in nature objectively. Rolston objects to Callicott's truncated intrinsic value theory on the grounds that it ignores the objective component of what is being valued in nature such that value is simply projected on nature as if there is no value there. Rolston's own account of natural beauty, using his echo model, is basically the same account. Yet, this account is fairly tenuous given, as noted above, that Rolston admits that the echoing is "most often working the other way around; the human valuer is reflecting what is actually there." Thus, on the basis of this model, Rolston could have easily concluded that natural beauty is a value in the human subject reflecting the aesthetic properties actually there, resulting in objective natural beauty in nature. Rolston's theory of natural beauty would be much improved if he supplied the reason for his reluctance to draw this conclusion.

NOTES

1 See, for example, Holmes Rolston, III, "Mystery and Majesty in Washington County," *Virginia Wildlife* 29, no. 11 (November 1968): 6–7, 22–23;:"Hewn and Cleft from this Rock," *Main Currents in Modern Thought* 27, no. 3 (1971): 79–83; "Lake Solitude," *Main Currents in Modern Thought* 31, no. 4 (1975): 121–126; "The Pasqueflower," *Natural History* 88, no. 4 (April 1979): 6–16; "Beauty and the Beast:

Aesthetic Experience of Wildlife," in Daniel J. Decker and Gary R. Goff, eds., *Valuing Wildlife: Economic and Social Perspectives* (Boulder: Westview Press, 1987), pp. 187–196; "Does Aesthetic Appreciation of Landscapes Need to be Science-Based?" *British Journal of Aesthetics* 35, no. 4 (October 1995): 174–386; "Aesthetic Experience in Forests," *The Journal of Aesthetics and Art Criticism* 56, no. 2 (Spring 1998): 157–166; "Aesthetics in Swamps," *Perspectives in Biology and Medicine* 43, no. 4 (Summer 2000): 584–597; "From Beauty to Duty: Aesthetics of Nature and Environmental Ethics," in Arnold Berleant, ed., *Environment and the Arts: Perspectives on Environmental Aesthetics* (Aldershot, Hampshire and Burlington, Vt.: Ashgate Publishing, 2002), pp. 127–141.

[2] Aldo Leopold, *A Sand County Almanac, and Sketches Here and There* (London, Oxford, and New York: Oxford University Press, 1949), pp. 224–225.

[3] Holmes Rolston, III, *Environmental Ethics: Duties to and Values in the Natural World* (Philadelphia: Temple University Press, 1988), p. 182.

[4] Ibid., p. 225.

[5] See Holmes Rolston, III, "Values in Nature," *Environmental Ethics* 3 (1981): 113–28; "Valuing Wildlands," 7 (1985): 23–48; and "Humans Valuing the Natural Environment," chap. 1 of Rolston, *Environmental Ethics*.

[6] Holmes Rolston, III, "Are Values in Nature Objective or Subjective?" *Environmental Ethics* 4 (1982): 127.

[7] Ibid., pp. 128–129.

[8] Rolston, *Environmental Ethics*, p. 113; quoted from J. Baird Callicott, "On the Intrinsic Value of Nonhuman Species," in Bryan G. Norton, ed., *The Preservation of Species* (Princeton: Princeton University Press, 1986), pp. 142–143, 156.

[9] Rolston, *Environmental Ethics*, p. 115.

[10] Ibid., p. 41.

[11] Ibid., p. 27.

[12] Ibid.

[13] Ibid., p. 131.

[14] Ibid., p. 233.

[15] Ibid., p. 38: "[N]ature is either amoral or immoral. Morality appears in humans alone and is not, and never has been, present in the natural scene."

[16] Leopold, *A Sand County Almanac*, pp. 224–225.

[17] Rolston, *Environmental Ethics*, p. 182.

[18] Ibid., p. 233.

[19] Ibid., p. 27.

[20] Ibid., p. 233.

[21] Ibid., pp. 208–213.

[22] Ibid., p. 209.

[23] Ibid.

[24] Ibid., p. 210.

[25] Ibid., p. 234.

[26] Ibid. (emphasis added)

[27] Ibid.

[28] *Webster's New World Dictionary of American English*, 3rd ed. (Cleveland and New York: Simon and Schuster, 1988), p. 1344.

[29] Rolston, *Environmental Ethics*, pp. 234–235.

[30] Ibid., p. 233.

[31] Ibid., p. 211.

[32] Ibid.

[33] Ibid., p. 100.

[34] Ibid., p. 104.

[35] Ibid., pp. 59–62. The principle of nonaddition of suffering permits "innocent suffering" and goes together with the homologous principle, which permits suffering comparable to that of animals living

wild in an ecological system ("ecologically functioning suffering"). Finally, there is a hedonist principle or "concern" to reduce pointless suffering. This last principle is superogatory in that it is "commended but not required."

[36] Ibid., p. 120.
[37] Ibid., p. 110.
[38] Ibid., p. 187.
[39] Ibid., p. 109.
[40] Ibid., p. 112.
[41] Ibid., pp. 141–142.
[42] Ibid., p. 112.
[43] Ibid., p. 115.
[44] Ibid., p. 116.
[45] Ibid., pp. 140, 142.
[46] Ibid., p. 112.
[47] Ibid., p. 234.
[48] Eugene C. Hargrove, *Foundations of Environmental Ethics* (1989; reprint ed., Denton, Tex.: Environmental Ethics Books, 1996), p. 180.
[49] Rolston, *Environmental Ethics*, pp. 98–99.
[50] Hume writes: "All the colors of poetry, however splendid, can never paint natural objects in such a manner as to make the description be taken for a real landscape. The most likely thought is still inferior to the dullest sensation." David Hume, *An Enquiry Concerning Human Understanding*, sec. 2.
[51] Rolston, *Environmental Ethics*, pp. 238–239.
[52] Allen Carlson, "Nature and Positive Aesthetics," *Environmental Ethics* 6 (1984): 5–34.
[53] Rolston, *Environmental Ethics*, p. 241.
[54] Ibid.
[55] Ibid., p. 243.
[56] Ibid., pp. 244–245.

BRENDA HAUSAUER

9. WORDS GONE WILD: LANGUAGE IN ROLSTON'S PHILOSOPHY OF NATURE

"Nature writing" has a long and rich tradition around the world. In the English-speaking world during the past two hundred years, for example, there has been an incredibly diverse number of essays, poems, and assorted manuscripts written about nature. Nature was the primary topic for all the English Romantic poets, the American Transcendentalists, the writer-naturalists who explored and charted America, and many recent writers. Even though most of these writers have not been professional philosophers, many of them wrote with a philosophical bent, combining observations and stories about nature with philosophical musings and conclusions.

The study of nature and reality in professional philosophy is also as old and important as the discipline itself. Environmental philosophy as a sub-discipline, however, has a shorter history. Although there are some important earlier texts, much of environmental philosophy's literature dates from the 1960's forward. Environmental philosophy differed from the earlier, more metaphysical explorations of nature because it often combined metaphysics (sometimes informed by recent ecological science) with ethics, and it sometimes had a more specific agenda: to convince and convert readers in the face of some natural problem or crisis.

Holmes Rolston, III, as one of the first to write in environmental philosophy, has had a substantial influence in shaping this emerging field. From the 1960's to the present, he has written four books and well over two hundred articles in a range of philosophical, scientific, and theological journals. Perhaps partly because environmental philosophy is a newer and multi-disciplinary topic in philosophy, Rolston's written methods have been very different from the straight argumentative norm in the philosophical genre. In fact, Rolston is virtually unique among environmental philosophers in the way he writes. He is a more "literary" philosophical writer; metaphor, rhetoric, and careful attention to language mark his writing. Rolston has been criticized by some philosophers for not producing solid arguments, and for retreating into stories and descriptions instead of arguing. But even those critical of Rolston tend to agree that his texts are provocative.

Rolston's writing style and methodology of presenting his ideas can be more clearly examined when they are compared to the style and methodology of a more "philosophical" literary writer. Annie Dillard is such a writer. She also has received acclaim for her writing, though her praise comes from literary scholars, not philosophers. She has written books that vary in their genres; several are primarily about nature, including *Pilgrim at Tinker Creek* that won the Pulitzer Prize. *Pilgrim at Tinker Creek*, like other texts by Dillard, is a personal narrative, but it is also a metaphysical exploration of nature, an attempt to relate metaphysics to meanings for the world and for humans,

C.J. Preston and W. Ouderkirk (eds.), Nature, Value, Duty, 143–165.

and a series of thoughts and discoveries about what has value. Compared to much other "nature writing," Dillard's books are more "literary"—they use imaginative, figurative language; their style is highly original and personal; their devices and rhythms are often poetical; and they include a progression in which early parts are not always well understood until later parts are read. Dillard also is more philosophical than many other nature writers in the sense that she explores metaphysics and ethics, and she examines general claims and advances general conclusions.

Holmes Rolston, III, and Annie Dillard are contemporaries and have quite a few surface similarities in their topics and writing styles, which makes a study of Dillard's styles useful in illuminating Rolston's. For example, they often wish to convince readers of similar points (e.g., there is intrinsic value in nature; nature points toward God); they both spend much time trying to understand well what is going on in nature, both through personal experiences and understanding science; they both rejoice in and are amazed by nature; and they sometimes use similar writing approaches. Despite these surface similarities, their styles are so different that they give readers quite different views of the topic. Some of those differences will be explored below.

Both Rolston and Dillard add something unique, something that has been missing, to their respective genres. There are some important philosophical points in Dillard's books; there is some artful writing in Rolston's works. But, because their works fall into different genres, readers actually view the philosophical insights and writing methods in different contexts. Readers approach texts with sets of preconceptions and expectations associated with genre differences. For example, texts read in the past help form readers' expectations for other texts in the same genre. The context created by those preconceptions and expectations molds the meaning of the text. Texts that depart from the traditional writing methods and topics of their genre can begin to change readers' preconceptions and expectations of the genre. Rolston's texts fall into this category, and this may help explain why Rolston's arguments are criticized by some philosophers.

Rolston's philosophical genre and Dillard's literary genre each have strengths to contribute to the other. For instance, philosophers may be able to most easily focus questions about many topics which literature undertakes, and thus clarify and enlarge upon literary works' presentations of the topics. And, because philosophers excel at testing arguments for validity and defending them against alternatives, philosophers could use their skills to comment on the (sometimes implicit) claims made by literary writers. On the other hand, there may be some insights that can only be stated through certain writing methods; because writers and literary scholars have long studied writing methods, they likely have much to say to philosophers on these topics. This essay is a small step in the direction of using a literary text to inform and illuminate a philosophical one with respect to writing style.

In the sections below, I will outline some of the more "literary" writing methods Rolston uses and take a brief look at how they impact his texts. And, I'll compare those methods with examples from Annie Dillard's texts to show more fully how a literary writer uses similar writing methods to address similar topics, though with rather different results.

GETTING THE WORDS RIGHT: WORD-EXPLORATIONS AND METAPHOR

Word-explorations

Any reader of Rolston's texts will notice how important specific words are to his topics. He often uses a word in one context, only slightly to change the word or the context later. Exploring etymologies, denotations, and connotations of words also enters into this process. These typical methods are found on many of Rolston's pages. Note how in the first example below, Rolston is aware that he is "playing with words."

> *Humans* are cognate with the *humus*, made of dust, yet unique and excellent in their aristocratic capacity to view the world they inhabit. They rise up from the earth and look over their world (Greek: *anthropos*, to rise up, look up). . . . Humans want to be inhabitants who appreciate (in double senses of *finding value in* and *adding value to*) their habitats. Humans are value-able, *able to evaluate* the world, able to discover (as well as invent) value there. They can keep life wonderful because they have the capacity for wonder. . . . Humans are of capstone value because they are capstone evaluators. . . . In this sense, humans are or can be superior, superb, supervenient, even (in a provocative sense) supernatural, super-to-the-natural. They transcend the spontaneous environment because they have oversight of it; they have transcendence in immanence. To continue to play with words, humans are spectacular because they emerge to see the spectacle they are in (Rolston 1988, 338–339).
>
> Genesis is linked with genes, and these genes in "the wise species," *Homo sapiens*, produce a brain that sponsors a mind with what we will call, provocatively, a unique "genius." That reconnects the word "genius" with its etymology, recalling the Latin *genius*, spirit. Other derivatives are found in such words as "ingenuity" and "engineer;" the Latin root, *gigno*, is the same, interestingly, as that of "gene," to generate, involving now the generation of a procreative animating spirit. In German, our choice of words would be *Geist* (Rolston 1999, 140).

An important mark of Rolston's word-exploration is that the same or similar words (or ideas) continue to appear throughout the paragraphs and chapters, often building up a very rich, complex texture of words, meanings, and contexts.

Rolston's word-explorations reveal layers of meaning that readers might be unaware of, and thus can offer insights in unique ways. In addition, the word-explorations ask a lot of readers: to read very carefully, to notice nuance, to be patient!, and to remember and reflect upon the words while reading forward. Thus, instead of directing attention only toward the meanings and implications of concepts, Rolston's readers are asked also to pay close attention to the richness of specific words and the built-up layers of meanings (a more literary task). In fact, the meanings and implications of the words are a vital part of the concepts Rolston argues for.

Word-explorations sometimes have benefits, and sometimes may be detrimental. The word-explorations can reinforce the arguments, and render them more forceful and convincing. Rolston has pointed out that word-explorations *can* be part of a logical sequence in an argument, insofar as they unfold novel layers of meaning (Rolston 1992). But when word-explorations are used to excess or are overdone, they detract from or help to bury the main points of the text. Critics may feel word-explorations divert attention from a weak point. At the very least, some of Rolston's word-explorations divert the readers' attention from the current topic to the intricacies of language.

Metaphor

Just as Rolston consciously uses specific words to help make philosophical points, he also is concerned about how others use words. This is evident in his texts when he spends considerable time addressing how society talks about certain topics, or what metaphors we use. For example in *Genes, Genesis and God*, he spends a good deal of time discussing the use of the term "selfish genes," exploring whether "selfish" is the right metaphor to use with "genes." The following are some other examples:

> When scientists speak of ant wars, or queen bees and their slaves, or immunoglobulins as carrying on a "battle within" us against invading microbes, they borrow words from one domain of experience and transfer them to another. A careful analyst needs to be cautious about overtones also transferred. A great deal depends on the metaphors one chooses, since these so dramatically color the way we see the natural world. One must be careful not to let negative moral words, borrowed from culture, discolor nature. Something like this happened before in Darwinism when "survival of the fittest" was the paradigm, interpreted as "nature red in tooth and claw," but biologists now prefer to restate this as "adapted fit," a better description, since fitness takes place in various ways, only one of which is combative or aggressive. "Adapted fit" colors events differently than "survival of the fittest" (Rolston 1999, 48).
>
> We could say that, in sexual reproduction, the genes "throw away half their investment," or that they get "corrupted" ("adulterated"!) with every reproduction. We would prefer to say "distributed," or "divided out" or "shared"—the metaphors are important; they color how we perceive what is going on. "Sex," says Michael Ghiselin, is "synonymous with 'mixis'—literally 'mingling' " (Rolston 1999, 98).

Pinning down the perfect metaphor, as well as elaborating the full and layered meanings of certain words, is very important in Rolston's texts; so much so, in fact, that "getting the words right" *is* the argument sometimes, or at least a step toward the conclusion.

Annie Dillard's texts also are thick with careful attention to specific words, considerations of different metaphors, and a layering of metaphors throughout the texts. But notice in the examples below how she does not take word games to the same depth as Rolston, but merely chooses, consciously, the metaphor or word she wants, and uses language to evoke a sense that this is the correct choice.

> Van Gogh, you remember, called the world a study that didn't come off. Whether it "came off" is a difficult question. . . . But, Van Gogh: a *study* it is not. This is the truth of the pervading intricacy of the world's detail: the creation is not a study, a roughed-in sketch; it is supremely, meticulously created, created abundantly, extravagantly, and in fine (Dillard 1974, 134).
>
> The planet is less like an enclosed spaceship—spaceship earth—than it is like an exposed mangrove island beautiful and loose. We the people started small and have since accumulated a great and solacing muck of soil, of human culture. We are rooted in it; we are bearing it with us across nowhere (Dillard 1982, 152).

Dillard's use of these methods is often not as meticulous and explicit as Rolston's. For example, Dillard does not consider many possible word meanings or metaphor candidates as Rolston does. In this way, Rolston's texts ask readers to think more carefully about words than Dillard's. However, both writers often choose one metaphor as a conclusion (or a step toward a conclusion), and use many other paragraphs to convince or show readers that this is the correct metaphor.

PLAYING WITH LANGUAGE: REPETITION, QUESTIONS, AND EXCLAMATIONS

Repetition

Another signature method of Rolston is re-describing his points in new ways, often in order to reinforce ideas. His texts almost always show some measure of repetition—but not literal repetition. Rather, the texts skillfully repeat an earlier concept (perhaps while adding new bits of tangential information) in language which is novel, pleasing, or interesting—as in the last two sentences of this passage.

> Willy-nilly, life is symbiosis. Would-be imperialists cannot dominate the world; they can gain only situated environmental fitness. Would-be maximizers can be no more than optimizers (Rolston 1986, 91).

Consider two other passages where repetition plays an important role:

> Reptiles can survive in a broader spectrum of humidity conditions than can amphibians, mammals in a broader spectrum of temperatures than can reptiles. Once there was no smelling, swimming, hiding, defending a territory, gambling, making mistakes, or outsmarting a competitor. Once there were no eggs hatching, no mothers nursing young. Once there was no instinct, no conditioned learning. Once there was no pleasure, no pain. But all these capacities got discovered by the genes (Rolston 1999, 4).
>
> In this forest I am the lone exhibitor of what nature could not achieve in these paintbrushes and columbines, pipers and porcupines. Transcending

> these embodiments of herself, she has gifted this human son with what she could not attain elsewhere: self-conscious reflection. The beaver knows I am here, and slaps his tail—but I know that I know I am here. The chickaree notices me, and chatters, but I notice that I notice him. The doe halts to think by what route she may avoid my camp and yet reach the meadow. But I think about her thinking so. If in the prehuman mind, causes are already being supplanted by reasons, the conclusion of that replacement is a mental incandescence: this novel iteration of knowing, and seeing, and thinking. The loneliness here exceeds the absence of a companion; what is wanting is the intentional, the fashioned, the cognitive. The lake surface has an analog in my skin, but the lake has an "antilog" in my pondering mind. *Cogito, ergo sum solus* (Rolston 1989, 225–226).

In the second excerpt (from the essay "Lake Solitude"), the main point is that humans have attained something the rest of nature has not: the ability to self-reflect. Notice that by the second sentence, this point has been stated. The rest of the paragraph re-describes the main point in new ways, by giving repeated examples and fleshing out the main point, making it richer and more complex. This method, in fact, is similar to Rolston's method of exploring word meanings and etymologies: both examine a word or a concept from myriad angles.

This method of repetition, like the use of word-explorations, can enhance or weaken Rolston's texts. Repetition, a more literary style, requires readers to attend closely to words and piled-up meanings. But, if this type of repetition occurs when readers expect reasons or are not prepared to attend closely to words, then the effects of the method are detrimental. If the method is judiciously used to enhance the topic, however, then the main points of the text may be reinforced to the readers. This method redirects philosophical readers' attention by stressing that a new understanding of a concept can be gained simply by expressing it in different words. Thus, the method of repetition, like word-explorations and metaphor, encourages philosophical readers to look at and experience the text more artistically.

Questions and exclamations

Rolston's texts are also good at using different types of sentences, including exclamatory and interrogative sentences. Consider the following passages:

> Here humans value the wilderness or the park non-economically for its unbuilt characteristics. Is this only some sort of escape value? Or is there some more positive characteristic in wildness that re-creates us? (Rolston 1988, 8).

> But science is never the end of the story, because science cannot teach humans what they most need to know: the meaning of life and how to value it. The sciences are as practical as theoretical; science has evident survival value, teaching us how to gain benefits that we desire. But what ought we to desire? Our enlightened self-interest? Our genetic self-interest? More children? More science? The conservation of biodiversity?

> Sustainable development? A sustainable biosphere? The love of neighbor? The love of God? Justice? Equity? Charity? (Rolston 1999, 161).

> But how bizarre this surrounding nothingness, which breeds me to slay me, yet slays me to breed me! . . . Is this wildness only unfeeling and unloving? Has it not also granted me life, surrounded me with beauty, and even when wolven, does not its howling threat stimulate me into a higher competence? Wildness is the pressing night, but it is wildness too that with me kindles fires against the night (Rolston 1989, 231).

> A fair landscape indeed! What wonders and pleasures lie in the realm your boundaries enclose, from the majesty of Abrams Falls to the mystery of Ebbing Spring, from limestone sinks to green pastures and templed hills. *De profundis*! What abundance and splendor are compacted here! Lord, bid time and nature gently spare these hills that once were home (Rolston 1989, 247).

The first three passages present questions, and as readers we know how the writer wants us to answer them. In this way, the questions point us toward a certain conclusion. In fact, Rolston almost always uses questions rhetorically in his texts. Rolston has pointed out that although the questions in his texts are often rhetorical, they once were autobiographical questions that were important to him in the development of his ideas (Rolston 1992). Thus, Rolston uses questions to reconstruct the process by which he reached certain conclusions, perhaps hoping to lead readers along the same paths toward his conclusions.

The use of exclamatory sentences in the last two quotes above, and in most of Rolston's texts, almost always present instances of amazement, excitement, or surprise on the writer's part, and a wish to convey that feeling to the readers. In fact, these exclamatory sentences give readers a vehicle for experiencing with the writer a certain feeling for nature, instead of discussing or describing an emotion intellectually. Exclamatory sentences are a good vehicle for this because they do not explain, but express or evoke. Rolston has claimed that they mix experience with argument (Rolston 1992). And combining arguments or explanations with these kinds of experiences or evocations can create a very convincing impression if both are strong.

Dillard also uses questions and exclamations sometimes, although her usages are perhaps more moderate than Rolston's. As the passages below illustrate, Dillard usually uses questions rhetorically, pointing readers toward a certain conclusion. And, exclamations are used to help evoke a feeling for the readers. But Dillard does not ask quite so many questions of her readers as Rolston, or ask them to share her excitement as often.

> Nature is, above all, profligate. Don't believe them when they tell you how economical and thrifty nature is, whose leaves return to the soil. Wouldn't it be cheaper to leave them on the tree in the first place? This deciduous business alone is a radical scheme, the brainchild of a deranged manic-depressive with limitless capital. Extravagance! Nature will try anything once (Dillard 1974, 65).

> Like any out-of-the-way place, the Napo River in the Ecuadorian jungle seems real enough when you are there, even central. Out of the way of *what?* I was sitting on a stump at the edge of a bankside palm-thatch village, in the middle of the night, on the headwaters of the Amazon. Out of the way of human life, tenderness, or the glance of heaven? (Dillard 1982, 53).
>
> Now the point of all this is that there are a million nephrons in each human kidney. I've got two million glomeruli, two million Henle's loops, and I made them all myself, without the least effort. They're undoubtedly my finest work. What an elaboration, what an extravagance! (Dillard 1974, 134).

DRAWING ON THE PARTICULAR: EXPERT INFORMATION, STORIES, AND DESCRIPTIONS

Expert information

In addition to a deep appreciation and knowledge of words and meanings, Rolston also has an appreciation and knowledge of science and other disciplines, which he shares with his readers. In every chapter or essay, there are references either to specific scientific theories or to experts in a certain scientific field. Some examples show how Rolston uses this method:

> Laden with my pack, moving briskly along, I turn my thoughts to respiration. Present in every cell containing a respiratory chain—from microbes to humans—is an electron carrier called the cytochrome-c molecule that evolved over 1.5 billion years ago. Given that I plainly value respiration for myself, and that evolution has conserved this molecule since before plants and animals diverged, it seems some sort of wild type in value. If I become winded, my body is facing another problem. The citric acid cycle, which follows glycolysis in the processing of food molecules and is a more recently evolved skill, is not generating enough ATP . . . Such sophisticated insights, reached in biochemistry laboratories and ecological field studies, reveal the extent to which wild values surround and lie within us (Rolston 1989, 122–123).
>
> A porphyritic basalt float block, dislodged from Fork Mountain above, or perhaps left by the Doe in a forgotten flood, rests now in Cambrian terrain, a quarter of a mile northeastward. Earlier I hammered off a chip to see how the world was made, and how I was made. The secret was there. It crystallizes. It polymerizes. Set in the black groundmass of this chip, rotated now between my thumb and my forefinger, is a gray plagioclase phenocryst, a polysynthetic twin, larger than my thumbnail, the faces 110, 010, 001, and 101 especially well developed. With a lens I make out the lamellae. The scintillations of sunlight cast off these surfaces hint at

> the crystalline order beneath: silicon, oxygen, aluminum, calcium stacked and inlaid on a lattice that shames the finest arabesque (Rolston 1989, 234).

> Contemporary geneticists are insisting that we misperceive this process if we think of it as blind. It is not deliberated in the conscious sense; but it is cognitive, somewhat in the way in which computers, likewise without felt experience, can run problem-solving programs. . . . John H. Campbell concludes, "Cells are richly provided with special enzymes to tamper with DNA structure," enzymes that biologists are extracting for genetic engineering. But this "engineering" is already going on in spontaneous nature. . . . So far from disparaging the blind groping of genes, computer scientists may deliberately seek to imitate a similar process on their unconscious computers (Rolston April 1992, 262).

In the first passage, we see Rolston himself hiking and thinking about his respiration; a setting is provided for the scientific discussion, and that discussion quickly moves to conclusions about the philosophical meanings of the scientific facts. The second passage is from "Meditation at the Precambrian Contact," an essay which is so full of geological, technical, and foreign terms that average philosophical readers will have trouble understanding it unless they possess extra knowledge in science and languages. The final passage displays some of the repetition discussed previously, only here it is slightly different. The text again supports the same concept with new language, but this time it uses concepts and vocabularies from experts in genetics and computer science. The text argues in this section that nature is not blind, if understood rightly, and the theories and practices of geneticists and computer scientists are used as evidence in this argument.

Rolston's environmental ethic is closely related to or based upon science. Expert scientific descriptions are vital parts of his texts, because many of his prescriptions appear to follow easily from his expert descriptions. (See the section on Refutations for more on moving from "is" to "ought.") So, if the choice of experts is biased, if their theories are over- or under-estimated, if problems with their theories are not discussed—then there will be problems for the prescriptions as well. Since many readers will not have the breadth and depth of scientific knowledge as Rolston, though, they will have to take his expert information mostly on faith.

Rolston frequently asks a lot of his readers, plying them not only with scientific and linguistic details, but also religious, literary, and historical allusions. In one essay, for example, Rolston speaks in a two-paragraph section about Freud's and Marx's doctrines, Matthew Arnold's poem *Dover Beach* (with specific quotations), two of Sartre's and Camus' books, a Buddhist doctrine (with specific Buddhist terms), and a phrase of Virginia pioneers (Rolston 1989, 251–252). This passage is a good example of how Rolston's texts often make such allusions without explaining them in detail. Readers who are not familiar with all of those expert allusions will still understand the point of the two paragraphs, but their understanding will not be as deep or insightful as a reader who is familiar with them. Rolston has suggested that perhaps his method of using expert knowledge is a "shotgun approach" and "what doesn't hit one reader will hit another" (Rolston 1992). And, "if there is anyone who can absorb all the allusions, he or she will get an interdisciplinary or integrated view" (Rolston 1992).

Still, a kind of tradeoff occurs when genres begin to include new methods and disparate allusions. While the "shotgun approach" may attract readers in different fields, broaden philosophical readers' expectations about their texts, and encourage interdisciplinary treatments of the topic, it can also discourage some philosophical readers who do not want to be burdened with allusions or information that is not completely described, and disappoint readers who would prefer a more extensive treatment of the science topics in their own disciplines' languages.

Annie Dillard also uses scientific information and allusions to other thinkers and writers widely in *Pilgrim at Tinker Creek*. Her uses, though, are somewhat different from Rolston's. She often discovers or remembers a scientific fact, experiment, or other interesting expert opinion, describes it in detail, often as a story, and relates it to her on-going experiences and conclusions about nature.

> Well, the New York Botanical Garden put a dried *Ibervillea sonorae* on display in a glass case. "For seven years," says Joseph Wood Krutch, "without soil or water, simply lying in the case, it put forth a few anticipatory shoots and then, when no rainy season arrived, dried up again, hoping for better luck next year." That's what I call flying in the teeth of it all (Dillard 1974, 163).

> When I reach home, I turned first to the bookshelf, to see if I could possibly have seen what I thought I had. All I could find was this sentence in Will Barker's book, *Familiar Insects of North America*: "The bite of the female [Mosquito, *Culex pipiens*] is effected with a little drill that can puncture many types of body covering—even the leathery skin of a frog or the overlapping scales of a snake." All right then; maybe I *had* seen it. Anything can happen in any direction; the world is more chomped than I'd dreamed (Dillard 1974, 227).

Dillard's appeals to experts, used in this more story-like format, involves readers in an on-going, somewhat chummy dialogue with the experts quoted and Dillard. By contrast, Rolston's use of this method often piles up scientific terms and information more quickly, sometimes without full explanation. When they appeal to experts, both writers usually choose only expert information that supports their experiences or conclusions, and do not discuss information from a range of experts to make the discussion more comprehensive.

Stories and descriptions

Another important feature of Rolston's texts is the use of scientific and non-scientific descriptions, anecdotes, and stories, used as illustrations or evidence for his conclusions. Rolston has pointed out that although some frown on the use of anecdotes or stories in arguments because they do not prove anything, still "anecdotes do illustrate," and they can be "a context of discovery even where they are not a context of justification" (Rolston 1992). Furthermore, anecdotes "force the reader to ask whether the anecdote is anomalous or revelatory," and thus the method is a "use of the particular to discover the universal" (Rolston 1992).

In his texts, Rolston sometimes does appear to want to forcefully describe a personal encounter with nature and describe his own reaction to it. But instead of evoking that experience with a vivid description, he often stays on a higher level of abstraction, perhaps in order to draw a more general conclusion from it. Consider the following examples:

> One steps into the abandoned tunnels enroute here, lower in the gorge, with an initial shudder. He enters the stone bowels of the Earth as though they were haunted with the jinn of Hades. The darkness is lonesome and alien. Intuitions of the savage persist, modern as I am. But the shudder passes, and, as is the case with one's initial encounter with the sea, there follows a fascination born of the intuition of connection, of reconnection (Rolston 1989, 233).

> Sunday once, again in new fallen snow, early out and on the way to church, I chanced upon a compacted covey of quail, ringed tail to tail, just aside the gravel road. Surely that was a curious place to huddle together for warmth and protection. Suddenly the close knit covey, glimpsed but moments, became forever one of the pictures that I will not forget. A dozen pairs of eyes met mine in an encounter that somehow went right to the nerve of life itself (Rolston 1989, 243).

In the first passage, Rolston seems to struggle with pronouns, using "one," "he," "I," and then "one" again. (I discuss more about Rolston's use of pronouns later.) Here, though, notice simply that using "one" and "he" creates a distance between the speaker and the event, creating the impression that everyone who stepped into the tunnels would have a similar experience. This passage shows, in a short space, the tension in Rolston's texts between particular, personal experiences, and universal interpretations of the meaning of those experiences. On the one hand, readers feel as though Rolston wants to intrude in the text, to tell about something specific and moving that happened to him; on the other hand, readers often feel as though he wants to intrude in a formal way, to say that his experience, or a similar one, is what everyone would have experienced had they been there.

Rolston's story about the tunnel does not give readers a great deal of material to evoke the experience in their minds. For example, readers receive little description of what the tunnels looked like or how the writer approached them; thus, only a general setting is placed in the readers' minds before the subsequent reactions of the writer occur. The setting is given in the phrases "abandoned tunnels," "lower in the gorge," and "stone bowels of the Earth," but then the text moves on to Rolston's reactions to the setting. Thus, the readers are not led through and asked to join in the experience as they might be encouraged to do in a more literary text, but to fill in the details themselves that fit the writer's reactions. Here, Rolston does not so much bring readers along with him on the experience, as ask readers to imagine it or a similar experience themselves.

In the second passage above, the text uses the first person singular "I" throughout, a step toward giving an exclusively personal experience, instead of one mixed with universal interpretations of meaning. But notice that direct description of the event is still not the predominant method here, except for parts of the first and last sentences. The second and third sentences are thoughts about the event of "seeing the quail." The first part of the last sentence, "a dozen pairs of eyes met mine," is an evocative description; but the last part, "in an encounter that somehow went right to the nerve of life itself," moves to a more abstract, universal level. It is describing Rolston's interpretation of the meaning of the experience, but it is not directly describing the event itself. Again, there is little description of the setting and the experience itself.

By comparison, Annie Dillard brings readers along on her experiences. Consider how she describes looking into the eyes of a weasel.

> The sun had just set. I was relaxed on the tree trunk, ensconced in the lap of lichen, watching the lily pads at my feet tremble and part dreamily over the thrusting path of a carp. A yellow bird appeared to my right and flew behind me. It caught my eye; I swiveled around—and the next instant, inexplicably, I was looking down at a weasel, who was looking up at me.
>
> Weasel! I'd never seen one wild before. He was ten inches long, thin as a curve, a muscled ribbon, brown as fruitwood, soft-furred, alert. His face was fierce, small and pointed as a lizard's; he would have made a good arrowhead. There was just a dot of chin, maybe two brown hairs' worth, and then the pure white fur began that spread down his underside. He had two black eyes I didn't see, any more than you see a window. The weasel was stunned into stillness as he was emerging from beneath an enormous shaggy wild rose bush four feet away. I was stunned into stillness twisted backward on the tree trunk. Our eyes locked, and someone threw away the key.
>
> Our look was as if two lovers, or deadly enemies, met unexpectedly on an overgrown path when each had been thinking of something else: a clearing blow to the gut. It was also a bright blow to the brain, or a sudden beating of brains, with all the charge and intimate grate of rubbed balloons. It emptied our lungs. It felled the forest, moved the fields, and drained the pond; the world dismantled and tumbled into that black hole of eyes. If you and I looked at each other that way, our skulls would split and drop to our shoulders. But we don't. We keep our skulls. So.
>
> He disappeared. This was only last week, and already I don't remember what shattered the enchantment. I think I blinked, I think I retrieved my brain from the weasel's brain, and tried to memorize what I was seeing,

> and the weasel felt the yank of separation, the careening splash-down into real life and the urgent current of instinct. He vanished under the wild rose. I waited motionless, my mind suddenly full of data and my spirit with pleadings, but he didn't return (Dillard 1982, 13–14).

With Dillard, obviously, it is almost all setting and personal story. Though both Dillard and Rolston had similar experiences of suddenly and unexpectedly gazing into the eyes of wild animals, and feeling that their experiences went to the "nerve of life," Dillard's methods are more evocative. Readers can *see* the weasel, can *see* Dillard sitting on the log, and can visualize the locking of eyes, and the breaking of the spell. Later in the essay, Dillard does draw "conclusions" or "lessons" from the encounter, but not till the experience itself is fully explored. With Rolston's experience of seeing the quail, readers are asked again to fill in the extra details not given in the text, or perhaps to imagine a similar experience they might have had at looking into the eyes of a wild animal. In this way, Rolston's text can be a context of discovery, for those who are primed for it. But Dillard's text is likely more successful as a context of discovery for those who cannot fill in the details or have not had a similar experience of their own.

Evocative description seems to play an important role in making the case for the "revelatory" nature of a certain experience. If Rolston's anecdotes are meant to be revelatory to readers, as he suggests, they require more evocative description in order for many readers to recognize the experiences as revelatory. Consider a final example. In "The Pasqueflower," an essay similar to a nature meditation, a personal encounter with a field of pasqueflowers gives rise to various musings. However, the actual encounter with the flowers is not portrayed as a story, or even a vivid description. The essay begins in this way:

> Earliest among the rites of the western spring is the blossoming of the pasqueflower, which, like the eastern arbutus, precedes by a month the rest of the vernal flora. Its precocious beauty accounts for its name, a flower of the *Pasque*, Easter; and its loveliness, size, and season led Aldo Leopold to introduce his *Sand County Almanac* with the plea that "the chance to find a pasqueflower is a right as inalienable as free speech." Recently, just after the equinox, hiking a meadow in the foothills of the Rockies, I delighted in thousands in finest bloom, with nothing else out save the aspen catkins.
>
> Its finding is a joy immediately in the aesthetic encounter, but beyond that, this windflower is a cherished symbol of the wild for reasons that run deeper. In its annual renewal as the first spirited flowering against the blasts of winter, it is a sign against the eternal storm (Rolston 1989, 256).

Here, details which hint at a description of an experience simply create a setting for a more abstract discussion. Notice that we have no direct description of the flowers, but only of Rolston's reaction to them. We do learn from the text that the flower blooms early in the spring, that it has a "precocious beauty," that there is something remarkable about its size, and that Rolston saw "thousands in finest bloom;" but readers who have never seen a pasqueflower do not learn what color and size it is, what it looks and smells like, how Rolston came upon the field, and other such details. Thus, when the text immediately moves on to discussion of the aesthetic encounter and the cherished symbols of the flower, many readers have no image in their minds of what is being encountered and symbolized. If an evocative description *had* given readers such an image, then the thoughts which come in the rest of the essay would have a stronger effect, and the essay as a whole would have a much better chance of pointing toward a revelatory experience. Rolston is again, perhaps, asking readers to fill in the details of this experience, or if they are not familiar with a pasqueflower, to remember an experience of their own that was similar. For some readers, this will be a successful technique, and for others it probably will be asking too much.

METHODS OF PERSUASION: PRONOUNS, REFUTATIONS, AND CONCLUSIONS

Pronouns

As discussed above, Rolston often uses stories and descriptions to point to universal claims. Another, more subtle way he does this occurs in his use of pronouns. He often uses "we," which seems to ascribe a measure of universality to a claim. The following passage illustrates this method:

> Yet the bacterium values the magnetite, the coyote the orienting scent post, as surely as I value the compass. The wilderness is full of cleverness that we do not understand, of signals that we do not hear, of values that go on over our heads. We abandon our prejudices about how things are, start from scratch, and learn new scales of what ought to be (Rolston 1989, 127).

The first sentence uses "I," creating a personal claim. But the second and third sentences switch to "we" and "our," changing the particular claim to universal ones. The second sentence is a description of wilderness and of the way humans as a group relate to it; it uses "we" for "all humans." The third sentence, however, is slightly different. Here, "we" doesn't seem to be the writer's interpretation of how all humans act, because the writer probably does not believe that *all humans* abandon prejudices, and so on. If this is so, then "we" could mean "some humans," or more likely, "those humans, including myself, who have come to the realization of the above claim." Then again, the sentence could mean "we (all humans) *should* abandon our prejudices, etc." Finally, "we" in the third sentence could be a "royal we," meaning "I," which is a possibility because Rolston uses it as such in other places. Readers, then, can interpret the meaning of this sentence in several ways, depending on the way in which they read "we." And because the claim changes when the interpretation of "we" changes, this passage is confusing with regard to which claim is actually being advanced. Moreover, there is a movement in this passage from a personal, more factual claim to a description of what humans should do, all in a tone that sounds just as factual.

Another passage illustrates the potential trickiness of "we."

> If feeding some humans is good, feeding more is better. And more. And more! Feeding all of them is best? That sounds right. We can hardly bring ourselves to say that anyone ought to starve. But we reach a point of diminishing returns, when the goods put at threat lead us to wonder (Rolston 1996, 260).

The first time "we" is used, it could mean that "I" or "most people" don't believe anyone ought to starve. But in the last sentence, it suddenly becomes more important to know how "we" is used; if "we" and "us" mean that "I" start to wonder whether some people ought to starve, then Rolston is saying what he believes; but if they mean that "you and I" or "most people" start to wonder whether some people ought to starve, there is a problem. Suddenly I, the reader, am ascribed an opinion I may not hold.

One final passage illustrates the problems with ambiguous pronoun usages. This passage appears near the end of the first chapter of *Environmental Ethics*; notice that the paragraph begins with interpretive description, moves to the royal "we want," then to "one needs," then back to interpretative description, and finally to "our conduct ought."

> Certainly there is struggle in nature. But there is also, and even more, adaptedness. There is resistance to life, but there is a conductance of life. The account we want—and will be seeking in subsequent chapters—contains both elements, and not merely as a nonsensical mixture of goods and evils. What one needs is a nature where the evils are tributary to the goods, where natural resistance is embraced within and made intelligible by natural conductance. Not death but life, including human life fitted to this planetary environment, is the principal mystery that has come out of nature. For several billion years the ongoing development and persistence of that life, culminating in human life, have been the principal features of eco-nature behind which the element of struggle must be contained as a subtheme. Our conduct ought to fit this natural life story (Rolston 1988, 43).

The use of the phrase "What one needs" in the fifth sentence is especially interesting here. Notice that if the text instead read "What we need," it could easily sound as if the writer was still using the "royal we," as he did in the previous sentence; then the sentence would mean "What I need." To avoid this (and it must be avoided for the claim to sound more universal), the writer must switch to another pronoun or subject, and he chooses the formal indefinite pronoun "one." "One" usually means "everyone" or "a person (i.e. *any* person)." But if "one" is replaced by either of these words, the claim sounds weaker or more implausible. "One needs" in this context could mean "everyone ought to need," or "those of us who want the right account need." But neither of these meanings is normally implied with the pronoun "one."

Furthermore, the words "our conduct ought" in the last sentence suggest that Rolston *has* been talking about "everyone" or "anyone," not just "some of us." And the use of "ought" in the same sentence might indicate that he was speaking of prescriptions in the earlier usage as well.

What this illustrates is that it's not always crystal-clear whether Rolston is saying what he believes, what he thinks most people believe, what he thinks most people ought to believe, etc. And this can be not only confusing, but at times coercive.

Annie Dillard, like Rolston, desires to move frequently from personal experiences, using "I," to more universal ones, using "we." But her usages seem more straightforward; it seems clearer that her uses of "we" mean "the human race" as in the following paragraph that follows a personal description of watching sharks:

> We don't know what's going on here. If these tremendous events are random combinations of matter run amok, the yield of millions of monkeys at millions of typewriters, then what is it in us, hammered out of those same typewriters, that they ignite? We don't know. Our life is a faint tracing on the surface of mystery, like the idle, curved tunnels of leaf miners on the face of a leaf. We must somehow take a wider view, look at the whole landscape, really see it, and describe what's going on here. Then we can at least wail the right question into the swaddling band of darkness, or, if it comes to that, choir the proper praise (Dillard 1974, 8–9).

Refutations

Rolston's method of presenting and refuting alternate points of view is also distinctive in its somewhat casual, story-like manner. The following passages contain such language: note especially the phrases, "The better answer" and "But the more one studies . . . the less evidently."

> Plainly, humans have expanded their territories all over the globe. But what is an appropriate lifestyle for residing in this globally occupied territory? To maximize the high intrinsic value of their kind? Nothing more? The better answer, found in environmental ethics, is that humans ought to be ideal observers . . . (Rolston 1988, 338).

> One could label all this so much resource use, and then stipulate that values necessitate sentient awareness. Objective organic processes form roots, precursors of value, but valueless in themselves, becoming of value only when experience is superadded. But the more one studies organic bodies, the less evidently this is the most plausible route for mapping value. It starts with a psychological or hedonic result of the biological processes, values this experiential effect, and devalues the productive causes except in terms of a late conclusion, in which, subjectively, we happen to stand. It takes a derived thing as the only thing that really counts (Rolston 1989, 125–126).

The phrases mentioned above are the key turning points for these passages. In the first passage, Rolston presents the alternative point of view in the third and fourth sentences, and then rejects that point of view in the next sentence. But he does not strongly insist that the other point of view is wrong—he merely suggests that there is a "better answer." In the latter passage, the third sentence gives the reason for rejecting the alternative point of view, and again, the view is not rejected strongly. Instead, the text says, "the more one studies organic bodies, the less evidently this is the most plausible route."

This appeal to consider "what is" in reaching conclusions is another of Rolston's methods. Before the passage below, for example, Rolston talks about whether to use "blind" as a metaphor for how genes act. Note how the paragraph below starts with an alternate point of view, moves to a question that suggests Rolston does not believe that point of view, and ends with an appeal to consider more details of genetic processes.

> Well, it may be replied, the stored information is not so blind, but the method of discovering any new genetic information is blind. Because genes do not "see" where they are going, the variations are accidental and groping, and for this "blind" is a convenient metaphor. As organisms move from earlier genetic achievements to the discovery of later ones, there are certainly elements of random exploration. But is that all there is to be said? Consider how these elements of trial and error are incorporated in a larger generative process (Rolston 1999, 28).

Rolston's movement illustrated in the above paragraph gels nicely with his view that descriptions and prescriptions can emerge simultaneously. He has said, "Much of the moving from 'is' to 'ought' . . . is not moral but logical. It is not that we have a moral duty to hold these recommended accounts, but that we have a rational 'duty'—we really *ought* to draw the conclusions this way . . . if we are to have the most plausible account" (Rolston 1992). This belief will affect the way Rolston gives "evidence" for some of his claims and refutations. That is, because he believes his accounts follow directly from or emerge with his descriptions of nature, the accounts do not always, to him, need as much as or the same kind of traditional philosophical evidence. Instead, their evidence *is* the descriptions of nature. Thus, refutations of alternate points of view sometimes involve, for Rolston, attempts to get the reader to "see" that his account follows directly from the description, while the alternate accounts do not follow. With this method, he asks readers to come along with him and trace either a thought process he once went through, or one he has created, almost as a set of musings-out-loud.

Conclusions

Rolston's texts are also very skillful at combining sentences to make interesting, varied paragraphs, and combining the paragraphs to lead to forceful conclusions. The movement of his texts often exhibits a rather casual progression, using many of the methods discussed above, leading to a firm conclusion. The following paragraph gives some sense of this movement in a short space. Notice how in this one paragraph, the text uses word-exploration, repetition of claims with different wordings, descriptions, the beginning of a story, expert knowledge in the form of Latin terms, a movement from the pronouns "I" to "he" to "we," a movement among existence verbs, action verbs, and prescriptive verbs, and claims that emerge from the kind of "seeing" discussed above. And notice especially the last four sentences, and the effect they have after the previous part of the paragraph.

> Myriads of larvae infest the lake. At first this annoyed, for it forced a regular walk to the Inlet for fresher water. But now it refreshes, as each walk prompts a moment of truth. This teeming aquatic life reveals an inexhaustible vigor, and I myself flourish, seeing the spontaneous vitality of the Earth. When I leave, the thrush will still spill its song into the forest, yet I will somehow be richer for the beauty that escapes my hearing. Spruce should rot where they fall, repay the elements, recompose in *Sedums* and *Calthas* appropriated from the humus—and so teach of death and life. If he rises to his name, *Homo sapiens* will leave this place unmutilated. There is a fragile fullness here, which, oddly, can be shared only when it is honored. We gain what we give; to be whole, I must leave the Earth whole. Who troubles these waters, troubles himself (Rolston 1989, 226).

These last sentences, which give readers the conclusion of the paragraph, are carefully worded and enjoyable to read. They are an interesting mix of repetitive claims and novel wordings, which combine to form a kind of twist at the end. And after the somewhat meandering paragraph, these sentences are forceful because they *are* a conclusion; even if it is not entirely clear how the conclusion was derived from the rest of paragraph for some readers. In addition to this kind of layering of methods within paragraphs, Rolston is very skillful at layering the paragraphs themselves to create a strong cumulative effect for the conclusions toward the ends of his chapters or articles. And the force of this kind of cumulative effect cannot be underestimated. For some readers, this force is all that they need to be completely convinced. Other readers, perhaps some philosophical ones, will have mixed feelings after arriving at the conclusions; the force of the skill of the writer may be admired, but the way at which it was arrived may result in some measure of ambiguity.

The methods Dillard uses to state conclusions in some ways are similar to Rolston's. She uses personal experiences, expert information, and layers of explanations to describe what "is," and attempts to get readers to see that certain truths follow from these. Her language surrounding conclusions is often casual, like Rolston's. But Dillard's statements of conclusions, while not always reducible to a single sentence, are in some ways more straightforward than Rolston's (perhaps because she does not need to argue!). Though both writers are stating implicitly in their texts "this is what I believe to be true," and "this is how humans should act," Dillard's texts do so explicitly as well. Consider these passages of Dillard's:

> What do I make of all this texture? What does it mean about the kind of world in which I have been set down? The texture of the world, its filigree and scrollwork, means that there is the possibility for beauty here, a beauty inexhaustible in its complexity, which opens to my knock, which answers in me a call I do not remember calling, and which trains me to the wild and extravagant nature of the spirit I seek (Dillard 1974, 139)

> Go up into the gaps. If you can find them; they shift and vanish too. Stalk the gaps. Squeak into a gap in the soil, turn, and unlock—more than a maple—a universe. This is how you spend this afternoon, and tomorrow morning, and tomorrow afternoon. *Spend* the afternoon. You can't take it with you (Dillard 1974, 269).

Rolston uses the methods related to plural subjects, refutations, and conclusions when very important explanations and claims need to occur in his texts. The plural subject usages almost always occur in passages with important explanations or conclusions. The refutations of alternate points of view are key explanations, because they represent vulnerable points in the developing theory; if the alternate points of view are not refuted strongly enough, readers who have been convinced up to this point may be lost. And of course, conclusions are very important in any text, because they represent the main points of the theory the writer has been working to support. In Rolston's texts, the conclusions often sound like explanations, or re-explanations or restatements of an earlier claim. Thus, the methods Rolston uses surrounding his explanations are important because those explanations play a large part in the effectiveness of the texts. Because the methods examined in this section are subtle ones, and are sometimes difficult to detect, they may contribute to the persuasiveness of the explanations without all readers' notice. Rolston has, in this way, added powerful tools to his explanations, which may be all the more powerful because they can go unnoticed.

THE LARGER CONTEXT OF ROLSTON'S METHODS

This investigation has explored some of the more "literary" methods in Rolston's texts; however, those methods are not the only ones in the texts. There are also many traditionally philosophical methods, such as a concentration on defining terms, drawing distinctions, placing arguments within the historical dialogue of ideas, and investigating issues and arguing for philosophical points. Because these more philosophical topics and methods combine with the literary topics and methods to create the total effect of Rolston's texts, and because the philosophical modes have not been explored in this investigation, the total cumulative effect of Rolston's texts cannot be evaluated here. Nevertheless, in order to begin to place the literary methods examined here within a larger context of the whole of Rolston's works, it will be helpful to examine briefly two further issues.

One of these is the relation between some of Rolston's "literary" methods and the content of the texts. For example, Rolston's implementation of story-telling is interesting in light of what his texts actually say about stories. In many texts, Rolston stresses strongly the importance of stories in the natural world and their contribution to an environmental ethic.

> Early science thought this nature to be lawlike and repetitive, but recent science has learned the evolutionary Earth history, indeed the history of our universe. And this story is still taking place. Life is still arriving. Earth is not so much a syllogism with premises and conclusions as it is a text to be interpreted. It is stories being told (Rolston 1988, 343).
>
> The rationality of the ethic, as well as the area to be mapped, will be historical. That is, logic will be mixed with story (Rolston 1988, 342).
>
> Out of physical premises one derives biological conclusions, and taking these as premises in turn, one derives psychological conclusions, which recompounded again, yield spiritual conclusions. This kind of logic seems more story than argument; the form of argument is not so much rational as, to use a religious word, incarnational, since each step has to be embodied. Story is a better category than unfolding law, much less random drift, or selfish defense of life, when one wants to get more out of less (Rolston 1999, 299).

Rolston's emphasis on stories in these passages matches his inclusion of stories in his texts, though he does not use stories in the same way as a writer such as Dillard. As discussed above, Rolston's stories often appear in an undeveloped way in the texts; therefore, the form of the text is not always consistent with the content, though an insistence on the importance of stories in environmental philosophy may not imply a developed use of stories in such texts. Still, the texts will not be as unified as texts which have a form that is completely consistent with the content.

The second issue which helps to place Rolston's more literary methods in a larger context is that of the overall arrangement of his texts. The topics and methods of the first six chapters of *Environmental Ethics* are fairly or very philosophical, but the topics of last three chapters are less traditionally philosophical: the chapters are about how the theory relates to environmental policy, how it relates to environmental business, and how people should live out an environmental ethic in an individual sense. *Philosophy Gone Wild* is organized in a similar way. Because the book is a collection of essays written over 20 or more years, the book contains some overlap, and does not present as unified and cohesive account as *Environmental Ethics*. Still, the essays are organized in a thoughtful manner; the first two sections of the book contain the more "philosophical" essays, and the last two sections, "Environmental Philosophy in Practice" and "Nature in Experience" move away from traditional philosophical topics and methods of expression. "Nature in Experience" is a collection of essays about personal experiences in nature that form the base for further reflections about nature. *Genes, Genesis and God*, Rolston's most recent book, starts with chapters on genes, and moves "up the ladder" to human culture, science, ethics, and religion, the latter being the area Rolston has not explored explicitly in many of his earlier works.

The arrangement of the chapters within Rolston's books gives readers a hint about how important the philosophical and literary topics and methods are for the texts. In *Environmental Ethics* and *Philosophy Gone Wild*, the *last* chapters or sections are the ones that depart most from the traditional philosophical methods and perhaps content. All three of the books exhibit a movement from scientific to cultural dimensions. Because the literary topics and methods occur mostly at the ends of the books, some readers may think these topics and methods are the most important ones; on the other hand, some readers may interpret the topics and methods as the least important, because the writer has waited till the end to introduce them.

THE IMPACTS OF ROLSTON'S LITERARY METHODS

Rolston's texts retain many traditional philosophical methods, but they join those methods with literary ones that clearly and boldly move philosophical methodology in a new direction. His texts are not only logical, "objective" arguments using traditional philosophical methods, but a complex interplay of argument, rationality, narrative, autobiography, emotion, and rhetoric.

All of Rolston's literary methods ask more of readers, more than philosophical readers are often expecting. With word-explorations, discussions of metaphor, repetition, questions, and exclamations, readers are asked to notice words, their origins, their ancient and modern meanings, their double meanings; to notice new ways to talk about the same concept, and to relate points to other contexts and meanings; to pay attention to sentence structure and how it contributes to explanations or arguments; and to appreciate questions and exclamatory expressions. Expert information gives readers facts they need to know for the developing theory, but also asks readers to appreciate many scientific, religious, literary, and historical allusions. Stories and descriptions ask readers to be open to discovering a new point via a story, and "ask whether the anecdote is anomalous or revelatory," in Rolston's words. And, they ask readers to fill in the details of a sketched-in story or to remember a similar personal story. Rolston's use of varying pronouns, and his casual methods for refuting alternate points of view and advancing toward conclusions may not "ask" as much of readers explicitly as the other methods. Still, readers must disentangle what is meant with changing pronouns, and be ready to follow along on thought processes to refute alternatives or reach conclusions.

Rolston's texts, in sum, are quite complex. They ask readers to do a great deal, just with the literary methods; add to this what the texts ask of readers with the traditional philosophical methods, and the texts can take much work to fully appreciate.

Rolston's literary methods ask more of philosophical readers than they may be ready for, moving philosophy in a perhaps more literary direction. But Rolston often does not go "all the way" toward the literary genre, as I have discussed above. His uses of stories are an example of this; they usually do not take readers along on an experience, as Dillard's stories do, but give a rough sketch of a story, asking readers to fill in the rest. In this way, his texts ask more of readers than Dillard's texts in terms of the use of stories. In order for more readers to acknowledge a story as revelatory, if that is one of Rolston's aims, it is necessary for Rolston to go all the way or at least further toward the literary genre with stories.

Rolston's texts also do not make their methodology explicit, something that Dillard's *Pilgrim at Tinker Creek* does. This is another way in which Rolston does not go quite as far as Dillard. Dillard's book is made up mostly of personal experiences and discoveries, combined with references to scientists and forceful general conclusions. Her primary methodological category is the personal. In *Pilgrim at Tinker Creek*, Dillard explains what she sets out to do in the book, to "explore the neighborhood, view the landscape, to discover at least *where* it is that we have been so startlingly set down, if we can't learn why" (Dillard 1974, 11–12). While she explores her physical neighborhood, a valley in Virginia's mountains, she writes about it and explores the thematic neighborhood of certain topics through reading. And, her writing explores thematic concepts prompted by personal experiences. The strength which comes from making a methodology explicit to the readers in this way is that the text then provides an example of the methodology which, if it is well-done, will implicitly argue for that methodology.

Because Dillard's readers know that her primary methodological category is the personal, her texts avoid a kind of disunity that Rolston's texts cannot escape. With Rolston's texts, there seem to be a number of candidates for the primary methodological category (rational argumentation, personal experience, emotive followed by rational evaluations, etc.). Rolston's texts would be more unified if their methodology were made explicit to readers (e.g., giving reasons for using the methodology, an outline of how elements in the methodology relate). Readers would then be conscious about what the writer's methodology is, they would notice an example of the methodology in the form of the text, and they would have both forms of information when they evaluate how effective the methodology is.

But perhaps Rolston does not want to give so much away. In many ways, Rolston's approach is more Socratic than Dillard's, using language as a context of discovery and asking readers to draw the conclusions out of themselves and their own experiences. Rolston asks much of his readers; he asks readers to utilize more of their whole selves in thinking and feeling through philosophical ideas than many philosophers do. That is an advance toward broadening philosophical methodology to include the myriad ways humans experience and find value in nature. Rolston may not go all the way to the literary genre with some of his methods, but he boldly goes where few in philosophy have gone, providing readers with provocative, enjoyable language that challenges and gives insights in novel ways.

REFERENCES

Dillard, Annie. 1974. *Pilgrim at Tinker Creek*. New York: Harper & Row.

Dillard, Annie. 1982. *Teaching a Stone to Talk*. New York: Harper & Row.

Rolston III, Holmes. 1986. The Human Standing in Nature: Storied Fitness in the Moral Overseer. In *Values and Moral Standing*. Bowling Green, OH: Bowling Green Studies in Applied Philosophy.

Rolston III, Holmes. 1988. *Environmental Ethics: Duties to and Values in the Natural World*. Philadelphia: Temple University Press.

Rolston III, Holmes. 1989. *Philosophy Gone Wild*. Buffalo, New York: Prometheus Books.
Rolston III, Holmes. April 1992. Disvalues in Nature. *The Monist*, 250–278.
Rolston III, Holmes. 1992. Personal communication with the author. Fort Collins, Colorado.
Rolston III, Holmes. 1996. Feeding People versus Saving Nature? In *World Hunger and Morality*. Englewood Cliffs, NJ: Prentice-Hall.
Rolston III, Holmes. 1999. *Genes, Genesis and God: Values and Their Origins in Natural and Human History*. Cambridge: Cambridge University Press.

VICTORIA DAVION

10. CARING FOR NATURE: AN ECOFEMINIST'S VIEW OF ROLSTON ON EATING, HUNTING, AND GENETICS

This chapter examines several of Holmes Rolston's views through an ecological feminist lens. I argue that although some of Rolston's specific conclusions, particularly those regarding domesticated animals and sport hunting, are highly problematic, much of what he says is not only consistent with, but also highly relevant to, central views within ecological feminism. I shall conclude that given his other more general views, he should reconsider his earlier positions regarding domestic animals and sport hunting. I shall begin with a discussion of the aspects of Rolston's work that are clearly problematic from ecological feminist standpoints, and then show how other, broader aspects of his work fall right in line with key ecological feminist insights. I shall conclude with an argument for why Rolston should change his positions on both the treatment of domestic animals and sport hunting, and by pointing out what I think ecological feminists such as myself, and Holmes Rolston, can learn from each other.

Ecological feminism is a term that covers many different positions (Warren 1994). In fact, ecological feminists disagree on many key issues, making it problematic to call any one position or approach *the* ecological feminist view. However, most ecological feminists agree that dualistic thinking has permeated traditional western world views, and that it contributes to the domination of nature and the oppression of women and various men. A dualism is a kind of "super dichotomy" in which difference is transformed into radical difference, and in which one category is seen as morally superior to the other. Val Plumwood has argued that a reason/nature dualism undergirds traditional western frameworks, in which whatever is seen as falling under the category of "reason" is considered to be radically different from and morally superior to whatever is seen as falling under the category of "nature" (1993). This dualism forms a fault line and is the basis for the construction of a variety of other problematic dualisms such as reason/emotion, masculine/feminine, and civilized/primitive, to name a few. In each case, whatever represents reason is thought to be radically different from and morally superior to whatever represents nature. Hence, men are seen to be more "reasonable" and women more "emotional," and this justifies sexism, while the civilized/primitive dualism justifies imperialism. Justifications for the system of American slavery included the fact that Africans were supposedly more primitive, more like animals, and therefore less capable of reason than white northern Europeans.

EATING

Rolston's position regarding the treatment of domestic animals, particularly food animals, clearly invokes a problematic nature/culture dualism, although unlike many

C.J. Preston and W. Ouderkirk (eds.), Nature, Value, Duty, 167–181.

in the traditional western philosophical tradition, Rolston urges that non-human beings, or "natural" beings have intrinsic value and are worthy of moral respect. He reasons that each living being has a telos or a goal, and therefore can be seen as an evaluative system, valuing some things over others. Living beings have their own "good" that can be respected by humans, should humans choose to do so. This is squarely in the spirit of ecological feminism which urges respect for non-humans. Rolston maintains that in seeking an environmental ethic, we need one that understands how nature functions. Utilitarianism, for example, would be misguided since pain and suffering have important functions in nature, so pain is not necessarily always a negative, contra the traditional utilitarian view. Rolston maintains that pain is a negative in culture, because in culture the pressures of natural selection are relaxed. Hence, while utilitarianism might be an appropriate interhuman ethic, it is inappropriate as an environmental ethic.

My purpose here is neither to evaluate whether Rolston is right that the pressures of natural selection are relaxed in culture, nor to explore whether utilitarianism is an appropriate ethical theory for humans "in culture." Rather, I shall argue that given his position, one would expect Rolston to argue that the pain of domesticated animals, because it occurs in culture, is bad. Nevertheless, employing a strict yet vague nature/culture dualism, Rolston draws the opposite conclusion. He says of domesticated animals that they are actually no longer natural kinds but human artifacts. They fit no environmental niche. They are removed from the forces of natural selection through human breeding, and without human interests they would cease to exist at all. Sheep, cows, pigs, and chickens are oblivious to the purpose and cultural context of those who care for them. "Pets may eat a hot dog at the family picnic but pets are not in culture" (Rolston 1988, 79). And, Food animals live neither in nature nor in culture but in the peripheral rural world. Meanwhile they can suffer" (79). In answer to the question of how humans ought to respond to the suffering of food animals, Rolston responds:

> They ought to be treated, by the homologous baseline principle, with no more suffering than might have been their lot in the wild, on average, adjusting for their modified capacities to care for themselves. . . . (whether modern industrial farming introduces suffering in excess of ecological norms will have to be investigated elsewhere.) (79).

Leaving aside the question of how to establish how much suffering would be the lot of domesticated animals in the wild, given that they have never lived in "the wild," it would seem that the amount of suffering caused by some methods of factory farming would exceed it. Not only does Rolston leave this question open, indicating that quite a lot of suffering is acceptable, he also claims that "a cow is a meat factory, pure and simple" (83).

Given Rolston's view that domesticated animals are the products of culture, that they do not exist in nature, and that humans are responsible for them, he could argue that the pain of domestic animals is prima facie bad, and should be alleviated when possible. Instead, Rolston invokes his nature/culture dualism to argue that the pain and death of domesticated food animals is not bad in itself. Therefore, humans have no obligation to avoid eating animals, even when we have other viable options.

> Humans eat because they are in nature, not because they are in culture. Eating animals is not an event between persons but a human-to-animal event, and the rules come from ecosystems in which humans evolved and which they have no duty to remake. Humans, then, can model their dietary habits on their ecosystems, but they cannot and should not model their interpersonal justice or charity on ecosystems. (81)

Examples of practices not to be modeled on ecosystems include marriage, truth telling, and promise keeping. In response to the objection that if we do not and should not model these practices on ecosystems, then there is no reason to model eating meat on them, Rolston writes:

> Other matters of cultur—marriage, truth-telling, promise-keeping, justice, charity—these are not events at all spontaneous in nature. They are person-to-person events. (81)

Hence, some practices that humans engage in are engaged "in nature," and it is appropriate to model such practices on ecosystems, while others are purely "cultural."

Ecological feminists such as Carol Adams (1991), Greta Gaard (1993), Chris Cuomo (1998), and Lori Gruen (1993) have long been concerned about the plight of domesticated animals. Rolston claims that while human culture creates moral obligations to be concerned about human suffering, this does not require us to interfere in the suffering of wild animals interacting "naturally." There is however a weaker obligation to be concerned with the suffering of domesticated animals in that they shouldn't suffer more than they would have in the wild. Whatever that means, it doesn't clearly rule out factory farming, which is arguably unnecessarily cruel. The obvious question from an ecological feminist perspective is why do humans eat "in nature" and get married "in culture"? One could just as easily argue that marriage is simply the regulation of sexual activity in culture, and that one should look to nature for the appropriate way to guide sexual activity, since animals "do it" too. Similarly, one could argue that people eat "in culture." There are huge dietary differences between different cultures of humans. Food practices are as much a part of culture as anything else. In fact, Rolston himself argues that some of the slaughtering practices of Jews and Muslims are problematic in that they cause "pointless" pain and are "cultural" not "natural."

> [R]eligious methods of slaughter result in a degree of suffering and distress that does not occur in a stunned animal. Muslims and Jews have joined forces to defend their practices. But the additional pain that their methods impose, no longer necessary, cannot be interpreted in the context of ecology; it is pain inflicted for culture-based reasons. Unblemished animals make better sacrifices to God; they enhance religious cleanliness. This pain is ecologically pointless; it has point only culturally and by the account given here, is not justified. (84)

If we are to look to "culture" as a guide-post for ethical eating habits, then on Rolston's own argument, it follows that food animals' pain is bad, and unnecessary pain should be avoided. We would still need to determine how much animal pain is acceptable in order to provide food for humans, especially in situations where other food sources are readily available. This would not be an argument in favor of universal vegetarianism, as in many places it is necessary for people to eat meat either to survive or to have tolerable lives. In addition, some cultural traditions may provide people with morally justifiable reasons to consume meat. But, this is a cultural argument; precisely the type that Rolston appears to think should not be used in relation to human dietary habits. My point here is that the nature/culture dualism invoked by Rolston is by no means clear. If "nature" and "culture" are radically different realms, it certainly isn't clear which human practices should be evaluated according to standards provided by "nature," and which by standards provided by "culture." If human beings are required to use "cultural" standards when behaving "in culture," and if eating is arguably "cultural" as well as "natural," the ethics of the treatment of "food animals" is extremely unclear.

HUNTING

Rolston's views on sport hunting are similarly problematic from an ecological feminist perspective. According to Rolston, although some wild animals may "enjoy" a hunt, hunting purely for pleasure is nonexistent in nature. Hence, it would appear that humans are not modeling "nature" when we hunt for leisure rather than for food. And, Rolston quite rightly points out that eating what one kills doesn't solve the problem, if the main motivation for the kill is sport and one has other things to eat. Hence, it would seem that Rolston should be against hunting for sport, which would put him squarely in line with most ecological feminist positions. Yet, Rolston is unwilling to take a firm stance against sport hunting. He states:

> The best case that can be made for sport hunting is that it is not merely recreational but is a vicarious, therapeutic, character-building, re-creational event, where a visceral urge is vented in the sport hunt, carried forth in its ecological setting. The sport hunt sublimates the drive for conquest, a drive without which humans could not have survived, without which we cannot be civilized. The hunter is as propelled as the bullets he or she shoots. Civilization has required the modified expression of this atavistic urge, but dimensions of it are still nowhere better expressed than in this reversion from civilization to nature. . . . In this sense hunting is not sport; it is a sacrament of the fundamental, mandatory, seeking and taking possession of value that characterizes an ecosystem and from which no culture has ever escaped. (91)

And:

> The human being is uniquely unspecialized, a species in whom aboriginal genetic drives can be sublimated and primitive ecologies modified.

> Perhaps the hunting drive, like the sexual urge, is dangerous to suppress and must be reckoned with . . . some hunters need to be immersed immediately in the bloodletting. The hunter feels not "perfect evil" (Krutch), but "perfect identification" with the tragic drama of creation, the blood sacrifice on which sentient life is founded, which both is and ought to be. In ways that mere watchers can never know, hunters know their ecology. The hunter's success is not conquest but submission to ecology. It is an acceptance of the way in which the world is made. In all this nowhere should it be forgotten that there is also a nature to which humans by genetic disposition are attracted, and hence the congenital ambiguity toward a nature we need to slay and to love, surfacing as a sacrament of the way the world is made. (92)

Rolston's analogy of the "hunting drive" with "the sexual urge" which is "dangerous to suppress and must be reckoned with," is highly disturbing from ecological feminist perspectives. Rolston merely states that there is a "sexual urge" which is dangerous if suppressed, and bases his position on "the hunting drive" on the obvious existence of the "sex urge." If the hunting drive is like this urge, then the danger in suppressing it is supposedly obvious.

Feminists, including ecological feminists, have long been rightly suspicious about certain durable but highly problematic "insights" about male sexuality in western patriarchal contexts. The "dangerous" and "uncontrollable" male sex drive continues to serve at least as an excuse for, if not a justification of, a tremendous amount of sexual misbehavior on the part of men (Stoltenberg 1989; Sommers 1994; Schlessinger 1994). We hear little about such a drive in women. And, as he often does, Rolston leaves this gendered component totally out of his analysis. The "boys will be boys" excuse that justifies putting up with such behavior because it is based on a "biological drive" is highly problematic. It is unclear how one might go about proving such a claim. However, even if there is some biological drive that causes male sexual misbehavior, we can, should, and sometimes do expect men to control that drive. Hence, we do sometimes hold men responsible for extreme sexual misbehavior, such as rape.

Feminists have long argued that given social contexts in which boys and men are encouraged to see themselves as having strong, if not insatiable, sex drives, and in which they are teased if they admit that they do not have such drives, it would be difficult, if not impossible to show that such "drives" are somehow "natural," as opposed to being socially constructed. In addition, even if such a thing could be shown to be true, we can argue that men are not completely "propelled" by such drives, and can learn to control them. Clear evidence for this is that many men do not misbehave sexually, so whatever "biological drives" there may be, they need not result in such behavior.

In discussing sport hunting, Rolston again assumes that the "hunting drive" is basically male, while failing to make this explicit in his analysis. Although in referring to the hunter as "he or she" in exactly one of his sentences about hunting, which is quoted above, he often refers to the "human" urge to hunt, but his analysis makes it clear that he is envisioning the sport hunter with the dangerous "hunting drive" as male. In a section discussing moose hunting in Maine, he grants that "this particular form of hunting does seem more to flatter male vanities, not unlike jaguar coats flatter female vanities"(92). However, he concludes that not all sport hunting simply flatters male vanities:

> The satisfactions and skill at the hunt must run deep; they sometimes flatter male vanities, but we cannot always interpret them so. And the fact that men enjoy skilled hunting, mixed with unease about killing, is no embarrassment. (93)

That the problematic "drive to hunt" for sport is male, is clear.

Ecological feminists have long been concerned with leisure sport hunting. In response to the idea that sport hunting is somehow character building, and that the "connection" the hunter feels with "nature" during a hunt is somehow sacred, ecological feminist Marti Kheel says:

> Identification may, in fact, enter into this philosophy, but only to the extent that it flows from an existing connection with individual lives. Individual beings must not be used in a kind of psychological instrumentalism to help establish a feeling of connection that in fact does not exist. Our sense of oneness with nature must be connected with loving actions. (Kheel 1990, 131)

Unfortunately, in a move that some but by no means all ecological feminists have made, Kheel appeals to "women's unique sense of connection with the natural world," (1990) as a source for legitimate connected loving actions. I have critiqued this approach elsewhere, as being essentializing, and as glorifying "femininity" in highly problematic ways (Davion 1994). My point here is that ecological feminists have urged that we examine various types of so-called feelings of identification with nature, as such feelings may be highly problematic.

Even if sport hunting does answer to some "natural" drive in men, this certainly does not make it ethically acceptable. And, as in the case of sexual misbehavior, it is unclear how one might go about establishing that it is motivated by any "biological drive" at all. Given that sport hunting is socially encouraged among certain groups, and a dislike of hunting is seen as abnormal and unmanly in many such groups, the fact that many men (and some women) enjoy sport hunting by no means establishes that hunting for sport is somehow a "natural drive." And, even if such a thing could be established, it does not follow that we cannot be expected to control such drives. Like the case of sexual misbehavior, claims that such drives are uncontrollable benefit those who engage in problematic behavior at the expense of their innocent victims, which makes them highly ethically suspicious.

Rolston maintains that some of the "hunting drive" can be satisfied in ways that avoid actual killing, as in camera safaris (92). However, he claims that some hunters just "need to be involved in the bloodletting," in order to feel "perfect identification" (92). My own answer is "so what"? Various immoral acts may give certain people positive feelings, but this in no way serves to justify them! In addition, as with eating, it is far from clear that sport hunting occurs "in nature" as opposed to "in culture." Given the huge number of technological gadgets often involved in sport hunting, it is difficult to deny a strong "cultural" component. Driving into nature in an SUV equipped with the latest in hunting technology hardly seems to mimic anything at all that animals "in nature" engage in. It seems to me to be a lot more like going to a human-made amusement park. Rolston clearly admits that there is nothing at all like sport hunting "in nature." Therefore, it is unclear that hunting for leisure is in any way more natural than cultural, and it is unclear that the desire to do it is based on some "natural, biological, urge." In warning that attempts to suppress the drive for sport hunting may be dangerous, Rolston creates the kind of excuse that those who excuse male sexual misbehavior have used. We'd better tolerate some ethically questionable behavior because after all, those engaging it cannot control it. It would be both unfair and dangerous to demand that such urges be controlled. Hunters, to use Rolston's own words, are "as propelled" as their bullets. However, it should be noticed that those who are asked to "tolerate" such behavior are asked to accept abuse, and in the case of sport hunting, death. Hence, Rolston's justification/excuse for sport hunting is deeply problematic from an ecological feminist perspective. It posits problematic "biological drives" and fails to demand that even if they exist, men are morally responsible for controlling them. Adding the gender aspect reveals that it is women and animals that are asked to bear the brunt of this lack of control. I want to be absolutely clear that Rolston does not explicitly state that he endorses sexual misbehavior on the part of anyone. In fact, I take it that he would either deny that his position implies any such an endorsement or change his view so that it no longer has such implications. In my view, it is part of the strength of ecological feminist analysis to dig out hidden gendered components in positions, revealing the sexist aspects of positions that appear to be gender neutral.

POPULATION AND SOCIETY

Another main focus of ecological feminism has been to insist that all environmental problems are social, and that factors of race, class, ethnicity, geography, and a variety of others concerning social location, in addition to sex, are crucial for the analysis of any environmental problem. Here Rolston's discussion of population and conservation fare somewhat better than his analysis of sport hunting and the treatment of domesticated animals. In his well-known paper, "Feeding People Versus Saving Nature," (1996) Rolston asks the question of whether it can ever be justifiable to allow people to starve in order to preserve certain aspects of nature, such as endangered species. Ecological feminists have also been concerned about the so-called population problem. They have critiqued approaches that simplify the problem into one that pits poor starving people, usually in third-world countries, against various endangered natural ecosystems or beings. Ecological feminists have asserted that such simple approaches absolutely fail to take into account differences crucial between humans (Cuomo 1998). Such approaches ignore questions of wealth and resource distribution as well as questions of social location, which include race, sex, class, and geography. Hence, questions concerning why the poor are poor and the rich are rich go unexamined, as do questions about why women, who bear the brunt of the pain and work involved in childbearing, which is often extremely difficult, if not fatal in very poor places, have so many children to begin with.

In "Feeding People Versus Saving Nature," Rolston appears to be arguing that it is okay to let poor people die in order to save certain endangered species and natural places. However, a close reading reveals a powerful social critique. Rolston points out that we need to look at why poor people and nature become pitted against each other to begin with. He maintains that the problem is often one of unjust distribution rather than simply not enough resources. He reminds us that the rich do all sorts of things that clearly indicate feeding the poor is not really a particularly high priority. The rich spend on their militaries while letting poor people starve to death, the rich pursue leisure activities rather than aiding the poor, and this may not always be wrong. However, given that this is the situation, then it might be acceptable to protect an endangered species rather than feeding the poor, if the choice comes down to that. Rolston argues that the problem needs to be addressed as a distribution problem rather than one to be solved by expanding the consumption of precious natural resources.

> The welfare concept introduces another possibility, that the wealthy could be taxed in order to feed the poor. We should do that first, rather than cut into much else that we treasure, possibly losing our wildlife or wilderness areas, or giving up art, or underpaying the teachers. In fact, there is a way greatly to relieve this tragedy, there could be a just distribution of goods of culture, now often inequitably distributed. Few persons would need to go without enough if we could use the produce of the already domesticated landscape justly and charitably. (253)

Rolston refuses the simplistic dichotomy of "nature versus the poor," insisting on an analysis of why such a "conflict" occurs in the first place. After all, he notes, it is better for the poor not to have to ruin precious natural habitats. He quite rightly points out that the question of the wealthy over the poor undergirds the question of whether to conserve nature or feed poor people.

> The decision about social welfare, poor people over nature, usually lies in the context of another decision, often a tacit one, to protect vested interests, wealthy people over poor people, wealthy people who have exploited nature already, ready to exploit anything they can. At this point in our logic, en route to any conclusion such as let-people-starve, we regularly reach an if-then, go-to decision point, where before we face the people-over-nature choice we have to reaffirm or let stand the wealthy-over-poor choice. (254)

Rolston also clearly realizes the racial dimension of the problem: "Unjust sharing between the whites and the blacks is destroying the green" (255).

Although Rolston answers the question of whether it can sometimes be justified to let people die in order to conserve nature affirmatively, this is a highly reluctant response. He clearly thinks the question is misguided given issues of maldistribution. While he does not pursue many avenues that ecological feminists view as crucial in dealing with issues of population, particularly issues of gender and sexuality, he is on the right track by insisting that the problem is not really one of insufficient resources, but of unjust distribution. It is really a social problem more than a problem of the so-called "carrying capacity" of the earth.

GENETICS

Rolston's more recent work also falls in line with many ecological feminist insights regarding theory itself. In his most recent book, *Genes, Genesis, and God* (1999), Rolston takes on the issue of genetic determinism. His criticism is very similar to criticisms of scientific reductionism offered by ecological feminists such as Val Plumwood (1993) and Carolyn Merchant (1983). According to Plumwood, some recent trends in sociobiology demonstrate what she calls a "truncated reversal." In a truncated reversal, the devalued term in a dualism is given value, and the previously valued term is eliminated altogether. For example, Cartesian metaphysical dualism involves seeing "mind" as something totally different from "body." "Mind" and "reason" are valued, "nature" and "body" which are mindless, and lack reason, are devalued. Plumwood traces a reason/nature dualism from Plato through Descartes in *Feminism and the Mastery of Nature*. However, for my purposes, what is of interest here is her discussion of the turn from Cartesian rationalism to the complete mechanism demonstrated in views such as physicalism. In these views:

> Materialist positions, which have become popular and self-consciously modern positions, attempt to reduce the mental side of the dualism to the bodily, as in physicalism, which reduces mind to brain (in modern mind/brain identity theory), or to complex organizational machine states (functionalism.) But the original dualism remains in the wings in such a conception to the extent that an impoverished and polarized conception of the material or bodily sphere deriving from the original dualism is affirmed as the ground of reduction. What is granted reverse value status as the ground of the reduction is a conception of body or the physical sphere stripped of psychological and mind like attributes, considered to be part of the sphere of the 'non-scientific.' All talk of teleology, of agency, of goals, of striving, of choice and freedom is exorcised here. (Plumwood 1993, 121)

In approaches such as Dawkins (1989), humans are viewed as "survival machines—robot vehicles, blindly programmed to preserve the selfish molecules known as genes" (Plumwood 1993, 121).

According to Plumwood, such approaches affirm human/nature continuity, but at the expense of failing to accord value to nature. Instead, they simply devalue humans as well as nature. They see everything as basically the same, denying complexity and difference. When there appears to be something that doesn't fit into the theory they reduce it into something that does. So, for example, if all human behavior is defined as selfish, then any behavior that is seen as possibly altruistic must be reconceived, and apparent complexity is reduced in ways that fit the particular theory being employed. The result is to validate "a 'scientific' stripped-down, impoverished and subordinated conception of the object of knowledge through observational restrictions on what can count as a knowledge claim and the policing of a strict 'hard'-'soft' subject-object boundary" (121). In such strategies, complexity and difference—which is not reducible to dualistic difference—is denied.

Rolston's excellent critical discussion of recent trends in sociobiology also claims that it denies complexity and difference, even when doing so causes scientists to make highly implausible claims. He worries about theories that deny the possibility of real altruism. Such theories try to use evolution to argue that the only real motivation for either humans or any other animals is to propagate their genes into future generations. Examples of such theorists include E.O. Wilson, who states "the central theoretical problem for sociobiology is how can altruism, which by definition reduces personal (individual) fitness, possibly evolve by natural selection?" (1975, 3). According to George Williams, "Natural selection . . . can honestly be described as a process for maximizing short-term self interest" (1988, 385). Richard Dawkins states:

> The logic . . . is this: Humans and baboons have evolved by natural . . . selection. Anything that has evolved by natural selection should be selfish. Therefore, we must expect that when we go and look at the behavior of baboons, humans, and all other living creatures, we will find it to be selfish." (1989, 4)

Perhaps Michael Ghiselin sums up this view best:

> No hint of genuine charity ameliorates our vision of society, once sentimentalism has been laid aside. What passes for co-operation turns out to be a mixture of opportunism and exploitation . . . Given a full chance to act in his own interest, nothing but expediency will restrain [a person] from brutalizing, from maiming, from murdering—his brother, his mate, his parent, or his child. Scratch an "altruist" and watch a "hypocrite" bleed. (1974, 247)

Hence, cooperation is engaged in only for the purpose of either long-term or short-term self-interest. Darwin's theory of evolution is said to have brought a revolution, showing what ethics really is. As Dawkins puts the point, "There is such a thing as being just plain wrong, and that is what, before 1859, all answers to those questions were" (1989, 267).

If there is no such thing as genuine altruism, consciously deciding to put someone else's interests ahead of one's own, all intuitions that we witness such acts need to be reinterpreted. They are merely acts of "pseudo-altruism," the hypocrite is always there. Rolston points out that having to interpret all acts as "selfish" is actually quite difficult. Examples where individuals act at seeming costs to themselves in order to protect other family members might be plausibly interpreted as acting for the sake of "inclusive fitness," where the "selfish" goal of extending one's genes into future generations is stretched to include one's kin. However, it needs to be stretched even further than this to make sense of the claim that all actions are "selfish." People cooperate in extremely large numbers. Communities form. Hence, the "selfish" gain of cooperation must explain large cooperative schemes where benefits of reciprocity are extremely indirect. I may cooperate to help you, hoping someone else will cooperate to help me, if these are the norms in society. This explanation may work, but it hardly seems obvious that personal gain is the aim of someone who helps a stranger with whom he or she will probably have no further interaction. As Rolston notes, once we decide that all actions must be selfish, we can then attempt to reinterpret actions in this way, but this is not necessarily the best interpretation available. We reinterpret because we have already decided what the truth must be.

To demonstrate the difficulties with the reinterpretation of altruism into selfishness, Rolston offers a powerful critique of Richard Alexander's reductive account of the tale of the Good Samaritan (1987). According to Alexander, the only real motivation that people have is to insure the future propagation of their genes. In the tale of the Good Samaritan, The Good Samaritan helps a complete stranger who was robbed by thieves on the road to Jericho. This tale is supposed to be an example of genuine altruism, as the Samaritan does not expect to encounter this stranger again. However, as Rolston notes, "genetics allows only one explanatory framework for any and all human (or animal) behavior, and so the Good Samaritan must be fitted into that explanatory box" (Rolston 1999, 252).

Alexander suggests that there are benefits to such seeming "altruism," that allow it to be reinterpreted as merely "enlightened self-interest" or "selfishness." Alexander's explanation is that one enhances one's reputation by being known as an "altruist." And this can be seen as a long-term benefit. Rolston reminds us that before the "great revolution" of 1859, nobody even suspected that they had genes. The Good Samaritan, according to the tale, feels genuine compassion for a victim of thieves. Hence, he must have been unaware that he was actually motivated by the goal of propagating his genes into the future. This means that the Good Samaritan is self-deceived about the real motivating forces for his "kind" actions. According to Alexander, the whole transaction in which the Good Samaritan promotes his own genetic self-interest goes better if people are self-deceived, believing that they and others can be genuinely altruistic. Apparent sincerity helps guarantee reciprocity. If the victim knew that the Samaritan's true motives were simply to put his own genes into the next generation, the victim would be less inclined to feel indebted and offer help later. And, if the Good Samaritan was aware of his own selfish motives, his insincerity might slip out. So, the whole thing goes more smoothly if both self-deception and deception are involved. Alexander says of people's true "selfish" motives, "I mean that such information is not part of their conscious knowledge, and if you ask people what they think their interests are they would usually give you the wrong answers" (1987, 36). Therefore, "if the theory is correct humans could not have evolved to know it, and to act directly and consciously in respect to it" (38).

Rolston convincingly argues that Alexander's explanation is by no means obviously the best. "Curiously, the double deception (deceiving others and being deceived about the fact that one is deceiving others) forces positing a double negative to substitute for an apparent positive (that one helps another altruistically). The appearance of this as a real positive cannot be allowed because the theory does not allow such emergents" (Rolston 1999, 259). Ecological feminist Val Plumwood complains that reductionist theories validate "a scientific, stripped-down, impoverished and subordinated concept of the object of knowledge through observational restrictions on what can count as a knowledge claim" (1993, 121). In a similar vein, Rolston explains that reductive theories are problematic because "Since psychological, ethical, and experiential evidence is inadmissible, we could find it difficult to reach the conclusion that biological determinants are under-determining the outcome" (1999, 259). This fits beautifully with Plumwood's worry that reductionist theories offer only "a subordinated concept of the object of knowledge," in that nothing that the "object" feels or believes is taken as evidence of anything but self-deception.

Another facet of Rolston's critique that is important for ecological feminists is his discussion of the strict selfish/altruistic dichotomy used to describe human motivation. Actions either are selfish, putting one's own interests above the interests of others, or altruistic, putting the interests of others ahead of one's own. Feminists have critiqued this dichotomy, arguing that many actions promote the interests of several parties, and if one cares about another, protecting the other's interests can become an interest of one's own. Hence, there are actions that are not neatly described as either "selfish" or "altruistic," at all. Rolston questions the firm selfish/altruistic divide in several ways. He challenges whether the so-called "selfishness" of other animals, is correctly called "selfishness" at all.

> Altruism in the ethical sense applies where a moral agent consciously and optionally benefits a morally considerable other, without necessary reciprocation, motivated by a sense of love, justice, or other appropriate respect of value. Selfishness applies where a moral agent exceeds the bounds of legitimate self-interest and is so concerned with the self that the appropriate motivation in love, justice, and respect for the interests of others fails (Rolston 1999, 217).

If a being is incapable of moral understanding, such a being cannot be selfish in any way that involves moral condemnation or altruistic in any way that deserves moral praise. Along these lines, Rolston questions the analogy with other animals used in saying that all other animals have only "selfish" motivations, and so we should expect to find that all human motivation is "selfish" too. He rightly wonders whether such terms are stretched too far when applied to non-moral animals, and even to genes. The idea of genes as "selfish," in any moral sense is an analogy which Rolston finds ultimately too simplistic and unhelpful. "words break when stretched too far" (1999, 279). For example, "When a bee flies to the hive and does the waggle dance, it communicates correctly. The bee is not to be commended for telling the truth; it does not have the capacity to lie, any more than a thermometer does" (1999, 279). Hence, Rolston rightly worries that moral concepts such as "selfishness" and "altruism" are stretched beyond the breaking point when applied to genes. He also worries, as I stated above, that the concept of moral selfishness blurs with the concept of moral altruism in situations where one identifies with and loves others. This is very similar to ecological feminist critiques of these concepts. Rolston, therefore, refuses to deny difference and complexity, and rejects a dualistic construction of "selfishness" and "altruistic" behavior, where they are seen as radically different.

Rolston's discussion of some trends in recent sociobiology not only echoes ecofeminist concerns about reductionist theories, it advances such concerns, making Rolston's work in this area highly relevant to ecological feminists. In what I take to be a brilliant move, Rolston applies the "selfish gene" theory to the theorists themselves. Strangely, Alexander claims that we can and should break away from our genetics, "To say we are evolved to serve the interests of our genes in no way suggests that we are obliged to serve them . . . Evolution is surely most deterministic for those still unaware of it" (1987, 40). And, he rather surprisingly maintains that despite his theory of behavior which shows we always act selfishly, " 'Thou shalt love thy neighbor as thy self' is an admirable goal" (1975, 96). Rolston correctly wonders how on earth, or from anywhere else for that matter, Alexander can claim to know that loving thy neighbor is admirable. How does he know that he is not self-deceived, and given his theory, why should we trust him. It doesn't appear that Alexander is producing science simply in order to insure that his genes survive in future generations, but since we are programmed for self-deception, appearances don't count! Altruism does not evolve on this view, so where will it come from? Rolston says that Alexander will have to cut the leash to biology and then "bring in by skyhook these non-evolved ethics, with their outside authority that has 'nothing to do with evolution' " (Rolston 1999, 246).

In a somewhat similar move, Richard Dawkins states, "Let us try to teach generosity and altruism because we are born selfish" (1989,3). Rolston again turns the theory on the theorist arguing that Dawkins can suggest this:

> only because he has found some vantage point from which he can recommend that altruism be taught in culture, educating us out of our beastly nature. If so, then he himself has reached a more comprehensive ethics—one in which he has escaped, or at least knows that he ought to escape, the biological legacy. He has found genuine altruism, ideal, if not yet as real as he wishes.

It is puzzling to say where he found this, since he has dismissed all ethics prior to Darwin as worthless, and all he can find in Darwinism is a disposition to selfishness, which is the wrong answer. (Rolston 1999, 265)

As Rolston points out, accepting these theories means we either have to override our biology or give up on teaching the genuine altruism these theorists endorse.

The idea that we have to "overcome" a "beastly biology" is nothing new to the classical western tradition. However, it is important that some kind of reason/nature dualism is reappearing within the works of some very prominent sociobiologists. This is very interesting from an ecological feminist perspective, as once again, reason is being constructed as opposed to and radically different from nature. Rolston resists this construction, offering a critique that is right in line with many ecological feminist insights. He maintains that we need to see altruism as something that has evolved from our biology, not something in stark contrast with it. Hence, he suggests that we see human beings as connected to and yet somewhat different from other animals. He therefore affirms complexity, similarity, and difference, resisting dualistic constructions. In his discussion of genetics, Rolston uses powerful arguments against reductionist theories, and argues persuasively that explanations including the possibility of an evolved but real altruism are at least as reasonable as those that deny such a possibility. He also resists explanations that posit "natural" drives to explain all of human behavior, particularly the drive to propagate one's genes. Hence, Rolston's work in this area in particular is highly relevant to ecological feminist concerns.

Rolston's work can also benefit from key ecological feminist insights. Ecological feminist analysis of his positions on the ethics of the treatment of food animals and the ethics of sport hunting reveal significant problems in both. His position that eating food animals is ethically permissible even when other food sources are readily available, because humans eat "in nature" rather than "in culture," is fraught with difficulties. His inability to condemn sport hunting, because he believes in a "biological drive to hunt," which is similar to the "sex urge" and therefore dangerous to repress is also highly problematic. However, Rolston's later work offers resources to escape these problems. With respect to food animals, even if many humans have the desire to eat meat when other food sources are readily available, it may be unethical to do so. If humans can evolve to care about other humans in a truly altruistic way, based on feelings of care and connection, there is no reason why humans cannot evolve to care about food animals, and particularly their pain, in a similar way. The fact that many humans may not always have done so "in nature," is irrelevant to where we are now. Rolston states, "Environmental ethics is . . . the most altruistic, global, generous, comprehensive ethic of all, demanding the most expansive capacity to see others" (1999, 288). The evolution of such abilities comes out of feeling of care, community, and genuine concern for others, not just selfishness, according to Rolston. So, it is unclear why this "generous" and "comprehensive" ethic should not be more concerned with the suffering of food animals than Rolston seems to allow. Rolston clearly believes we can evolve into genuine altruism, "acting out of moral conviction for the benefit of nonhuman others" (1999, 288). There seems to be no reason why such concern cannot or should not be applied to food animals. With regard to explaining human behavior through posited "biological drives," Rolston's analysis of genetics shows some of the problems in doing this. His discussion of the "hunting drive" which seems to excuse hunting for sport makes use of such problematic assumed drives, and an ecological feminist analysis reveals that the victims of accepting the existence of uncontrollable or dangerous "sexual urges," or "hunting drives" are often women and animals. It is here that I think Rolston can learn the most from ecological feminism. When we choose to "see the world" in terms of a particular theory, we should always ask who benefits from accepting that particular world view, and at whose expense.

REFERENCES

Adams, Carol J. 1991. "Ecofeminism and the Eating of Animals," *Hypatia*: A Special Issue. *Ecological Feminism* 6, no. 1 (Spring): 125–145.
Alexander, Richard D. 1975. "The Search for a General Theory of Behavior," *Behavioral Sciences* 20:77–100.
Alexander, Richard D. 1987. *The Biology of Moral Systems*. New York: Aldine de Gruyter.
Cuomo, Chris. J. 1998. *Feminism and Ecological Communities: An Ethic of Flourishing*. London and New York: Routledge.
Davion, Victoria. 1994. "Is Ecofeminism Feminist?" In Karen J. Warren, ed., *Ecological Feminism*. London and New York: Routledge.
Dawkins, Richard. 1989. *The Selfish Gene*. New York: Oxford University Press.
Gaard, Greta, ed. 1993. *Ecofeminism: Women, Animals, Nature*. Philadelphia: Temple University Press.
Ghiselin, Michael T. 1974. *The Economy of Nature and the Evolution of Sex*. Berkeley: University of California Press.
Gruen, Lori. 1993. "Dismantling Oppression: An Analysis of the Connection between Women and Animals." In Greta Gaard, ed., *Ecofeminism: Women, Animals, Nature*. Philadelphia: Temple University Press.
Kheel, Marti. 1990. "Ecofeminism and Deep Ecology: Reflections on Identity and Differences," In Irene Diamond and Gloria Feman Orenstein, eds., *Reweaving the World: The Emergence of Ecofeminism*. San Francisco: Sierra Club Books.
Merchant, Carolyn. 1980. *The Death of Nature: Women, Ecology, and the Scientific Revolution*. New York: Harper Collins.
Plumwood, Val. 1993. *Feminism and the Mastery of Nature*. London and New York: Routledge.
Rolston, Holmes, III. 1988. *Environmental Ethics: Duties to and Values in the Natural World*. Philadelphia: Temple University Press.
Rolston, Holmes, III. 1996. "Feeding People versus Saving Nature?" In William Aiken and Hugh LaFollette, eds., *World Hunger and Morality*, 2nd ed. Englewood Cliffs, NJ: Prentice Hall, pp. 248–267.
Rolston, Holmes, III. 1999. *Genes, Genesis and God: Values and Their Origins in Natural and Human History*. Cambridge: University of Cambridge Press.
Schlessinger, Laura. 1994. *Ten Stupid Things Women Do To Mess Up Their Lives*. New York: Villard Books, Random House.
Sommers, Christina Hoff. 1994. *Who Stole Feminism?* New York: Simon & Schuster.
Stoltenberg, John. 1989. *Refusing to be a Man: Essays on Sex and Justice*. New York: Meridian, Penguin.
Warren, Karen J. 1994. *Ecological Feminism*. New York: Routledge.
Williams, George C. 1988. Huxley's Evolution and Ethics in Sociobiological Perspective," *Biology and Philosophy* 1:114–122.
Wilson, E. O., 1975. *Sociobiology: The New Synthesis*. Cambridge, MA: Harvard University Press.

CLARE PALMER

11. RETHINKING ANIMAL ETHICS IN APPROPRIATE CONTEXT: HOW ROLSTON'S WORK CAN HELP

INTRODUCTION

Holmes Rolston has long been regarded as a leading figure both in environmental philosophy and in science and religion. In this chapter, though, I argue that Rolston's work also paves the way towards rethinking animal ethics. Given the well-known hostility between many forms of environmental philosophy and animal ethics, to turn to Rolston—a notorious champion of the former field—in order to advance work in the latter field, may seem singularly perverse. But, I will maintain, Rolston's arguments—whilst undeveloped and in some respects problematic—provide a better basis for advancing work in animal ethics than the advocacy or rejection of utilitarian or rights positions that have dominated animal ethics for several decades. In particular, I will suggest, Rolston's work provides tools for thinking through the complicated location of domesticated animals both conceptually and ethically. So, at the end of the chapter, I make some initial moves in outlining how Rolston's position might be developed to contribute to new thinking with respect to animals and ethics.

In focusing on the place of animals in Rolston's environmental ethics, I will touch only obliquely on the area of his work that has caused most controversy: his endorsement of a theory of objective intrinsic value in nature. That topic has been exhaustively discussed elsewhere, and I will not revisit that discussion here. The place of animals in Rolston's work is not, though, entirely virgin territory either. Some attention has been paid to it before (notably by Peter Wenz [1989], to which Rolston responded, and later by Ned Hettinger [1994] and Moriarty and Woods [1997]). But my interest takes a somewhat different trajectory to that of existing debates, concentrating on Rolston's understanding of nature and culture as part of the architecture of a context-oriented approach to animal ethics. In order to develop this argument, I will begin by outlining—in a very basic way—what seem to be some central problems in what we might call "philosophical animal liberation" approaches to animal ethics. Then I will move on to draw out key aspects of Rolston's understanding of "nature" and "culture". I will consider how animals are located within these categories, and then make some suggestions as to how Rolston's position might contribute to a more contextual approach to animal ethics.

Two further initial comments should be made for clarification. First, in using the term "animals", I intend to confine my discussion to non-human mammals and birds. Second, I will be assuming—as does Rolston—that, on grounds of sentience at least, it makes sense to talk about these animals as being morally considerable (no stronger claim, such as that animals have rights, is intended). I will not be putting forward arguments to defend the moral considerability of these animals here.

C.J. Preston and W. Ouderkirk (eds.), Nature, Value, Duty, 183–201.

PHILOSOPHICAL ANIMAL LIBERATION: OUTLINING A PROBLEM

Animal liberation is the common—although inaccurate and unhelpful—catch-all term for a range of different kinds of arguments concerning human ethical obligations towards animals. These arguments arise within both utilitarian and rights traditions and are usually associated in the case of utilitarianism with the work of Peter Singer (1979, 1983), and in the case of rights theory with Tom Regan (1984, 2001). What binds these different arguments together as advocating "animal liberation" is that they hold in common the view that animals make strong moral claims on humans; often so strong as to maintain that the interests of humans and animals should be taken equally into consideration in moral decision-making. Exactly what such claims entail in practice varies in different versions of utilitarian and rights theories, but it is commonly argued that some ways of treating animals—such as eating them—widely thought to be morally acceptable are, rather, morally unacceptable. (It is not necessary for me to pursue these arguments in more detail here.)

There are a number of standard objections to philosophical animal liberation arguments. Some are based on the view that the conception of animal capacities and cognitive abilities usually adopted in animal liberation positions is too high-level, sophisticated and anthropomorphic. Others focus on problems in the relevant ethical theory (see Lockwood [1979],Frey [1983], Leahy[1993]). Interesting though these objections are, I will not linger on them here. I want, rather, to consider objections that focus on what we might call the "class system" manifest in philosophical animal liberation. Here, I use "class" in two senses. In the first sense, class is about procedures of classification. Animals are grouped on the basis of their membership of a putative class; the class is defined by the possession of a particular keystone capacity or ability, or cluster of keystone capacities and abilities. The second sense of class concerns the hierarchical value of these classes. The selected characteristics that divide are also value-bestowing. Indeed, the keystone capacity or ability does not just pick out the animals in the class as being morally considerable. That certain basic capacities (such as the ability to feel pain) are central in determining a baseline of moral considerability does not seem especially problematic. But in philosophical animal liberation, such capacities solely and equally determine the degree and kind of moral significance of all animals in the relevant class.

The central notion, then, of philosophical animal liberation seems to be that animals possess certain innate capacities, and that these determine—and solely determine—moral significance. This thesis, though, is a vulnerable one in several ways. The way that interests me here stems from the kinds of criticisms made of philosophical animal liberation within environmental ethics. Examples of these criticisms are, for instance, that philosophical animal liberation has no grounds for thinking that a member of an endangered species should be of any more value than one of a well-represented species; and, depending on what "class" the rare individual is in, may be worth less. Another is that philosophical animal liberation positions seem to have a problem with predation that might mandate heavy human management of the wild for animal welfare reasons.

But, because of the "class system" I have already outlined, there is no easy solution for animal liberation to these kinds of criticisms. An endangered wild cat, a domestic house cat and a non-native feral cat all have (roughly) the same innate capacities. There are no grounds for treating wild animals one way while regarding domestic animals with similar innate capacities differently. This is because contextual differences of this kind are not of ethical significance to philosophical animal liberation. Yet it seems implausible that a thoroughgoing approach to animal ethics can be developed without attention to context. Thousands of different species, some not yet discovered by humans and many with members that will never encounter humans are involved. And, in contrast, also to be taken into account are animals shaped and formed by human actions in terms of genetic make-up, susceptibility to disease, reproductive capacity, bodily form, temperament and cognitive abilities. Focusing on capacities such as the ability to feel pain alone cannot, in particular, capture anything about the dramatic transformation in animals wrought by domestication. From a philosophical animal liberation perspective there is nothing of direct ethical interest to be said either about human intervention in the processes of bringing into being, selectively breeding and shaping the natures of domesticated animals; or about the human independent, ecological embeddedness of wild animals.

It is in recognizing the significance of context and relationships such as these to humans that, it seems to me, Holmes Rolston's work can assist in the development of animal ethics. His account—while I think it is problematic in some respects—provides important building blocks for a more contextual and relational animal ethics. In order to consider his contextual position more fully, though, it is essential to have some understanding of the place the contexts of nature and culture play in Rolston's work.

ROLSTON ON NATURE AND CULTURE

Understanding Rolston's approach to ethics in general, and to environmental and animal ethics in particular, requires an overview of his interpretation(s) of nature, and how nature may be distinguished from culture.[1] This is by no means straightforward. In the first chapter of *Environmental Ethics*, and in earlier chapters on the subject, Rolston outlines a number of different ways in which humans might be said to "follow nature." This inevitably involves Rolston in offering a range of different possible interpretations of the meaning of "nature" and "natural" behavior. However, Rolston's discussion does not map very well onto his own usage of the terms "nature" and "the natural." It is the ways in which these terms operationally contribute to Rolston's arguments that are significant here in trying to understand Rolston's position. I will begin, then, by outlining different ways in which an analysis of Rolston's writing suggests that he is using the terms "nature" or "natural".[2]

> (N0) That which most broadly "obeys natural laws" (Rolston 1986, 31). Although this may be interpreted to include "astronomical nature", Rolston prefers to restrict this to a global sense, and in particular to the physical and biological processes that form the system that "gives birth to life."

(N1) That which is not of human origin; the wild understood as "nature outside human control"(Rolston, 2003, 108); processes that are spontaneous, non-reflective and independent of humans, e.g. evolution and speciation.

(N2) Behaviors and practices that are common across species (including humans), for instance behavior that enables survival and reproduction e.g. sleeping, grooming, eating.

(N3) The expression of distinctive (not shared) species-specific characteristics and behaviors manifested by "natural kinds" e.g. coyote nature, warbler nature, human nature; all species, Rolston maintains "live in the world with some degree of uniqueness" (Rolston 1994, 1).

Complex relationships obviously exist between these senses of "nature" and "natural". N0 is the broadest sense; the N1 and N2 senses of nature are wholly subsets of the N0 sense. Some N3 behaviors and practices are also subsets of N0 and N1; but N2 and N3 behaviors are exclusive of one another. An example might help to explain these distinctions. Suppose a wild rabbit is eating grass on a mountain-side. In what senses is this natural behavior? It is natural in the N0 sense—part of the biological systems of earth; it is natural in the N1 sense—spontaneous and independent of humans; it is natural in the N2 sense—eating is a common cross-species behavior. It is not natural in the N3 sense, as expressing the specific, exclusive nature of "rabbitness" (unless the eating was being described in a particular, fine-grained way to emphasize its rabbit-distinctiveness as opposed to, say, the eating habits of prairie dogs).

These distinctions may seem tiresome, but they are necessary in order to understand where Rolston locates human beings. Humans are, inevitably, included in the N0 sense of nature; they emerged as part of the physical and biological systems that produced life on earth and are still dependent on these systems. But Rolston seems to have a more particular interpretation of what part of human life falls under the description of N0 nature—that is, those processes that happen to humans and that are (currently) beyond human control; so for instance, blinking when grit gets in an eye or sneezing at too much chili. Humans are, in contrast, by definition excluded from the N1 sense of nature where "any deliberated human agency, however well intended, is intention nonetheless and interrupts these spontaneous processes" (Rolston 1991, 371). Humans can, though, be included under both the N2 and N3 senses of nature. In the N2 sense, humans do, of course, display behaviors that when broadly described, are cross-species behaviors (such as sleeping and eating). In the N3 sense, humans display distinctive species-specific characteristics that are exclusively human and do not fall under the N2 sense of nature. These human-exclusive characteristics correspond closely to Rolston's use of the term culture. There are, presumably, some exclusively human characteristics, products of N0 nature, that do not fall into the category of culture (perhaps no other species sneezes at too much chili) but Rolston is not particularly interested in these. What he does argue is that culture falls wholly within the human sense of N3 nature; culture is exclusively human and species-specific.

What, then, does Rolston understand by culture? He does not precisely define cultural behaviors and characteristics, but stresses intention and deliberation (as opposed to the spontaneous action of nature in N1); the dramatic and deliberate rebuilding of the environment (Rolston 1988, 1, 5; 1989: 132) and "moral and spiritual sensibilities" including value invention and recognition (Rolston 1988, 40). In later work, Rolston repeatedly emphasizes that culture is distinguished by the transferal of information neurally rather than genetically, resulting in human education and knowledge-accumulation, and by the deliberation and reflectiveness involved in the inevitable process of environmental transformation (Rolston 1994, 2–3). Rolston's view of what qualifies as culture, then, is quite high-level, given the multiplicity of ways in which culture can be understood. He requires such an idea because he identifies culture with that which is species-specific to humans and is not shared by non-human animals.[3] Although culture is dependent on N0 nature in the sense that it still "requires nature in life support" (1988, 40), Rolston insists that culture is radically discontinuous with it (1989, 32); it "did evolve out of nature, but has evolved out of it"(1998, 3). Although some human behaviors (blinking, sneezing) remain N0 natural, human culture, according to Rolston, no longer falls into this category. One way Rolston conceptualizes this is to say that nature (understood in an N0, N1, N2 and non-cultural N3 sense) is primarily about causes (chemical and genetic for instance) while culture, as an expression of human-unique nature, is about reasons.[4]

However, we need to be careful here, for there is yet one other way in which Rolston deploys the term nature:

(N4) Cultural, deliberate human choices to behave "more naturally" or to use Rolston's term, "follow nature".

This understanding of "natural" is rather different from N0–N3; the N4 sense of nature is not about origin. In terms of origin, N4 behavior is not natural in any of Rolston's senses; it is cultural; it flows from culture and involves reason and deliberation. However, what is pursued in this deliberation, the object of the reasoning, is nature, in one or other of Rolston's senses. A deliberate choice might be made to act sparingly, hold back, withdraw or leave alone to allow N0 processes to operate (for instance, to use one of Rolston's examples "natural childbirth" is a decision to refrain from a series of alternative medical practices). Deliberate attempts to imitate, parallel, fit in with or approximate more closely some perception of wild, spontaneous N1 nature, perhaps by restoring land to look like wilderness or by growing crops to fit local soil and climate are natural in this sense. Rolston also suggests that humans may act naturally (in an N4 sense) by deliberately choosing to do what some other species do unreflectively—that is, by culturally imitating either N2 behavior or N3 behavior specific to another species.

ANIMALS, NATURE AND CULTURE IN ROLSTON'S WORK

Sorting out different senses in which Rolston uses "nature," "natural" and "culture" allows us to consider where Rolston locates animals. Most fundamentally, Rolston rests on distinctions between N1 nature and culture. Animals are divided into two groups. One group falls wholly into the category of N1 nature (and also, by definition, N0 nature; but the former is what is significant here). Members of this group are part of, and indeed partly comprise, wild, spontaneous nature. They exist independently of humans, though they may encounter humans when humans move or act in wild nature (by hiking, or constructing roads or pipelines, for instance); and they may be unintentionally affected by other human actions (in terms of, for instance, pollution or the introduction of exotic species). Both the N2 and the N3 behaviors of these animals are subsets of N1. The second group of animals Rolston considers are those that have been, as he puts it, "captured for food, domestication, research or other utility". He does not explicitly distinguish between different kinds of "captured" animals (for instance between zoo animals and household pets). His (relatively brief) discussion of these animals focuses on domesticates, in particular domesticates kept for food. He locates domesticates in the following way in relation to his nature/culture distinctions:

(a) Domesticates are outside N1 nature: Domestic animals are not part of spontaneous wild nature; they are "living artifacts"; they have been "transformed by culture" (1988, 79); they require human action to survive and sometimes to reproduce.
(b) Domesticates are subject to N0 natural processes: Even though their minds and bodies may have been transformed by human culture, they themselves are entirely subject to "natural laws" in Rolston's sense (to disease, injury and so on);
(c) Domesticates are not natural kinds and so have no N3 behavior: Because humans have selectively bred domestic animals, they are now breeds and not natural kinds. Thus domestic animals cannot behave naturally in an N3 sense; N3 behavior can only be manifested by natural kinds.
(d) Domesticates are not part of culture: Culture is the manifestation of deliberated N3 species-specific human behavior. Domesticates, not being human, cannot manifest these characteristics and so are outside culture.
(e) Domesticates can behave naturally in an N2 sense: Domesticates are not part of wild N1 nature; they are not natural kinds, and so cannot act naturally in an N3 sense; they are not human so cannot behave naturally in a deliberative N4 sense. They are, however, part of N0 nature in the sense that they are subject to natural processes; and they can manifest N2 natural characteristics in the limited sense of sharing in common cross-species survival behaviors.

Rolston maintains that both pets and agricultural animals have been removed from N1 nature, but that neither can be part of human culture. Agricultural animals, Rolston suggests, inhabit (literally) the no-man's land of the "peripheral rural world" (1988, 79). These categorizations are central to Rolston's discussion of human ethical responsibilities towards animals.

ROLSTON, ANIMALS AND ETHICS

Rolston argues forcefully in *Environmental Ethics* that all living beings carry intrinsic value. (As I have already said, this argument has been exhaustively discussed elsewhere, and I will avoid further comment on it here.) He also accepts (rather than argues for) the view that higher animals are sentient and that this sentience is of significance in ethical decision-making.[5] But although sentience is significant in ethical decision-making it is, crucially, context that gives guidance on duties towards animals (1988, 61). Duties towards similarly sentient animals vary according to how animals are placed in relation to N1 nature and culture, and what kind of human actions are, or have been, involved in their situation. Scrutiny of Rolston's work seems to suggest that three key contextual factors need to be taken into account when considering ethical obligations towards an animal:

1. How the animal is placed in relation to N1 nature and human culture;
2. Whether previous human actions have had an impact on the animal's current situation;
3. Whether humans are acting naturally towards the animal (and if so, in which sense).

Of particular concern for Rolston is human treatment of wild animals—that is, sentient animals wholly located within N1 nature. Such animals come into being and live independently of humans. Whilst he objects to gratuitously harming wild animals, he also argues against assisting them by, for instance, attempting to reduce their pain. Human intervention both changes N1 patterns of evolution and speciation (perhaps preserving individuals with weaknesses that otherwise would not have survived and allowing these weaknesses to be genetically inherited) and humanizes wild N1 nature. "Pain in nature is situated, instrumental pain; it is not pointless in the system, even after it becomes no longer in the interests of the pained individual" (Rolston 1988, 60).

That animal pain can be outweighed by other, ecological and evolutionary, values relating to the systems in which wild animals are embedded is one sense in which context might be important in making ethical decisions about animals. But Rolston hints at another sense. Humans have, he says "no obligations to help wild animals; we are obliged to leave them alone" (Rolston 1989, 134). Wild animals are self-sufficient; they can and do provide for themselves. There is no reason why they have any claim on humans, and to assist them is to create a relationship that did not previously exist. The basic argument here is, then, one I shall call the no relation/ no obligation argument. That humans have no relations to wild animals means that they also have no positive obligations to them. It is only where human actions have already created some kind of effect or relationship—where the suffering, even though it may be part of nature in an N0 sense, is not genuinely N1 in origin—any ethical obligations come into play. In this regard, Rolston comments: "If human intervention, and not just the forces of natural selection are causing the deaths, that does seem to make a difference to the welfare claim" (Rolston 1988, 56). He approves, for instance, of a case where a rancher was required to flatten portions of extensive new fencing in order to allow starving antelope to reach their traditional winter grazing areas. Again, although Rolston does not make this explicit, two kinds of reasons seem to hold here. Human actions have already created some kind of relation or interaction with the antelope which means they are no longer independent and self-sufficient; and wild ecological processes have already been disturbed by the separation of the antelope from the grazing; no further disturbance, and some diminution of its effects would result from removing the fencing.

There is, however, one context where, even without the justification of prior human activity, Rolston maintains that humans can act on sentient animals in N1 nature. This is when humans are acting naturally in an N2 sense—that is, when they are acting in ways common across species for survival. Some forms of hunting, Rolston maintains, fall into this category. When hunting is carried out in an N2 sense Rolston views it as a form of human predation. In this category he is inclined to include all hunting where the hunted is eaten or used in some other significant way (e.g. for fur, if the fur is for warmth rather than fashion), even if the food or fur is not hunted as a necessity for the hunter (as we would expect in the case of subsistence hunters). Trophy hunting, on the other hand, does not fall into this category. I will discuss some problems concerning this classification of hunting later.

We can, then, extrapolate something like the following principle from Rolston's work:

> (P1) Humans should not harm, but have no duty to assist, wild animals in N1 nature *unless* the action is in response to some prior potentially or actuallly harmful human activity *or* humans are behaving naturally in an N2 sense.

Based on the trumping significance of other ecological values, and the no relation/no obligation argument, Rolston creates a contextual principle, P1, of non-intervention in N1 nature for animal welfare reasons. This principle certainly avoids some of the difficulties arising out of philosophical animal liberation—such as intervention in predation.[6] But it's a contextual principle that solely concerns animals with which humans do not interact. What about ethical obligations towards animals with which humans do interact—the domesticates who are not part of N1 nature, but also, according to Rolston, not part of human culture either?

Rolston argues that domesticated animals should properly be considered in ethical terms against the background of the N1 nature from which they originally came. Unlike the philosophical animal liberation positions outlined earlier, the comparison class is not human beings who share the same valuable capacities as animals, but rather wild animals for whom humans should not relieve suffering. The origin of domestic animals in wild nature, for Rolston, still determines appropriate ethical treatment. This grounds what Rolston calls the "homologous principle":

> (P2): Humans should not cause inordinate suffering to domestic animals, beyond those orders of nature from which the animals were taken (1988:61).

Rolston calls this a principle of "non-addition": Humans should not add to the suffering an animal would have endured in the wild, but, Rolston points out, non-addition does not mean subtraction from suffering; there is no obligation to cause or allow only lesser suffering. Elsewhere, he comments "where culture captures value in nature, there is only a weak duty to subtract from the pain" (Rolston 1988, 61). (He does not explain, though, what distinguishes a strong duty from a weak duty, nor what kind of obligation this entails). To this, however, Rolston adds another principle:

> (P3): The suffering of domestic animals should be removed, as far as it can, where it has become pointless because the animals are no longer in the environment of natural selection, though this is "commended rather than required" and is "benevolence not justice" (1988, 61).

The homologous principle, P2, then, supposes that wild animals are the appropriate comparison class for domestic animals. If we accept the homologous principle, domestic animals should not be caused more suffering than occurs in the wild. But we already know that any amount of suffering should be permitted among wild animals in N1 nature, since we should not assist wild animals by relieving suffering. So P2 alone, as a principle, potentially permits humans to cause high levels of suffering in domestic animals, since high levels of suffering may occur without relief in wild animals. But P3 changes this situation significantly, because P3 introduces a strong contextual concern. It is the case, on Rolston's principles, that high levels of suffering are acceptable without relief amongst wild animals. But one of the reasons for accepting this is because wild animals are part of valuable N1 evolutionary processes. P3 notes that at least some domestic animals are outside the environment of natural selection, and that therefore a good reason for accepting suffering in a wild context does not exist in the domestic context. Therefore suffering should be removed as far as it can, though this is benevolence not justice.

Indeed, once P3 has been taken into account, it isn't clear why P2 is necessary, since there are no domestic animals that fall outside the scope of P3, but to whom P2 applies. No domestic animal is in the environment of natural selection, so (according to P3) the suffering of all domestic animals should be relieved, as far as it can, as benevolence not justice. It's also worth noting that what is entailed in P3 is somewhat stronger than that which is entailed in P2. P2 concerns human actions as the cause of suffering; P3 concerns human actions as removing suffering (which presumably incorporates not only the concern in P2 not to cause it, but also to relieve it when it arises from non-human N0 sources such as disease or accident).

A minor reformulation, P3* should serve instead of P2 and P3:

> P3* The suffering of all domestic animals should be removed, as far as it can, since it has become pointless because the animals are no longer in the environment of natural selection, though this is "commended rather than required" and is "benevolence not justice" (1988, 61).

In dismissing P2 and moving to reformulate P3, though, I perhaps moved a little too swiftly. Rolston does suggest another argument that might be used in support of P2. Even though domestic animals are not part of N1 nature, nonetheless, he argues, the process of domestication in agriculture parallels the actions of wild ecosystemic processes; agricultural animals are in quasi-ecosystems. The killing of animals for food then, parallels predation. So the same rules about suffering are appropriately applied to both wild and domestic animals. This seems to be based on a broader view that agriculture is a way in which "humans step back from culture into the wild"

(Rolston 1988, 79) and that agricultural systems should be ethically appraised just as wild ecosystems are ethically appraised.[7]

This suggestion puts domestic animals into a quasi-N1 category. But the argument presents problems in a number of ways. First, it is an argument that could apply only to some domestic animals, primarily agricultural animals (not to pets, nor to experimental animals, for instance). Second, we need to get clear about what Rolston means by "quasi-ecosystems". Is the suggestion a literal one, that domestic food animals, in roaming on pastures and mountains, become quasi-ecosystemic members? This is, I suppose, an empirical question (which raises questions about Rolston's division between N1 nature and other environments). Of course, this reading of quasi-ecosystems could only apply to a minority of agricultural animals in any case; animals intensively farmed, in particular those kept indoors, could hardly be described as members of ecosystems in ways comparable to wild animals. I shall argue below, however, that such a comparison—even of a minority of domestic animals—reveals a problematic understanding of domestication.

However, it may be that Rolston does not intend "quasi-ecosystems" to be taken in such a literal sense; rather he means that just as wild animals prey on other animals in ecosystems, so humans prey on animals in agriculture. The processes are parallel and therefore domestic animals should be regarded, ethically, in the same way as wild animals. Rolston certainly does seem to suggest that agricultural meat-eating is a form of predation and is thus natural in an N2 sense. But the objection to this argument must be that agricultural animals are selectively bred, controlled, sustained and killed as part of a deliberated cultural practice. Since this is not N2 human behavior, it cannot count as one of the P1 exceptions, even though it could be argued that animal agriculture is natural in one form of the N4 sense (that is, where culture mimics nature). This makes the continued maintenance of P2 problematic.

The elimination of P2, of course, does not mean the rejection of animal agriculture, nor entail vegetarianism. All that is established is that a good reason for accepting suffering in the wild context, which informed P1, is not a good reason for accepting suffering in the domestic context, as maintained in P3 (and P3*). It would be perfectly possible to accept P1 and P3* and to maintain animal agriculture. Indeed, P3* is not a strong principle. If P3* is taken to be compatible with intensive farming, removing suffering "as far as it can" may still leave a considerable amount of suffering (and the removal of the suffering is, anyway, not required by justice, only by benevolence). If P3* is interpreted as being incompatible with intensive farming, it is still compatible with other forms of animal agriculture and with meat-eating; since, as Ned Hettinger rightly points out elsewhere, Rolston offers no principle here that taking an animal life is itself bad, if no suffering is involved.[8]

This discussion has taken no account, however, of the no relation/no obligation argument that appeared to be part of Rolston's case for non-interference with wild pain. Introducing it here, though, would not change anything. Since domestic animals are not only outside the environment of natural selection, but also outside the context of no relation, neither of the reasons Rolston suggests for accepting wild pain apply in the domestic case. Domestic animals are, plainly, in a situation of relation with human beings. That this is a situation of relation rather than no relation, does not of course in itself entail any obligation. But Rolston does make a gesture towards affirming a possible form of a relation/obligation argument (and later in this chapter I will also be attempting to construct such an argument). Rolston comments of domestic animals "In taking an interest in them [relation] humans have assumed a responsibility for them [obligation]" (1988, 79). But his work makes little attempt to flesh this claim out, either in explanatory or practical terms. Primarily interested in the context of the wild, concern for the context of domestication is slight and engagement with it minimal (and there is a persistent sense that the domesticated and rural, tainted by human contact, are inferior contexts to the wild, as are the animals within them).

Indeed, Rolston simultaneously exaggerates some senses of the difference between domesticates and wild animals, whilst underplaying other senses. In Rolston's scheme, domestic animals are artifacts of human culture. They are, he maintains, breeds, not natural kinds, and so do not have species natures (we can, for instance, talk about wolf nature, but we cannot talk about dog nature; we can only talk about airedales or poodles). Yet there clearly are characteristics that all dogs have in common (for instance, that they are social animals with dominance hierarchies; that they have a highly sensitive sense of smell, and so on). Many of these characteristics are shared with wolves—albeit in neotonized form. Dogs are not fully artifactual; certainly they bear the imprint of human influence in their bodies and temperaments; but also they carry many characteristics of their wild forebears. Here, Rolston seems to take the transformations effected by domestication too far. But, on the other hand, Rolston does not seem to take other effects of domestication far enough. Artifactual though domesticates may be, as we have seen with the original P2, he nonetheless considers that their comparison class is the wild, and that we should think about "what would have been their lot in the wild, on average, adjusting for their modified capacities to care for themselves" (1988, 79). But how should this be interpreted? The domesticates concerned would not have existed if humans had not bred them. Without humans there would be no life at all, not an alternative possible wild life. And what sort of "adjustment" does Rolston mean? Does he mean us to imagine (say) that a poodle has the capacities to survive in a wild that a wolf has, and to compare suffering in the poodle's life with suffering in the life of an (average) wolf? Or to imagine the kind of life this particular poodle would have if released into the wild, adding a few extra claws and some extra aggression as an adjustment?

That Rolston thinks it possible merely to "adjust for modified capacities to care for themselves" is an indication that he takes the changes of domestication insufficiently seriously. Many domestic animals would not survive in the wild at all. Some would survive as scavengers. A few might "resettle homeostatically into environmental niches" (1988, 78–9)—though as animal breeding and genetic modification of animals progress, this possibility becomes increasingly unlikely.

It is, in general terms, the loss of the ability to survive well, or to survive at all, independently that significantly separates domestic from wild animals; indeed the creation of domesticates is the deliberate creation of dependence.[9] Dependence,

I will suggest, is an important contextual characteristic (amongst others) in reconsidering human ethical relationships with animals.

AN ASIDE: THE MISCLASSIFICATION OF HUNTING

Before moving on to consider how Rolston's animal ethics may be developed, I want to make one further point concerning an apparent misclassification in his work. Rolston maintains that some kinds of human practices (as we have seen with animal agriculture) fall into the category N2 natural—shared across species. Where human behavior towards animals is concerned, one main instance of this he cites is hunting. This claim that is attacked both by Wenz (1989) and by Moriarty and Woods (1997). All three question whether human hunting practices can be conceived of as natural (though Wenz allows that hunting by indigenous peoples might be thought so). Moriarty and Woods emphasize that hunting, in the US at least, even where the kill is consumed, is characterized by hunting codes, technological gear and equipment and reliant on automobile use. Therefore, they argue, it is a cultural practice misclassed by Rolston (and once it is classed in culture rather than nature, the reclassification has consequences for its ethical acceptability).

These criticisms can be explained in a more careful way using the nature/culture distinctions I have outlined. Rolston suggests that humans can behave naturally in different ways: in N0 ways when acted upon by the "laws of nature"; in N2 ways when manifesting shared, cross-species behavior; in N3 species-specific cultural behavior; and in an N4 way by deliberated cultural behavior that in some sense follows nature. Rolston maintains that hunting is natural in the N2 sense. But for this to be the case he would have to argue that hunting was a non-reflective, non-deliberate practice, since it is reflection and deliberation that, for him, are central in distinguishing N2 natural behavior from N3 cultural behavior. But studies of hunting throughout human cultures indicate that this is not the case. All forms of human hunting are deliberate in this sense. Some refinement to Rolston's spheres of nature and culture might, however, allow him to accommodate hunting as natural in an N2 sense. Suppose one took a controversially high-level view of what went on in some forms of hunting by other animals; that is, that something analogous to deliberation or planning took place. Then the deliberation and reflection involved in human hunting would look more like a shared, cross-species behavior. As currently constituted, though, attributing these higher-level abilities to non-human hunters would lead to more permeability at the boundary of animals and culture than Rolston would want to accept. It might be possible, though, for a finer definition of culture to be constructed such that animals are excluded even if they are able to plan or deliberate at some level; but more work would need to be done before this revision could stand.

While it may be problematic for Rolston to suggest that hunting is natural in an N2 sense, it is entirely plausible, though, that, in many Western non-indigenous cultures at least, hunting is frequently viewed as deliberately natural in an N4 sense. Hunting codes designed to give animals an opportunity at escape may be a deliberate attempt to hold back and not to use all the technological possibilities at hand (one possible interpretation of N4 naturalness). Further, the commonly-made case that hunting emulates or reminds of simpler, more traditional ways of living, or provides ways of obtaining food that are closer to the wild than those of mechanized farming also fits this idea of N4 naturalness. Rolston's account, suggesting that hunting is "harmonious with nature" seems to echo this view. He may also be thinking of hunting as humans imitating what animals do instinctively (Rolston 1988: 92, 94); but he never suggests that it is not deliberate or intended. In all these cases, naturalness is naturalness in an N4 sense, a cultural, deliberate attempt to follow nature. On Rolston's current account, then, all human hunting is misclassified as N2 behavior, since it is always deliberate (this is not to say that there are no N2 human activities, but rather that hunting is not one of them). This conclusion undermines the ethical case Rolston makes for hunting, by maintaining that Rolston has misclassified hunting, and that the change in classification means it can no longer be seen as a permitted exception under P1. (It is not intended to imply that therefore there is no ethical case to be made for any form of hunting, however; merely that this case is not it).

ROLSTON, CONTEXT AND ANIMAL ETHICS

As we have already seen, Rolston maintains that not only capacity (primarily sentience) but also context is important in ethical decision-making about animals. In understanding context, it is useful to distinguish between the levels of origin and prior contact, process and place. For instance, in terms of origin N1 nature is that which does not originate from humans and has not been altered by human prior contact; in terms of process it is all those processes (evolutionary, ecological) that operate independently of humans; in terms of place it is the land that issues from this origin and these processes—that is, the wilderness.

Although Rolston does not make these distinctions explicitly in the context of animals, nonetheless applying them to animals allows for the possibility of some finer discriminations. For instance, the separation of the context of origin from the context of place allows for distinctions between animals constitutionally independent of humans (that is, not having been bred by humans) and animals that live independently of humans (whether or not they are constitutionally independent). Questions about prior contact allow for the discussion of special claims. It might be argued that domestication itself provides grounds for a form of special claim. But alongside this, there are also more specific special claims, where an individual human or groups of individual humans have chosen to establish relationships with particular individual animals that otherwise would not have existed.

Rolston's own account does not develop such distinctions. Indeed, they set up some difficulty for his principles P1–P3. For instance, human commensals or scavengers may be wild in their context of origin but not in their context of place, since they may live alongside humans and rely on them for provision; do they fall under P1 or P2? Feral animals, in contrast, may not be from N1 nature in their context of origin—they have been domesticated—and they may affect N1 processes; but on the other hand, they may be independent of human provision and largely out of human control (which makes applying P2 and P3 awkward). Rolston's categorizations are problematic inasmuch as they fail to fit the more specific contexts in which many animals are located and therefore provide us with no guide as to how to how such animals should be treated ethically. A more discriminating categorization than the one Rolston offers—though using the kind of distinctions that can be drawn from his work—is required to lay out fully the different contexts in which animals are located. Such a categorization (though certainly not exhaustive) might go something like this:

1. A wild context, where animals originate, subsist and are located independently of humans, though sometimes encountering humans (N1 nature in Rolston's sense).
2. A domestic context, of animals that have been, at some time, bred by humans and are now dependent on them for food and shelter, where this has been deliberately encouraged by humans. This may be a relation of positive affect (e.g. pet animals).
3. A feral context of animals that have been, at some time, bred by humans and are now dependent on them, but as scavengers rather than as intentionally encouraged by humans (e.g. urban feral cat colonies).
4. An exotic context of animals, released by humans into an area, and now living independently of humans (e.g. mink released from fur farms in the UK).
5. A scavenging context of animals that are constitutionally wild (have never been bred by humans) but that live alongside humans, manifesting various degrees of dependence on humans for food or shelter. This dependence has not been deliberately intended by humans and in the case of animals viewed as pests, may be positively discouraged and be a relation of negative affect (e.g. rats, raccoons, pigeons).
6. A commensal context of animals that are constitutionally wild but that live some or all of the time alongside humans, and are partly or wholly dependent on humans. This dependence is intentionally encouraged by (some) humans, and the animals are regarded positively though rarely with individual affect (e.g. wild birds fed at bird tables).
7. An agricultural context of animals highly bred by humans, kept for a functional role, highly dependent on humans, and deliberately created by humans in this way (experimental animals fall into a similar category).
8. A captive context of animals that are constitutionally wild, but that are kept confined by humans for a variety of purposes (often for display) and are dependent on humans inasmuch as they are captive and outside their native habitat.

If these contexts are tabulated, differences can be clearly highlighted:

	Wild Constitution (Origin)	Wild Living (Process and place)	Dependent on Humans	Intended/Encouraged by Humans
Wild	√	√	X	X
Pet	X	X	√	√
Urban Feral	X	X	(√)	X
Invasive Exotic	X	√	X	X
Pest/scavenger	√	X	(√)	X
Commensal	√	(√)	(√)	√
Agricultural	X	X	√	√
Captive Wild	√	X	√	√

This account is, of course, not exhaustive, the boundaries here are fluid, and the categories rather generalized. Some scavenging and commensal animals, for instance, may only be partially dependent on humans; some pets or agricultural animals might manage to survive independently of humans (though they would then begin to move into the feral category). Some people may encourage animals that others despise; other animals, introduced at a period in the past and encouraged at that time—such as red foxes in Australia—later become widely regarded as pests to be destroyed. These complications would have to be taken into account in considering any particular case, and in any more detailed examination of animal contexts. However, this categorization suffices to pick out particular kinds of relations between humans and animals. But why are these kinds of relations—developed from Rolston's discussion—to be regarded as significant? Let's think a bit further about just one of these as an example: domestication, and in particular, the dependence it entails, in the context of a possible relation/obligation argument.

ONE VERSION OF A RELATION/OBLIGATION ARGUMENT

Rolston suggests that "taking an interest" in animals by domestication generates some kind of moral responsibility for them that does not apply to wild animals. He does not, however, fill out this claim at all, nor indeed does his account of actual obligations to domesticated animals (not to make their lives worse than those of wild animals) seem to support it. While the claim is intuitively attractive, much more work would need to be done to make it convincing; I will attempt here only to make some very preliminary moves.

First, we should distinguish between two different kinds of dependence, which I will call external and internal dependence. Captive wild animals, for instance, might be able to provide for themselves in their native species habitat but in confinement are circumstantially dependent on humans to provide food and shelter. Their dependence is an external, humanly-imposed effect of captivity. Domestication, though, is a process that usually produces internal dependence, dependence that is constitutional and permanent. Domesticated animals are, usually, dependent on humans to provide for their vital needs in terms of food and shelter (at least).

It is this internal dependence of domestication in which I am interested here, and it throws up a number of questions that turn out to be really rather complicated. Humans are, in some sense, deliberately responsible for the existence of dependent domesticated animals. Whatever one might think about the possibility of animal "collusion" in the initial relations that led to domestication, it is clear that humans actively and deliberately engaged and still do engage in ever more specialized forms of selective breeding and more recently genetic modification. Animals are thus brought, by humans, through deliberate human choices, into dependent relations. The resulting loss of animal independence (with respect at least, to wild forebears) while sometimes a side effect of selective breeding for other ends (e.g. the breeding of cats without claws) is also deliberate, since an ability to live independently may jeopardize the plasticity of domestic animals to human intentions.

But pursuing this notion of responsibility more closely raises a number of standard philosophical difficulties about responsibility. Who is causally responsible, and in what ways, for the creation of dependent domesticated animals? Was it the original domesticators? Those individuals (farmers, breeders, experimental scientists) who deliberately act to produce domesticated animals in the present? Or those who voluntarily demand the products of domestication, from pets to pastrami? How are individual and collective responsibilities to be understood, in such a complicated and ongoing situation as this?[10] These questions are interesting in themselves, but their significance is increased if, by the attribution of causal responsibility, some sort of moral responsibility is thought to come into play.

Of course, that someone or some group has played a part in creating *x*, with the deliberate intention of creating a dependent *x*, does not in itself show that they have any duties to take *x* into account in moral decision-making. *X* must have some further claim to moral consideration. But the assumption that animals are morally considerable was made at the beginning of the chapter, and is not at issue here. Given this, there does, as I have already suggested, seem to be a plausible argument along the line that the deliberate creation of a dependent morally considerable being brings obligations to provide for that being. This would be one particular form of a relation/obligation argument where importance is based both on deliberate creation of the relation and the kind of relation (dependence for vital needs) as the basis for obligation.

Some of those philosophers who have considered relations not dissimilar to these but in the human situation have, however, found this kind of analysis unsatisfactory. Goodin (1985) maintains that moral obligations in situations that involve one party's vulnerability to another—such as children to parents—are not related to previous actions voluntarily undertaken and self-assumed agreements. So, he insists, such obligations solely derive from whether the individual concerned is depending on us, is particularly vulnerable to our actions and choices (Goodin 1985, 11) and/or whether we are the best or most obvious person to meet their need or protect them in their vulnerability. In some cases this would not result in any difference in terms of who has the duty; but the reason for the duty is different.

As Goodin himself accepts, these two kinds of explanation for relational obligations in the context of vulnerability and dependence are not necessarily exclusive, although he argues that many relational duties to the vulnerable cannot properly be subsumed into the voluntaristic model, since they were never voluntarily assumed by anyone. This may be so, but equally some of the issues discussed here may raise questions over the broadness of application of his thesis. Goodin (1985, 181) maintains that humans have special duties to animals on the grounds that animals possess morally significant interests and that they are vulnerable to humans, both individually and collectively. But if the presence of vulnerability and being the most able person to assist are sufficient to generate moral obligations, then (for instance) the park-keepers in Yosemite National Park should treat the bighorn sheep's natural diseases. Of course, Goodin might respond that his position could be overridden by other environmental values. But he would not be able to accept Rolston's sense of no relation/no obligation because, in Goodin's scheme, encounter is enough to generate the relation. If the vulnerability presents itself, and you are in the best position to relieve it, then the relation comes into being and along with it, the obligation. That you had no active involvement or causal role in the situation is irrelevant. As with utilitarianism and rights approaches to wild animals, though, this position seems not to work well with wild animals, even if it were thought to work well in the case of people.

CONCLUSION

In this chapter, I've suggested that Rolston's work provides a way forward in addressing some, at least, of the difficulties posed to philosophical animal liberation by some forms of environmental ethics. Rolston's contextual emphasis (although problematic as it stands in his own account) alongside his acknowledgment of the value of animals as sentient individuals, begins to suggest some new approaches to animal ethics. In particular, his contextual account opens up new ways of thinking about domestication. Domestication can be seen as a process such that, in creating relationships that close down domestic animals' abilities to live independent lives, creates special human responsibilities to provide for them.[11] Indeed, it seems likely that further development of this position would deliver some, at least, of the protection for domesticated animals that philosophical animal liberation advocates set out to achieve in the first place. And this can be achieved without implausible consequences in the wild to which environmental ethicists so strongly object. This approach to animal ethics, then, may be able to bring environmental and animal ethics closer together, whilst at the same time providing tools to develop a more complex animal ethics, sensitive to human-animal relations and to the contexts in which different animals are located.

ACKNOWLEDGMENTS

I am grateful to Erica Fudge, Emily Brady and Jose Luis Bermudez for reading and commenting on earlier versions of this manuscript. Helpful comments were also made when versions of this chapter were given at Washington University in St Louis in February 2004 and at an APA-Pacific session honoring Holmes Rolston in March 2004. I am particularly in debt to Andrew Brennan and Yeuk-Sze Lo who, *inter alia*, argued that the N0 category was needed to fully work through Rolston's position.

NOTES

1 Indeed, in his response to Wenz, Rolston explicitly says that the chapter of *Environmental Ethics* on animals and ethics should be read in the light of his earlier chapter on following nature.

2 It would be spatially extravagant to explain how I have drawn these different uses of nature from Rolston's work; so for now the accuracy of these uses will have to be taken on trust.

3 Questions about individual humans who do not manifest these distinctive species-specific human characteristics are sidestepped by Rolston's focus on the norm for the species. Some humans may not manifest these distinctive characteristics; but they do not represent what is normal for the species and are not species-representative. The emphasis on species norms is also intended to protect Rolston against versions of the Argument from Marginal Cases.

4 Andrew Brennan (personal comm. 2004) drew my attention to this distinction in Rolston's work—see Rolston (1988: 34).

[5] I am not going to raise questions here about Rolston's take on sentience and moral considerability, nor am I going to quibble at his use of the term "suffering" where one might prefer to use pain. This issue has been widely discussed elsewhere, and is not especially interesting in Rolston's account. Hettinger rightly argues that, as constructed, Rolston's principles here favor plants over animals, since plants, unlike animals, are governed by a principle of "non loss of goods". Hettinger's account here is interesting and useful; but I want to take a different direction in this chapter, so I will not be pursuing Hettinger's particular arguments more closely.

[6] P1 is not a principle without difficulties, though it is less problematic than P2 or P3. One legitimate worry (expressed in questions on each occasion when I have given this chapter) might be as to whether there really is any N1 nature in existence; without N1 nature, P1 has no application.

[7] Rolston (1994: 72) develops the idea of "agricultural integrity," where agricultural areas are "managed sustainably" such that their operation does not disrupt the surrounding natural systems" and they are "enveloped by natural systems". Nonetheless, nothing about this account suggests that he would evaluate agriculture by any other standards than those he adopts for wild nature.

[8] Hettinger rightly notes that this seems to privilege plants over animals, and suggests a revision to Rolston's principle by extending his Principle of Non Loss of Goods to animals from plants. See Hettinger (1994) for the detail of this argument.

[9] By indicating that humans deliberately created dependence, I am not meaning to rule out the thesis, most prominently argued by Budiansky, that (some) animals in some sense "colluded" with domestication. See Budiansky (1992)

[10] Brennan (personal communication 2004) quite rightly raises the question how far I am responsible for things that are problematic because of the behavior of other members of my species, not of me. I began to consider some issues of this kind in Palmer (2003) with respect to the feeding of feral pigeons in Trafalgar Square; but plainly this is a major area for further research.

[11] Of course, it might be argued that the moral problem lies in creating dependent animals in the first place, rather than in the ways they are treated once in existence, but I do not intend to propose this argument here.

REFERENCES

Budiansky, Stephen. 1992. *The Covenant of the Wild*. New York: Weidenfeld and Nicholson.

Frey, R.G. 1983. *Rights, Killing and Suffering*. Oxford: Blackwell.

Goodin, Robert. 1985. *Protecting the Vulnerable*. Chicago: University of Chicago Press.

Hettinger, Ned. 1994. "Valuing Predation in Rolston's Environmental Ethics." *Environmental Ethics* 16:3–20.

Leahy, Michael. 1993. *Against Liberation: Putting Animals In Perspective* (Revised). London: Taylor & Francis.

Lockwood, Michael. 1979. "Killing and the preference for life." *Inquiry* 22: 157–190

Moriarty, Paul Veatch, and Mark Woods. 1997. "Hunting is not Predation." *Environmental Ethics* 19: 391–404.

Palmer, Clare. 2003 "Placing Animals in Urban Environmental Ethics." *Journal of Social Philosophy* 34/1: 64–78.

Regan, Tom. 1984. *The Case for Animal Rights*. London: Routledge.

Regan, Tom. 2001. *Defending Animal Rights*. University of Illinois Press.

Rolston, Holmes. 1986 *Philosophy gone Wild*. New York: Prometheus.

Rolston, Holmes. 1987. "Beauty and the Beast: Aesthetic Experience and Wildlife." In Decker and Goff, eds., *Valuing Wildlife: Ecological and Social Perspectives*. Boulder CO: Westview.

Rolston, Holmes. 1988. *Environmental Ethics: Duties to and Values In the Natural World*. Temple University Press.

Rolston, Holmes. 1989. "Treating Animals Naturally?" *Between the Species* 5/3: 131–137.

Rolston, Holmes. 1991. "The Wilderness Idea Reaffirmed." *The Environmental Professional* 12: 370–377

Rolston, Holmes. 1994. *Conserving Natural Value*. New York: University of Columbia Press.

Rolston, Holmes. 1998. "Technology versus Nature: What is Natural?" *Ends and Means* 2/2 online at http://www.abdn.ac.uk/philosophy/endsandmeans/vol2no2/index.shtml

Rolston, Holmes. 2003. "Life and the Nature of Life—in Parks." In Harmon, David and Putney, Allen, eds., *The Full Value of Parks: From the Economic to the Intangible*. Lanham MD: Rowman and Littlefield, pp. 103–113.

Singer, Peter. 1975; 1983 ed. *Animal Liberation*. Wellingborough: Thorsons.

Singer, Peter. 1979. *Practical Ethics*. London: Cambridge University Press.

Wenz, Peter. 1989. "Treating Animals Naturally." *Between the Species* 5/1: 1–10.

JOHN LEMONS

12. NATURE DIMINISHED OR NATURE MANAGED: APPLYING ROLSTON'S ENVIRONMENTAL ETHICS IN NATIONAL PARKS

INTRODUCTION

Many people whose work deals with science, public policy and environmental decision making believe their work must take into account moral responsibilities and duties to nature and human beings. In other words, those working in these areas know they must advocate for public policies and decisions that are morally justified (Barry and Oelschlaeger 1996; Shrader–Frechette 1996). At the same time, the value–laden aspects of scientific methods and techniques, the use of scientists' work in value–laden policy and decision making, and the ethical implications of much public policy making and decision making are often under–recognized and understudied (Lemons and Brown 1995). It is particularly the case that when scientists, public policy makers, and decision makers want to take into account environmental ethics more fully, they often are unsure of how to do so because of their lack of training in environmental philosophy.

Holmes Rolston, III has been one of the most helpful voices at this intersection of ethics, public policy, and science. As demonstrated by this volume, he has had a profound influence on the development of environmental ethics; and other philosophers continue to critically assess his ecocentric theory. Rolston recognizes that justification of an environmental ethic rests, in part, on good science to inform the ethic. For this reason, he serves philosophers, scientists, public policy makers, and decision makers well by combining good philosophy with good science. (See, e.g., Rolston 1988, 1989, 1991, 1994, 2004.)

Philosophers such as Goodpaster (1978), Regan (1983), Taylor (1986) and Singer (2001) developed theories that in their view justify the moral considerablilty of a wide range of nonhumans as moral patients. Generally speaking, they have argued that (at least some) nonhuman individuals are moral patients because they are sentient or have a "good of their own." Attempts to extend the theories of these philosophers beyond individuals to, e.g., species or ecosystems, generally have proven more difficult. The (most well-known) approaches that have come closest are those of J. Baird Callicott and Rolston.

Callicott (1991) develops his theory by drawing on Leopold's *The Land Ethic* to locate value in biotic communities or ecosystems. An important aspect of Callicott's position is his contention that value cannot exist in nature without a self-conscious human doing the valuing. He also argues that some concepts of wilderness are based on a faulty dualism between putatively non–natural human behavior and actions on the one side, and nonhuman nature on the other. Accordingly, he argues that although

C.J. Preston and W. Ouderkirk (eds.), Nature, Value, Duty, 203–219.

some wilderness preservation is justified, greater emphasis should be placed on sustainable development, biodiversity preservation, and the maintenance of ecosystem health.

Alternatively, Rolston (1994, pp. 167–202) argues for an ecocentric ethic based on the intrinsic value of species and ecosystems ("natural wholes"). He also recognizes the intrinsic value of individual organisms. More specifically, Rolston emphasizes four concepts: (1) intrinsic value of and derivative duty to respect all living organisms, (2) intrinsic value of species as natural groupings of life, (3) intrinsic value of abiotic nature, and (4) holism as an essential part of ecology and environmental ethics. Rolston derives all of these duties from the intrinsic value of nature in a manner that denies the naturalistic (is-ought) fallacy.

Scholarly objections to Rolston's ethics focus around several main areas (Palmer et al. 2001): (1) some social scientists and ecofeminists argue that Rolston's notions of intrinsic value and evolutionary complexity are too hierarchical; (2) some philosophers and pragmatists reject the concept of intrinsic value and prefer to focus on other types of values while others argue that value is by necessity subjective (by humans) and, therefore, value does not exist if no humans are doing the valuing; (3) Rolston's rejection of the legitimacy of the so-called naturalistic fallacy is regarded as only weakly detailed and argued; (4) some philosophers and others hold that Rolston's environmental ethic is merely the beginning of a fuller theory and that many conflicts involving particular cases as well as broader practical and theoretical issues still need to be resolved.

In this chapter, I discuss this latter objection to Rolston's work, namely, that to develop his theory more fully he needs to root it in or demonstrate it by greater use of particular cases. I approach Rolston's work from my background as a scientist who has tried to integrate science with public policy and environmental ethics, not as a formally trained environmental philosopher. For this reason, and because I assume that readers of this volume are familiar with Rolston's environmental philosophy, I focus my comments on some of the practical scientific and public policy questions and implications arising from his work. Specifically, I mostly use concrete examples from U.S. national parks so that my discussion is less abstract and theoretical. The first set of examples concerns the use of science to inform policy making and decision making in national parks, and the second set concerns the interpretation of national park legislation. Understanding whether or under what circumstances Rolston's theories are of practical use can help inform public policy and decision making and, in particular, assist people based in one discipline who seek familiarity with another. Without combining ethics and science, environmental ethicists will be unable to undergird their philosophy with science where relevant; and scientists, public policy makers, and decision makers will be limited in contributing to ethical environmental policies and decisions. Although I conclude that Rolston's ethic needs to be developed more fully with the use of practical cases to help inform difficult policy problems, I also believe that it nevertheless holds great promise to motivate and guide decision making in places such as our national parks.

MANAGEMENT CHALLENGES IN NATIONAL PARKS

Some of our nation's most important ecological and scenic resources are protected in national parks, whose fundamental purpose is commonly stated as threefold: (1) to conserve the scenery, natural and historic objects, and wildlife; (2) to promote the enjoyment of scenery, natural and historic objects, and wildlife; and (3) to provide for that enjoyment in such a way that the scenery, natural and historical objects, and wildlife are unimpaired for future generations (16 United States Code [USC] Section 1). The National Park Service (NPS) manages over sixty-three natural area units ranging from 30,000 to over 3,000,000 million acres in size (Sellars 1997). Included in this category are parks such as Yosemite, Yellowstone, Glacier, Grand Canyon, Great Smoky Mountains, Everglades, Kings Canyon–Sequoia, and Denali. Twelve of the parks are Biosphere Reserve Parks dedicated to long–range ecosystem monitoring under the UNESCO Man and Biosphere program. Importantly, the U.S. National Park Biosphere Reserves are considered ecological standards for the network of International Biosphere Reserves.

Management of national parks has been difficult and controversial because of problems associated with external threats to parks, i.e., threats originating outside of parks' boundaries and over which the NPS has little control, and because of internal threats to parks resulting from visitor use and facilities over which the NPS has direct control. Although more recent detailed data are lacking, research has identified over seventy different kinds of threats in the categories of aesthetics, air pollution, physical removal of resources, encroachment by exotic species, visitor physical impacts, water quality pollution and water quantity changes, and park operations and facilities development, and over 2000 internal threats and over 2400 external threats have been identified for biological, physical, aesthetic, cultural, and park operations resources (NPS 1980). Because of the value–laden scientific and public policy issues inherent in their management, national parks provide a good context for discussing and evaluating some concrete aspects of Rolston's ethics.

ROLSTON'S ETHICS APPLIED TO PARTICULAR CASES

Rolston uses some specific but limited kinds of examples to demonstrate decisions in national parks that are consistent with his ethics. For example, he believes there are no rights in wild nature and that nature is indifferent to the welfare of particular animals. He cites an example of a bison that fell through the ice into a river in Yellowstone National Park (Rolston 1994, p. 113). The NPS decided neither to save the animal nor kill it as an "act of mercy" to reduce its suffering. Rolston (1988, p. 182) also cites another example from Yellowstone National Park, where bighorn sheep caught pinkeye (*Chlamydia*), which occurs naturally in them, and which ultimately resulted in the death of about 300 individuals. Had wildlife veterinarians treated the sheep, presumably some of them would have been saved. For Rolston, the actions of the NPS with respect to the stranded bison and the infected bighorn sheep were consistent with his environmental ethic because they respected "nature taking its course." According to Rolston, the NPS decision makers knew that death of individuals from natural causes operates functionally. The dead bison became part of the ecosystem's nutrient cycle and the energy contained in its biomass nurtured the animals that fed upon its carcass. The death of the bighorn sheep from pinkeye served not only *Chlamydia* (also a part of nature) but in an adaptive sense might increase the biological fitness of the bighorn sheep because those individuals carrying genes more resistant to *Chlamydia* could be favored by differential reproductive success.

Rolston cites a third example from Yellowstone where a female grizzly bear and her cubs made their way over winter ice on Yellowstone Lake to an island to feast on a dead carcass but became stranded when the ice melted and they could not return to shore (Rolston 1994, p. 113). In this instance, the NPS saved the bears by relocating them from the island. Rolston maintains that on ethical grounds this intervention was justified because grizzly bears are an endangered species, and that fact imposed a duty not so much to save individual bears but rather to help save an endangered species. According to Rolston, saving species is important because "life on Earth cannot exist without its individuals, but a lost individual is always reproducible, a lost species is never reproducible." Saving species is important for Rolston because of their role in the evolutionary development of other species and increasing or at least maintaining the complexity of life.

In these examples and arguments, Rolston's conclusions disagree with norms that might favor saving the stranded bison or infected bighorn sheep in order to mitigate suffering and death (e.g., arguments from Singer [2001] to minimize suffering or from Regan [1983] to prevent the death of conscious beings). According to Rolston, such norms conflict with what goes on in nature (death from falling through ice on a river or from natural predation) and therefore conflict with ecological processes on which species depend and through which they have evolved.

Before proceeding to more complicated problems, I want to provide some examples that can be used to question Rolston's conclusion that duties toward individuals of other species should rest on whether their species are "endangered." The first issue is that scientifically it is difficult to know when many species are, in fact, endangered, and even more difficult to know when their populations are on a trajectory that in the future might lead to their becoming endangered despite present indications that they are "healthy." Populations of species can proceed down a slippery slope as the species locally, regionally or globally declines from a healthy status to a threatened status to an endangered status and possibly to extinction. Furthermore, the reasons for the decline or even that the decline has begun often are not known or not known with full scientific certainty (Lemons 1995).

Consider the case of grizzly bears within the Central Rocky Mountains in Alberta, Canada (Herrero 1994). Although the scientific facts are uncertain, there is concern that in only a few generations the grizzly bears in this region are being confined to about five ecosystems surrounded by increasing levels of human development and activities. (The Canadian government has not concluded that its nation's grizzly bears are "endangered."). While some scientists and preservationists argue that the risks of extinction are significant, others argue that the bears have characteristics that make them sensitive to human activities and natural events such that they are "threatened with becoming threatened with becoming extinct-a rather remote threat." Such an argument raises the question: Should we intervene to save individual animals when there is uncertainty over whether or to what extent risks to the species exist? Further, consider a more extreme version of this question: Do we have an obligation to save individuals who are not members of a currently endangered species, based on a vague and general concern about the effects of humans' unprecedented detrimental activities on the future status of species, even though for the time being the species appear to be "healthy" and a scientific basis for the concern is impossible to establish? For example, one might be concerned that although species such as the grizzly bear in Canada (or for that matter bison and bighorn sheep in Colorado) are not endangered today, they very well could be in the future, given continued human alteration of the Rocky Mountains. Does such a vague and general concern warrant saving individuals in an attempt to stave off possible threats to the species from future human activities, which we know have resulted in serious declines in biodiversity? Of course, these latter two questions could be viewed as "begging" some of the problems Rolston is concerned about, namely, that saving individual bison or bighorn sheep would interfere with "letting nature take its course" or would interfere with ecological processes on which individuals of other species are dependent. However, his view assumes there is a clear, distinct difference between endangered and non-endangered species. The less clear the distinction, the more we might see a responsibility to save individuals whose species are not "officially" endangered.

Consider too that the under the Endangered Species Act (ESA) the US Fish and Wildlife Service (USFWS) recently "delisted" the population of grizzly bears in the "Greater Yellowstone Area" (GYA) (New York Times 2005). According to the USFWS (2000), the rationale is that the following population characteristics of the bears meet guidelines to support delisting: the number of bears has increased to about 600 individuals from 175–200 in the late 1970s; there is a four-percent mortality rate in the bears' population; and females and cubs now populate most bear management areas of the GYA. Yet, other biologists question whether delisting the population makes scientific sense on the basis of only 600 individuals. This number is very close to an estimated minimum viable population. All estimates of minimum viable populations contain significant scientific uncertainty as well as value-laden assumptions; and basing delisting on data the USFWS has collected over a short time period (late 1970s–present) is questionable because the ecology of populations operates on time periods much longer than a couple of decades (Lemons 1996a).

My concern with Rolston's position about such matters is that, when he writes of obligations to save endangered species, his expressed view is unclear about how ethical deliberation should take into account scientific or other uncertainties about the status of species whose endangerment is uncertain or subject to change under undetermined conditions. Is his view about duties to endangered species based on the best available science concerning the status of the species or on whether the species has been legally listed as an endangered species? Because of the considerable politics surrounding the ESA these are two very different questions.

A significant complication for protection of grizzly bears in the GYA is the habitat fragmentation expected to occur as a result of President Bush's recent repeal the US Forest Service's "Roadless Areas Conservation Rule," originally promulgated to protect the last remaining wildlands in our national forest system. The repeal of this rule likely will exacerbate external threats to parks because adjacent lands will experience increased road building, timber harvesting, and other uses such as gas, oil and mineral extraction and thereby subject grizzly bears to increased threats both directly and indirectly through habitat fragmentation (CCNPR 2004). If grizzly bears were delisted prior to the manifestations of the ecological consequences of the repeal of the "Roadless Areas Conservation Rule," this ultimately could threaten the long-term status of the bears, especially given continued opposition to listing species for protection under the ESA. Rolston's guidance to following nature or to allow nature to run its course within the parks may well become impossible as a result of such complications.

The complexity of the question of when to let "nature take its course" is illustrated by Rolston's evaluation of a recent rare event. In 1986 the Hubbard Glacier in Alaska accelerated its movement and closed off Russell Fjord from the sea, thereby trapping large numbers of porpoises and harbor seals, whose lives were then threatened as fresh water replaced salt water in the fjord (Rolston 1988, p. 50). Shortly after the animals were trapped, animal rights advocates argued that the animals should be rescued, while scientists argued that nothing should be done because of the opportunity to conduct scientific research. In my reading of Rolston's discussion of this event, I infer that because the porpoises and harbor seals were not members of endangered species they should not have been rescued from Russell Fjord because its closure presumably was a natural event. This is consistent with his conclusion about letting the bison stuck in the river and the bighorn sheep infected with *Chlamydia* in Yellowstone National Park die.

While I am sympathetic to Rolston's view that in these kinds of cases we let nature take its course, I wonder how easy it is to know factually whether certain events are "natural" or whether they are triggered or exacerbated by humans. Scientifically speaking, there is a possibility that many Alaska glaciers were affected by human-induced global climate change in the 1980s; and, although this is speculative, the Hubbard Glacier event could have resulted from such change. For purposes of discussion, if we assume some probability that such climate change was responsible for that event, would Rolston recommend that the animals be rescued, despite the fact that they were not members of endangered species? If there is intrinsic value in nature, it would seem that rescuing non-endangered animals from human–induced or possibly human-induced changes could be ethically mandated. Scientifically, it could be argued that we should save not only members of endangered species for the reasons stated by Rolston, but we should also save individuals from non–endangered species because the latter possess the potential to contribute to future evolution and increasing the complexity of life, probably more so than endangered species, due to the former's greater numbers and ecological importance compared to rare species (Myers 2002).

THE COMPLICATED REALITY OF PARK SERVICE DECISIONS

Although Rolston's examples of the bison, bighorn sheep, and grizzly bear demonstrate how his ethics can inform some kinds of decisions, my larger concern is that these kinds of examples are not really typical of the more difficult problems confronting the NPS, where science is needed for informed decision making (Lemons 1995, Lemons and Junker 1996). Further, the examples deal with relatively uncomplicated internal threats to species and ecosystems in national parks, i.e., those that the NPS has more autonomy and authority to manage, as opposed to comparatively more difficult external threats arising from outside the parks' boundaries and which therefore the NPS has much less authority to manage. Because the kinds of examples Rolston uses are not typical of the more complicated practical problems confronting public policy makers and decision makers where science and ethics are relevant, the examples demand too little from both science and ethics to be of practical use. Consequently, those concerned with developing practical environmental ethics might require more from Rolston than the kinds of examples he typically has used. My concern is that there is a need for more collaborative work by philosophers, who seek to develop a coherent and ecologically sound ethic, and scientists, who can articulate the kinds of complex problems faced by policy and decision makers. Then the framing and solutions of such problems can be informed by ethics that are practical, motivational, and inspirational.

Nationally, areas where "natural conditions" exist are rapidly diminishing and are at risk from accelerating anthropogenic forces (Graber 2003). These anthropogenic forces include but are not limited to habitat fragmentation and concomitant biodiversity loss, invasions of alien species, loss of natural fire regimes, acid precipitation and contamination from other chemicals, and global climate change. Whereas national parks once were relatively buffered from many prevailing anthropogenic forces, the acceleration of such forces has now made parks' ecosystems ever more threatened. Compounding the threats to parks' resources are those stemming from high levels of visitor use and facilities development in parks.

Rolston's emphasis on maintaining the "wild" in nature seems to resonant with the NPS's management goals to identify the parks' resources and values that are subject to "no impairment" (NPS 2001):

> The park's scenery, natural and historic objects, and wildlife, and the processes and conditions that sustain them, including, to the extent present in the park: the ecological, biological, and physical processes that created the park and continue to act upon it; scenic features; natural visibility, both daytime and at night; natural landscapes and smells; water and air resources; soils; geological resources; paleontological resources, archaeological resources; cultural landscapes; historic and prehistoric sites, structures, and objects; museum collections; and native plants and animals.

In support of this management standard, the NPS guidelines state:

> Natural processes will be relied on to control populations of native species to the greatest extent possible. Unnatural concentrations of native species may be controlled if the activities causing the concentrations cannot be controlled. Nonnative species will not be allowed to displace native species if this displacement can be prevented.

While a reading of Rolston's ethics suggests he would support these guidelines, nevertheless troubling questions remain (see, e.g., Westman 1990, Lemons 1996a): What is the scientific definition of "natural concentrations" of native species and how is this concentration to be determined? How are native species to be controlled by natural processes while nonnative species are to be actively managed? What scientific principle or goal allows for this differentiation? Would active management also affect native species? What is a "native species" and how arbitrary is the definition? What do the terms "natural" and "unimpaired" mean? Is or should the answer to these types of questions to be found in NPS legislation and/or policy? In the methods and tools of science? In the principles of environmental ethics?

The NPS states: "Whenever possible, natural processes will be relied upon to maintain native plant and animal species, and to influence natural fluctuations in populations of those species." Despite that statement, policy also permits management intervention to restore parks' species and ecosystems. In fact, due to the rate at which parks' ecosystems are becoming compromised, restoration increasingly constitutes a larger part of management. Although NPS management is more scientifically informed than in the past, the threats posed by pervasive anthropogenic threats to parks' resources present serious problems that likely will not be resolved satisfactorily. These problems transcend in seriousness and complexity the recommendations made by Rolston to let nature take its course when he presents the examples discussed above.

Other, yet more complex kinds of national park management challenges also demonstrate this difficulty. Rolston recommends the use of some natural–caused or prescribed fires to help restore natural fire regimes to ecosystems that for over a century were altered due to human suppression of fires. He writes: "A forest fire harms an individual aspen tree, but it helps *Populus tremuloides* because fire restarts forest succession without which the species would go extinct." (The view that *Populus tremuloides* is a successional species dependent on fire is not unequivocal scientifically but the reasons are beyond the scope of this paper; see, e.g., Grahame and Sisk [2002]. Regardless, Rolston's observation that fire suppression has greatly altered ecosystems is well taken.) Although the NPS has used natural–caused and prescribed fires in an attempt to restore more natural fire regimes, their use has neither significantly reduced the huge fuel buildups resulting from fire suppression nor regenerated ecological structure or function to conditions that existed previous to fire suppression policies (Caprio and Graber 2000). Scientific constraints on the use of any kind of fire for management purposes include uncertainty about such things as the natural intervals between fires, the fires' intensities, the ecological endpoints or trajectories resulting after fires, the potential of fires to facilitate in some instances the invasion of alien species, and a recognition that on the scale of centuries to millennia climate variation has imposed large changes on vegetation. Political constraints on natural-caused or prescribed burning stem from the need to ensure that fires do not get out of control, air quality regulations, availability of fire monitors and other personnel, proximity of human settlements to fires, and increasing political support for mechanical thinning of forests. So although fires might in some sense be a "natural" management tool, there is no simple and direct way to apply Rolston's ethical guidance ("let nature take its course") to their use.

Consider too some of the problems of atmospheric contamination in national parks. Aquatic ecosystems in eastern parks are threatened by acid precipitation, including the loss of fish and amphibians. In western parks, atmospheric deposition of pesticides is an important factor in the continued decline of frog populations, and atmospheric deposition of nitrates and ammonium in this region provides through fertilization a competitive advantage to alien herbaceous plants compared to native species. In some parks experiencing serious problems of atmospheric pollution, it is possible to identify the individual and local sources of the pollution. For example, in Grand Canyon National Park, the coal-fired power plant in nearby Page, Arizona, is a significant contributor to declining air quality in the park. However, there is no clear scientific evidence that this power plant is the major cause of atmospheric pollution the majority of the time. In Big Bend National Park, although air quality continues to decline, no single pollution source can be pinpointed. Undoubtedly, smelters and coal-fired power plants from the U.S. and Mexico contribute to Big Bend's air quality problems; but so too does global industrialization because some atmospheric pollution there also stems from areas as far away as Asia (NAS 1993). Mitigation of such effects within the parks may well require decision makers not to follow Rolston's admonition to let nature take its course.

Finally, consider the problems global climate change poses to national parks' ecosystems and species illustrated by the issue of the Hubbard Glacier raised above. Ecosystem and species-level changes from global climate change already are being seen in Alaska (Hassol 2004). Based on climate models for the Sierra Nevada in California, warmer winters, reduced snow packs, and more rain relative to snow suggest that some biological communities and species presently found in the region's national parks may not be able to persist due to the changing conditions (Jayko and Millar 2002).

These kinds of large–scale environmental problems demonstrate that no national park is large or isolated enough to avoid the impacts of ecological change and political decisions beyond its borders. In essence, almost any ecosystem management undertaken by the NPS is in reality experimental, because the outcomes cannot be predicted with very much precision due not only to the limitations of ecology but also to confounding political decisions for other jurisdictions that have consequences to parks. How should the NPS manage parks under rapidly changing, uncontrollable, and uncertain conditions? Obviously, NPS managers must work with others in an attempt to overcome barriers to establishing effective partnerships with diverse agencies and institutions extending over much larger areas than a particular park or group of parks. However, with few exceptions, such barriers have been insurmountable, given the fact that the NPS has lacked legislative authority and political influence to control external threats. Thus, the limitations of ecology and influences beyond its control constrain the NPS both in achieving its goal of "no impairment" as well as in using natural processes to maintain plants and animals (Lemons 1996b, Shrader-Frechette and McCoy 1993).

Given the complexity of external problems confronting national parks, combined with scientific uncertainty about the threats to parks, it appears that NPS policy makers and decision makers must choose between two unsatisfactory alternatives. One, allow parks' ecosystems to develop under the influence of anthropogenic influences and threats, or, two, engage in intensive and intrusive intervention and restoration. The first alternative, because it involves less active intervention, will provide the appearance of "wildness," while the second alternative will turn parks into intensively managed areas as a kind of hopeful experiment to save whatever species and ecosystems as we can. Neither of these alternatives would seem suitable to Rolston yet it is not clear how his ethics can yield practical solutions to the problems.

THE VALUE OF ROLSTON'S SOFT ETHICS IN GUIDING NATIONAL PARK POLICY

What the above considerations reveal is that asking Rolston (or anyone) to provide an ethical framework capable of yielding practical solutions to complicated environmental problems is probably asking too much. From an environmental professional's perspective, there is neither an environmental ethic that is fully developed nor one that has been used extensively to inform policy making and decision making. Perhaps Rolston's goal in developing his ethic has been to motivate and inspire rather than to impart a system of ethics capable of supporting precise solutions to controversial and complex environmental problems. The chance for developing an environmental ethic relevant to NPS policy making and decision making might be increased if legislation and ethics converge (Caldwell 1996). For example, a clearer preservationist legislative mandate would make things simpler for both the management professional and the ethicist. Although the goals and purposes of national park legislation are ambiguous, a strong but understudied argument can be made that the fundamental purpose of national parks is to more strictly preserve them in as "pristine" state as possible, or alternatively, that NPS administrators have discretion to do so. Rolston's ethic clearly articulates frameworks and guidelines that can help motivate and inspire those interested in preserving parks' ecosystems and species and, in particular, those with strong sentiments or intuitions about the need to do so but who are unable to express their views clearly in ethical terms. Accordingly, Rolston's ethics is latent in the sense that, although it is "soft" and general, it could become more useful if there existed a clearer mandate to preserve pristine nature in the national parks.

PRESERVATION OR CONSERVATION?

National Park Service policy and decision making continue to be extremely controversial and in this sense reveal that the parks' authorizing legislation embodies a host of competing values. Controversies stem from different interpretations of the purposes of establishing parks and of the forces influencing their management. Further, some controversies stem from the lack of a firm underlying predictive and explanatory science to inform policy and decision making. Strictly speaking, park management must conform to the stated goals and purposes in national park legislation. The most often quoted legislation mandating the purposes of national parks is 16 United States Code (USC) Section 1, referred to as the NPS "Organic Act," which was passed by Congress in 1916 authorizing the creation of the NPS. However, the NPS is not entirely consistent in its stated administrative policies regarding its own interpretation of Section 1; in various documents it has emphasized differing interpretations (Lemons and Junker 1996; Sellars 1997).

There are several possible reasons for different interpretations of Section 1 (and other 16 USC sections to be discussed). First, despite Section 1's being prescriptive and normative because it establishes the goals and direction of NPS policy and decision making, it contains key words such as "natural" and "unimpaired" (to use only two examples) that have not been defined by the NPS. No scientific, legal, or ethical criteria or standards have been developed that can be used to indicate whether NPS decisions are consistent with the promulgated normative language. A review of the relevant literature demonstrates the confusing variability of the term "natural." A short list of representative examples include Devall and Sessions (1984), Smith and Theberge (1986), Westra (1994), Parsons et al. (1986), Bonnicksen and Stone (1985), Jorling (1976), Callicott (1991), Sagoff (1995) and Rolston (1991). All of these writers propose different definitions and criteria for the meaning of "natural."

Second, NPS legislation is imperfectly bounded in the sense that it presumably requires some ecocentric values but at the same time implicitly reflects an anthropocentric ethic designed to promote human use and enjoyment of parks. Third, a more cynical reason for the NPS's apparent ambiguity regarding the purpose of national parks is that even the apparently ecocentric values protected by Section 1 can be interpreted as strictly human values that differ only slightly from the values protected by common laws of nuisance but which are cloaked in an appealing veil of protection for nature. Complicating the quest for clarity in all of this is that there have been few NPS judicial interpretations and little case law and, therefore, discussions of NPS policy and goals have necessarily been based on stipulative arguments as opposed to constitutive meanings of legislative language.

The ambiguity or vagueness of NPS legislation has also resulted in conflicting political views about park management, ranging from managing according to strict preservation ideals, to managing for increased visitor use and facilities to satisfy public preferences, to managing for the economic interests of park concessionaires. From the standpoint of preservationists, the controversies have not been satisfactorily resolved and parks' resources have continued to decline. I believe the value of

Rolston's ethics is that it can help foster a greater preservationist public policy for national parks and related issues.

Typically (but not exclusively), the NPS interprets Section 1 to require a balance between preservation and use of parks. However, preservationists argue that the NPS, in attempting to implement that balance, has allowed significant deterioration of parks' ecosystems due to heavy visitor use and development. Mantell (1979), Chase (1986), Lemons and Junker (1996), and Keiter (1997) have argued that the legislative meaning of Section 1 interpreted in the context of other relevant national park legislation sets forth an impressive and unambiguous resource preservation mandate, including the preservation of pristine-as-possible ecological processes. Interestingly, the clause in Section 1 for the NPS to manage parks "for future generations" presents a standard different from that of the well being of many other stakeholders who claim to have an interest or power or influence with the NPS. In other words, while various stakeholders might focus on policy making and decision making that satisfies particular values, or provides short–term benefits, or reflects current political ideology, if this argument is correct, the NPS would be required to manage parks with long–term interests in mind. Even if strictly speaking an interpretation of Section 1 that requires a strong preservation mandate is assumed to be incorrect, the fact remains that NPS decision makers have wide discretion in how the legislation is interpreted so long as it is not applied in an arbitrary and capricious manner. In other words, NPS administrators can promulgate stricter preservation policies in the parks should they choose to do so.

In addition, legislation subsequent to Section 1 also requires NPS policies and decisions to promote preservation values. For example, in 1965 Congress supplemented Section 1 with 16 USC Section 20 (a–d) to help resolve conflicts between preservation and use in national parks that were deepening as a result of a larger national population, greater mobility, changing visitor expectations, and some NPS policies that had encouraged increased visitation and development of facilities in parks. In particular, one major goal of Section 20 was to limit concessionaire developments and facilities to places that were outside of important scenic, natural and ecological areas. Importantly, Section 20 also reaffirmed the importance of Section 1 in calling for the preservation of parks' resources. On several occasions, Congress has reaffirmed the primary responsibility of the NPS to preserve park's resources (16 USC Sections 1(a)–1, Section 55).

A stricter policy of preservation in national parks would protect better the attributes of biological diversity; the complexity and interactions of organisms, populations, and ecosystems; evolutionary processes; rare species; and species richness. All of these attributes are important to stakeholders in the NPS, including those in the scientific community who advocate for more stringent preservation policies. Yet, although many such stakeholders are well–intentioned, they are not well versed in formal philosophy and therefore often do not embed their arguments within a strong ethical framework. Rolston provides strong ethical support for a stricter policy of preservation because he has developed his ethics by focusing on: (1) biological diversity, including at different levels from the genetic to the landscape; (2) complexity of organisms, populations, and ecosystems; (3) processes of evolution and the emergence of diversity and complexity; (4) the significance of rare species; and (5) the significance of species richness. He also focuses on connections between concepts that link science and ethics. For example, Rolston attaches great importance to DNA, not only for its obvious biological significance, but also because he posits that the information it carries helps us to distinguish what "is" from what "ought to be." *Inter alia*, the genetic information coded in DNA and manifest in organisms and species drives genotypic potential to phenotypic expression and, therefore, results in every organism having a "good of its own." In this sense, each individual has intrinsic value. The diversity of DNA fosters greater competition between organisms, plant and animal succession, ecosystem development, and speciation and an ever-richer biota. Rolston's ethical system recognizes intrinsic value in all of these, and he claims that ethical obligations to respect them arise from such values.

The reliance of Rolston's ethics on attributes of nature and its intrinsic value can support an ethical justification for concepts of "natural" that require human intrusions in nature and national parks to be minimized. Similarly, application of his ethic to the understanding of national parks' purposes would require definitions of "unimpaired" to be closer to what is meant by "pristine nature," compared to an environment altered significantly by humans that, nevertheless, humans might find satisfactory. In essence, Rolston provides a systematic analysis needed to identify the values that may be implicit or explicit in NPS legislation and policy. If NPS legislation is to mean anything more than valuing human autonomy conceived as freedom to engage in economic or other activities with little or no constraint, Rolston's ethic assumes significant importance because it requires that we attend to intrinsic values, including those better suited for preservation of nature for future generations.

RESPONDING TO STAKEHOLDER VALUES

The NPS is not an expert–centered agency but rather a more responsive one that is concerned with questions of representation of various stakeholders in policy making and decision making, access to information, and responsiveness to public demands and perceptions (Freemuth 1999). Stakeholders for the NPS include the general public, elected officials, heads of other agencies, special interest groups, and to the extent possible future generations. The NPS expects its administrators to be responsive to the views of stakeholders while simultaneously being responsive to policy priorities. Given the vagueness and ambiguity of NPS legislation, the agency, as one might expect, often makes decisions based on the prevailing opinion and influence of its stakeholders. For those concerned with preservation of parks' resources, the problem is that all too often public preferences and commercial interests have dominated NPS policy, but so too have vested interests within the NPS itself, such as regional directors or higher administrators, whose political agendas might be very different from, say, park scientists or naturalists or others concerned with strict preservation (Sellars 1997). At the same time, there are people both within and outside the NPS who advocate stricter preservation policies. Mostly, the arguments are predicated on stipulative legal arguments discussed above, on scientific reasons, or on vague and somewhat implicit values that people hold about nature but cannot adequately express or put into a coherent ethical framework.

As more people interested in preserving national parks gain insight into Rolston's ethics, presumably they will become more effective at stating clearly and more comprehensively the ethical reasons for a stricter policy of preservation in national parks. In other words, Rolston's ethic is supportive of and congruent with scientific reasons as well as with arguments that the purposes of national park legislation are to more strictly preserve parks' resources based not on anthropocentric utilitarian reasons but on recognition of nonhuman nature's intrinsic value. Further, Rolston's arguments about the moral duty to maintain biodiversity because of its long–term importance to emergent diversity and complexity through the processes of evolution also are congruent with Section 1 which states that the NPS shall provide for the enjoyment of parks "as will leave them unimpaired for the enjoyment of future generations." To the extent current NPS policies and decisions allow harm to biodiversity or evolutionary opportunity, the very naturalness parks were established to protect will not be available for future generations.

Finally, despite the fact that NPS legislation mandates protection of parks' values, nowhere does it define the values. Consequently, NPS administrators are assumed to have wide discretion in identifying and giving importance to parks' values in their attempt to fulfill the legislative mandates. Some of the typical park values identified include: (1) market or commercial values; (2) life support values; (3) recreational values; (4) scientific values; (5) genetic and biodiversity values; (6) aesthetic values; (7) cultural symbolization values; (8) historical values; (9) character-building values; (10) therapeutic values; (11) sacramental values; (12) aspirational or option values; and (13) intrinsic values. (Runte, 1979; Sax, 1980; Lemons, 1987; Rolston, 1988) However, the values are usually only implicitly embedded in most policy making and decision making, which is to say most policies and decisions are not based on explicit identification, discussion, or analysis of the importance or wisdom of the values or how they might conflict with one another. One of Rolston's (1989) significant contributions is that he has analyzed these values and developed a set of principles that defend on moral grounds why some values should take precedence over others in such places as national parks. In particular, market or commercial values should be the least emphasized by NPS policy makers and decision makers because they pose the greatest conflicts or threats to all of the other values, which are less consumptive of parks' resources. Rolston's argument that market or commercial values should be least emphasized by the NPS is consist with Section 20 (described previously) calling for development to be limited to those areas in parks "outside of important scenic, natural and ecological areas."

CONCLUSION

My comments are an attempt to provide insight on the practicality of Rolston's ethic, mostly using selected examples and issues from national parks. Although he provides general ethical guidelines consistent with NPS policies and decisions based on relatively simple examples, those guidelines do not seem sufficiently prescriptive to deal with the more complex external threats to biodiversity and natural resources in such places as national parks. To be fair, an expectation that anyone's system of environmental ethics be sufficiently prescriptive to meet the demands posed by all external threats to parks is unrealistic. Most importantly, Rolston's ethic is motivational and inspirational and as such offers great potential to those attempting to advance the cause of preserving nature. His ethic is important because it dovetails with, and therefore can be used in conjunction with, understudied legal and public policy arguments to implement stricter preservationist policies on public lands such as national parks.

Rolston's call for obligations and duties to wild nature raises the question: How can we more effectively apply complex ethical positions to actual policies on the ground? For one thing, the general public, policy makers and decision makers need to be more educated in environmental philosophy and its relevance to public policy. Ecologists and other scientists can begin addressing value–laden environmental policy issues and decision making through public education, working with professionals and nonprofessionals in interdisciplinary settings, changing research agendas to focus on policies and decisions that have ethical implications but rely on science to inform the decisions, and finding other ways to demonstrate the importance of ethics as part of public policy making and decision making (Ehrlich 2001). Included in this is the need to adapt research agendas to yield as much information as possible about anthropogenic impacts on the future course of biotic evolution and attempting to find ways to ameliorate them.

Finally, those wishing to foster NPS or other environmental agendas in ways that take into account Rolston's ethic must be aware that their own training and expertise likely incorporate cultural attitudes that promote the domination of nature. Indeed, few people hold purely preservationist views. In the sense that Rolston's ethic is motivational and inspirational and we can understand it as a challenge to achieve a more enlightened public policy.

REFERENCES

Barry, D. and M. Oelschlaeger. 1996. Science for Survival: Values and Conservation Biology. *Conservation Biology* 10: 905–911.

Bonnicksen, T.M. and E.C. Stone. 1985. Restoring Naturalness to National Parks. *Environmental Management* 9: 479–486.

Caldwell, L.K. 1996. Science Assumptions and Misplaced Certainty in Natural Resources and Environmental Problem Solving. In J. Lemons (ed.). *Scientific Uncertainty and Environmental Problem Solving*, Cambridge, MA: Blackwell Science, p. 394–421.

Callicott, J. Baird. 1991. The Wilderness Idea Revisited: The Sustainable Development Alternative. *The Environmental Professional* 13: 235–247.

Caprio, A.C. and D.M. Graber. 2000. Returning Fire to the Mountains: Can We Successfully Restore the Ecological Role of Pre–European Fire Regimes to the Sierra Nevada? In D.N. Cole, S.F. McCool, W.T. Borrie and J. O'Loughlin (eds.). *Wilderness Science in a Time of Change Conference, Proceedings RMRS–P–15–Vol 5*, Ogden, UT: US Department of Agriculture, Forest Service, Rocky Mountain Research Station.

Chase, A. 1986. *Playing God in Yellowstone: The Destruction of America's First National Park*. Boston: Atlantic Monthly Press.

CCNPR (Coalition of Concerned National Park Retirees). 2004. *Analysis: Two Dozen U.S. National Parks Threatened by Bush Roadless Forest Rule Repeal*. http://www.ens.org/nws/2004/07/28/analysis_two_doz (access 08/03/05)

Devall, B. and G. Sessions. 1984. The Development of Natural Resources and the Integrity of Nature. *Environmental Ethics* 6: 293–322.

Ehrlich, P. 2001. Intervening in Evolution: Ethics and Actions. In *The Future of Evolution*. Washington, D.C.: Proceedings National Academy of Sciences, p. 5477–5480.

Freemuth, J. 1999. Absolutely American and Absolutely Democratic: National Parks and Policy Change. *George Wright Forum* 16: 63–76.

Goodpaster, K.E. 1978. On Being Morally Considerable. *Journal of Philosophy* 75: 308–325.

Graber, D.M. 2003. Facing a New Ecosystem Management Paradigm for National Parks. *Ecological Restoration* 21: 264–268.

Grahame, J.D. and T.D. Sisk (eds.). 2002. *Canyons, Cultures and Environmental Change: An Introduction to the Land–Use History of the Colorado Plateau*. Flagstaff, AZ: Land Use History of Northern Areas Project http://www.cplubna.nau.edu/index.htm (access 08/03/05)

Hassol, S.J. 2004. *Impacts of a Warming Arctic–Arctic Climate Impact Assessment. Cambridge*, UK: Cambridge University Press.

Herrero, S. 1994. The Canadian National Parks and Grizzly Bear Ecosystems: The Need for Interagency Management. *International Conference Bear Research and Management* 9: 7–21.

Jayko, A.S. and C.I. Millar. 2002. *Impacts of Climatic Change on Landscapes of the Eastern Sierra Nevada and Western Great Basin, U.S. Geological Survey Open File Report 03–202*. Washington, D.C.: U.S. Geological Survey.

Jorling, T.C. 1976. Incorporating Ecological Principles into Public Policy. *Environmental Policy and Law* 2: 140–146.

Keiter, R.B. 1997. Preserving Nature in the National Parks: Law, Policy, and Science in a Dynamic Environment. *Denver University Law Review* 74: 649–695.

Lemons, J. 1987. United States' National Park Management: Values, Policy, and Possible Hints for Others. *Environmental Conservation* 14: 329–340, 328.

Lemons, J. 1995. Ecological Integrity and National Parks. In L. Westra and J. Lemons (eds). *Perspectives on Ecological Integrity*, Dordrecht, The Netherlands: Kluwer Academic Publishers, pp. 177–201.

Lemons, J. 1996a. (ed.). *Scientific Uncertainty and Environmental Problem Solving*. Cambridge, MA: Blackwell Scientific, Inc.

Lemons, J. 1996b. The Conservation of Biodiversity: Scientific Uncertainty and the Burden of Proof. In J. Lemons (ed.). *Scientific Uncertainty and Environmental Problem Solving*, Cambridge, MA: Blackwell Science, pp. 206–232.

Lemons, J. and D.A. Brown. (eds.). 1995. *Sustainable Development: Science, Ethics, and Policy*. Dordrecht, The Netherlands: Kluwer Academic Publishers.

Lemons, J. and K. Junker. 1996. The Role of Science and Law in the Protection of National Park Resources. In R.G. Wright (ed.). *National Parks and Protected Areas*. Cambridge, MA: Blackwell Science, p. 389–414.

Mantell, M. 1979. Preservation and Use: Concessions in National Parks. *Ecology Law Quarterly* 8: 1–54.

Myers, N. 2002. Biodiversity and Biodepletion: The Need for a Paradigm Shift. In *Biodiversity Sustainability and Human Communities: Protecting Beyond the Protect*, S. Stoll–Kleeman and T. O'Riordan (eds.). Cambridge, UK: Cambridge University Press, p. 46–60.

NAS (National Academy of Sciences). 1993. *Protecting Visibility in National Parks and Wilderness Areas*. Washington, D.C.: National Academy Press.

New York Times. November 15, 2005. *Yellowstone Grizzlies to Lose Endangered Species Status*. http://nytimes.com (access 11/15/2005)

NPS (National Park Service). 1980. *State of the Parks–1980, A Report to Congress*. Washington, D.C.: Office of Science and Technology, US Department of the Interior.

NPS (National Park Service) 2001. *Management Policies* http://www.nps.gov/policy/mp.htm (access 08/03/05)

Palmer, J.A., D.E. Cooper and P.B. Corcoran. (eds.). 2001. *Fifty Key Thinkers on the Environment*. New York: NY: Routledge.

Parsons, D.J., D.M. Graber, J.K. Agee, and J.W.V. Wagtendonk. 1986. Natural Fire Management in National Parks. *Environmental Management* 10: 21–40.

Regan, T. 1983. *The Case for Animal Rights*. Berkeley, CA: University of California Press.

Rolston (III), H. 1988. *Environmental Ethics*. Philadelphia, PA: Temple University Press.

Rolston (III), H. 1989. *Philosophy Gone Wild*. New York, NY: Prometheus Books.

Rolston (III), H. 1991. The Wilderness Idea Reaffirmed. *The Environmental Professional* 13: 370–377.

Rolston (III), H. 1994. *Conserving Natural Value*. New York, NY: Columbia University Press.

Rolston (III), H. 2004. Life and the Nature of Life– In Parks. The *George Wright Forum* 21: 69–77.

Runte, A. 1979. *National Parks: The American Experience*. Lincoln, NB: University of Nebraska Press.

Sagoff, M. 1995. The Value of Integrity. In L. Westra and J. Lemons (eds.). *Perspectives on Ecological Integrity*. Dordrecht: The Netherlands, p. 162–176.

Sax, J. 1980. *Mountains Without Handrails: Reflections on the National Parks*. Ann Arbor, MI: University of Michigan Press.

Sellars, R.W. 1997. *Preserving Nature in the National Parks*. New Haven, CT: Yale University Press.

Shrader-Frechette, K.S. and E. D. Mccoy. 1993. *Method in Ecology: Strategies for Conservation*. New York, NY: Cambridge University Press.

Shrader-Frechette, K.S. 1996. Throwing out the Bathwater of Positivism, Keeping the Baby of Objectivity: Relativism and Advocacy in Conservation Biology. *Conservation Biology* 10: 912–914.

Singer, P. 2001. *Animal Liberation*. New York, NY: Harper Collins Publishers.

Smith, P.G.R. and J.B. Theberge. 1986. A Review of Criteria for Evaluating Natural Areas. *Environmental Management* 10: 715–734.

Taylor, P.W. 1986. *Respect for Nature*. Princeton, NJ: Princeton University Press.

(USFWS) US Fish and Wildlife Service. 2000. Strategy for Managing A Recovered Population of Grizzly Bears in the Yellowstone Ecosystem. Lakewood, CO: US Fish and Wildlife Service http://www.r6.fws.gov/endspp (access 08/14/05).

Westman, W.E. 1990. Managing for Biodiversity. *BioScience* 40: 26–33.

Westra, L. 1994. *An Environmental Proposal for Ethics: The Principle of Integrity*. Lanham, MD: Rowman and Littlefield.

JAMES W. SHEPPARD AND ANDREW LIGHT

13. ROLSTON ON URBAN ENVIRONMENTS

No one working in environmental ethics today can overestimate the landmark contribution of Holmes Rolston, III to our field. This volume and his many honors and accolades in and outside of the academy are ample testament to the importance of his work. Rolston, perhaps more than any other environmental ethicist in the last three decades, has provided a steady and wide-ranging stream of publications on all manner of subjects concerning how and why nature has value.

While truly impressive in its breadth and coverage, Rolston's body of work also omits certain subjects. Such omissions are not necessarily a problem; they can charitably be seen as helping to ensure that certain subjects in an author's work will be carefully treated even if it also means that other subjects will be left out. Nonetheless, certain kinds of omissions can be problematic if they are unjustified, and especially if the neglected subjects can be demonstrated as necessary for an overall project to be complete. In the case of Rolston's work, some omissions open the door to such criticism.

In this chapter, we consider one such largely excluded item in Rolston's work—urban environments. Our goal, however, is to do more than just identify this omission. We also will explain why this omission is unnecessary given Rolston's other philosophical commitments, many of which involve his understanding of natural value. One easily could wonder why it is important to point out this missing subject matter in Rolston's work. What, after all, is so important about urban environments for an environmental ethicist? Space does not permit us to make a positive case for the ecological importance of urban environments here. We have defended that thesis elsewhere for the skeptical reader.[1] Instead, in what follows we offer first a general explanation for why traditional environmental ethics largely has ignored human concerns, second, provide an overview of Rolston's theory of natural value, third, demonstrate an unfortunate consequence for the appreciation of urban environments following from Rolston's theory of value (whether intended by Rolston or not), and finally, fourth, demonstrate that Rolston's work can more easily accommodate an account of the natural value of urban environments than some may expect.

THE NONANTHROPOCENTRIC PREJUDICE OF ENVIRONMENTAL ETHICS

Over the last three decades the field of environmental ethics has come to organize itself around discreet sets of philosophical, political, and practical issues. At least in North America, but also in other new world countries like Australia, environmental philosophy has been dominated by a concern with more abstract questions of value theory,

C.J. Preston and W. Ouderkirk (eds.), Nature, Value, Duty, 221–236.

primarily focused on the issue of whether nature has intrinsic value, or some other form of non-instrumental value. If such value can be justified independent of human consideration then it is an instance of "nonanthropocentric" value as opposed to anthropocentric, or human-centered forms of valuation which have dominated the history of ethics in the West. If nature has such non-instrumental, nonanthropocentric value, then, so most of the major theorists in environmental philosophy would argue, a wide range of duties, obligations, and perhaps even rights obtain in our treatment of it.

Curiously, however, environmental ethicists have been largely silent about urban environmental issues, let alone the normative status of urban environments themselves. In one of the few articles on the subject in the last fifteen years, Alastair Gunn reports that in three top selling environmental ethics textbooks, out of nearly 200 readings between them, not one single selection deals explicitly with urban environments.[2] Some authors even appear outright hostile to the potential of finding value in humanly produced cultural landscapes.[3] Our first claim then is that this dismissal of cultural landscapes, the built world, and urban environmental problems can be traced to the foundations of the field in the search for nonanthropocentric forms of non-instrumental value. Why? One answer is that for the majority of environmental ethicists engaged in this traditional enterprise, the embrace of nonanthropocentric foundations for an environmental ethic entails a rejection of anthropocentric forms of value, and, we would argue, consequently, of anthropogenically created landscapes.

Since the beginnings of the field in the early 1970s, environmental ethicists have routinely rejected anthropocentric forms of moral consideration as both part of the cause of the current environmental crisis and as an impediment to any solution to those problems.[4] While the rejection of anthropocentrism has been far from univocal in the short history of the field, it is still arguably true that the majority of environmental ethicists reject anthropocentrism as a plausible foundation for an environmental philosophy in general or for a theory of the value of nature in particular. Due largely to the pioneering work of Robin Attfield, J. Baird Callicott, Arne Naess, Val Plumwood, Rolston, Richard Sylvan, Paul Taylor, and others, the burden of proof in the literature is clearly on the proponents of anthropocentrism to prove that it has any place in discussions of environmental ethics and not on the opponents. Nonanthropocentrism is thus often presumed to be an uncontroversial starting point for an environmental ethic. We agree with Gary Varner's argument that the predominance of nonanthropocentrism (and holism for that matter) is more of an accepted prejudice than a proven position.[5]

What explains this presumption as well as the continued rejection of anthropocentrism in environmental ethics? Though we will not fully defend the proposition here, a strong case can be made that it follows from the intuition that anthropocentric schemes of value are equivalent to the instrumental valuation of nature and that, as a consequence, anthropocentrism is incapable of providing the ground for a guaranteed rejection of certain cultural forms of valuing nature that tend to lead to ecological degradation. At bottom here is a worry concerning the adjudication of environmental debates solely in terms of human preferences, preferences which are based in the very cultural norms and social practices that are viewed to have led to ecological degradation in the first place. Some of the earliest critiques of anthropocentrism were embedded within a critique of the perceived cultural relativity of many contemporary environmental policies. At the end of his landmark 1973 article, "Is There a Need for a New, an Environmental Ethic?" Sylvan provided an example of just this sort of worry.

> it would just be a happy accident, it seems, if collective demand for a state of the economy with blue whales as a mixed good, were to succeed in outweighing private whaling demands; for if no one in the base class happened to know that blue whales exist or cared a jot that they do then "rational" economic decision-making [a form of anthropocentric instrumental valuation] would do nothing to prevent their extinction.[6]

Taylor took this claim for a culturally neutral foundation for natural value further, and more forcefully, by arguing that anthropocentric value is "entirely relative to culture: if any particular society did not hold ideals that could be symbolized in nature and wildlife (for example, if it happened to value plastic trees more than real ones), then there would be *no reason* for that society to preserve nature or protect wildlife."[7]

What then is the connection between the rejection by environmental ethicists of anthropocentrism and their curious silence about urban environmental issues? One answer involves the fact that since many if not most environmental ethicists see the principal goal of their inquiry to involve the identification of a culturally neutral basis for the nonanthropocentric value of nature, most theorists focus their work on what they perceive to be pristine exemplars of this type of natural value, such as wilderness areas. Stated in different terms, if nature is considered to be valuable in and of itself, however the ground of that value is metaphysically based, it is likely that it will be best identified in those areas relatively independent of human intervention and intentionality as opposed to humanly shaped areas. As a consequence, urban environments are either unintentionally left out of most systematic accounts of environmental ethics, or, worse, intentionally cited as a form of disvalue. In what follows, we will use Rolston's work to investigate this dynamic further. More specifically, we will try to ascertain whether the direct and indirect implications of Rolston's theory of natural value unintentionally or intentionally condemn urban environments.

ROLSTON'S WORLDS OF VALUE

As we suggested at the start, Rolston necessarily leaves certain subjects out of his work in order to do justice to those subjects he does address. While possibly representative of a tension in his overall view, these omissions could serve as a guide for important theoretical discussions in the field insofar as they could guide our thinking about how to address a broader array of environments in the field and hopefully address a wider set of problems than Rolston could ever address by himself. The best way to consider Rolston's tendency to leave certain subjects out of discussions and to capitalize on this opportunity is to consider carefully the natural value schematic he offers as well as its many implications. While a focus on intrinsic value arguably guides Rolston's work, it is important not to overlook how much attention he directs at developing a complete value schematic that also includes a detailed set of anthropocentric instrumental values. This attention to the plural values associated with the natural world, to the worlds of value, without a doubt, is one of the more admirable traits of Rolston's mature work. In Rolston's schematic natural values are organized hierarchically. There is a primary level that includes intrinsic values and a secondary level that includes instrumental values. On the primary level, the three types of value are spontaneous, systemic, and projective natural values.[8]

Spontaneous natural values correspond to natural properties and qualities that have existed for as long as the natural world has been evolving, such as the basic physiological characteristics of flora and fauna.[9] These properties and their values are autonomous, wild, and untrammeled. They are, more to the point, intrinsic to the natural X that exhibits these properties and qualities which means that they are also independent of any human use or interaction. Characterized as such, these independent and intrinsic values, these spontaneous values, these objective values will resonate with those whose aim is to offer an account for those natural Xs perceived to be untouched or pristine. An exemplar of this value type would be a pristine wilderness area, such as the Amazon rain forest. While not completely untouched by the actions of humans, the Amazon wilderness does exude a sense of wildness and, for some, autonomy. In addition to "untouched" wildernesses, certain entities also have spontaneous value. Species of wild flora and fauna, left to flourish on their own, also have the spontaneous properties that make them intrinsically valuable.

Systemic values, the second type of primary intrinsic value, emerge when a species' instrumental value within an ecosystem is mixed with its spontaneous value. Systemic value, according to Rolston, is "something in addition to intrinsic and instrumental value, a creative potential that steadily becomes actual."[10] The American alligator of the Florida everglades illustrates this type of value. Alligators have spontaneous natural value because they possess spontaneous natural properties such as aspects of their basic physiological structure that stem from their ongoing evolutionary development. In addition to these properties, alligators are also valuable instrumentally, or more properly, systemically, because they exist side-by-side or together with other species in an ecosystem. It is because they are, to use Rolston's terminology "good in a role, in a whole" that we can identify them as a type of "value-in-togetherness."[11] More specifically, they are systemically valuable because they carry out important ecological services and fill a specific ecological niche as a keystone species. To recite an oft-mentioned example, the footprints that alligators leave in the mud create indentations that, when filled with rainwater, provide water for numerous other creatures in times of drought. Without the alligator, these creatures would be without water and thus not be able to operate as effective members in the ecosystem. The removal of the alligator thus would send ripples throughout the ecosystem threatening not just the alligator, but by association, the whole ecosystem. In simplest terms then, alligators *qua* alligators are bearers of spontaneous value; alligators *qua* keystone species in an ecosystem are systemically valuable.

Projective values, the third type of intrinsic value, correspond with the ability of species and ecosystems to tell a story—to "pro-ject all the storied natural history," to use Rolston's phrase.[12] This value is associated with species and ecosystems by virtue of the fact that they are repositories of natural evolutionary lineages. Of the three types of intrinsic value, this is the least precise. Why? Presumably *Homo sapiens sapiens*, like other species, pro-ject a natural history. But, if we follow Rolston's guidelines, many of the anthropocentric and anthropogenic causes to which humans dedicate themselves to as well as the various ways humans necessarily, at times, run roughshod over spontaneous properties and spontaneous values should cause us to pause in recognizing the pro-jective value of the human story. If this is the case, is it because the story that humans are telling, indeed the evolution of the human species itself, is not valuable? Or, perhaps less severely, is it because certain parts of the story are less desirable than others. If it is the former, an explanation of why this is the case is needed. If it is the latter, one has to wonder if a veiled misanthropy is at work. Minimally one is left to wonder about this type of natural value. How useful is it as a category if human beings, far and away the most dominant species on the planet, lack this type of value? Supporters of Rolston might claim that we have exaggerated the potential problems with pro-jective value especially since the anthropocentric and instrumental natural values Rolston outlines more than suffice for dealing with the story of human beings and how human beings have interacted with and impacted non-human nature.

According to Rolston, these anthropocentric and instrumental values, the last category of natural value in Rolston's schematic, emerge when spontaneous natural properties are mixed with some human use, experience, action, or interpretation of those properties.[13] Humans and their creations, while valuable, have arrived on the scene relatively late—that is, relative to the rest of the already valuable spontaneous natural world. The earth is composed of valuable spontaneous natural properties that have existed for around approximately 4 billion years. Human beings that ascribe value to nature have existed for 30–35,000 years.[14] While the former (spontaneous natural properties) are necessary for the latter (value ascribing humans), the latter are not necessary for the former. This helps to explain why these instrumental values are less important than and dependent upon nature's intrinsic values.

Where then does this leave us in terms of determining which, if any, natural Xs or environments are left out of the Rolstonian equation? Fortunately, this is not a difficult question to answer especially if it is admitted, as we suggested in the last section, that behind many accounts of natural value are intuitions, assumptions, and judgments about which types of natural Xs and which environments are likely to be exemplars of that value. Again, the suggestion is that for every type of natural value we are able to identify, we can also identify a natural X that best reflects that conception of value. Said differently, accounts of natural value often explicitly or implicitly posit some natural X or some set of natural Xs as being the most appropriate focal point of a specific type of value attribution. It is useful to recognize from the outset what is and what is not included here for more than just the sake of intellectual curiosity. Why? When types of natural value are advanced, we arrive also at a normative focus for efforts to move from a description of natural value to ethical or policy prescriptions for our treatment of bearers of that value.

In very broad terms, Rolston's thinking over the years about value has corresponded to five basic focal points in the natural world:

A. The Genome
B. Individual members of species
C. Species
D. Ecosystems
E. The Earth[15]
F. Some combination of A, B, C, or D

This list is a helpful point of departure but it is also deceptive. Consider what happens when we hold this list up next to Rolston's value schematic. One of the reason's Rolston's approach is so clever has to do with the fact that everything seems to be covered. From the genome to the Earth, from spontaneous, systemic and pro-jective values to anthropocentric and instrumental values, Rolston seems to have covered all of the bases. This apparent coverage is deceiving though insofar as certain Cs (human beings) and Ds (urban ecosystems) are at worst typically excluded or at best provisionally included. When they are included, it is an inclusion based on the conviction that they belong, at best, in the category of things that have anthropocentric value or are of instrumental value. What we want to suggest is that Rolston's account of natural value underestimates how certain species and certain environments may be more valuable than heretofore thought.

At this point it is appropriate to recognize that Rolston's environmental ethic, at least in its early incarnations, does not appear to offer any specific proposals that might be applicable in or to urban areas. This early exclusion, as we have already suggested, is perhaps best characterized as an unintentional exclusion and as a necessary part of a broader philosophical methodology. In later, more recent formulations of his environmental ethic, Rolston seems to at least gesture in the direction of including urban environments in his theoretical account. Sandwiched in between these two periods was a period where there does appear to be a clear anti-urban bias guiding Rolston's work. Before concluding this analysis with a consideration of his most recent work, we will consider this anti-urban bias and some of its implications.

ROLSTON'S GEOGRAPHICAL BIAS

So far we have tried to articulate the theoretical basis and scope of Rolston's account of natural value. But to better contextualize the anti-urban bias which has characterized at least some of his work we need to further consider and appreciate the kind of landscapes which appear to motivate those value accounts. Any reader of Rolston knows of the focus in his work over the years on wilderness issues and most sympathetic readers regard this focus as entirely appropriate. In one of his more famous essays on the topic, "Values Gone Wild," Rolston exemplifies the intuition that a nonanthropocentric ethic starts in the realization of the value of wilderness and then moves on to reevaluate other spheres of cultural value:

> Only about 2 percent of the contiguous United States remains wilderness; 98 percent is farmed, grazed, timbered, hunted over, dwelt upon,

> paved, or otherwise possessed. (. . .) But . . . when the wildness is almost conquered, we begin to awake to error in the mastery theory. Not all value is labored for, assigned, or realized at our coming. The anomalous 2 percent that we will to keep wild, and then realize to be valuable without our will, reveals that the theory of value that has governed our handling of the 98 percent is flawed, only an approximation over a certain range.[16]

As was pointed out in the last section, for Rolston, nature, in its many guises, is the source of value. Therefore, a focus on wilderness issues makes sense as the most fecund repository of that value. But as Rolston's work on this topic evolved it seemed that cultural landscapes had to be primarily described as first, lacking the kind of value we find in pristine nature, and second, deserving of a thorough reassessment. In a later essay considering the possibility of the human improvement of wild nature, Rolston's nature-culture division is more evident: "The architectures of nature and of culture are different, and when culture seeks to improve nature, the management intent spoils the wilderness. (. . .) The cultural processes by their very 'nature' interrupt the evolutionary process: there is no symbiosis, there is antithesis."[17]

Before proceeding to interrogate these claims, it is important to acknowledge that Rolston does not argue that there is no value in human culture or in urban environments. In one account, after stipulating that Earth contains three environments, urban, rural, and wild, Rolston follows with the Aristotelian claim that humans are a political animal and that their "essence is to build a polis, a town." "The city is in some sense our niche; we belong there, and no one can achieve full humanity without it."[18] From our history of living in urban environments comes literacy and advancement, many things which make us human.

But the existence of wilderness, as suggested above, brings us something else: the recognition that all values are not simply the human values cultivated in urban environments. While in itself such a claim does not represent much of a problem for the appreciation of urban environments, Rolston goes much further in reversing the priority of forms of valuation over that which we would find in a more conventional Arisotötelian account:

> No one can form a comprehensive worldview without a concept of nature, and no one can form a view of nature without evaluating it in the wild, deliberating over spontaneous nature and whether and how it can have value. In that sense, one of the highest of cultural values, an examined worldview, is impossible to achieve without wild nature to be evaluated as foil to and indeed source of culture.[19]

What we find most worrisome about a view like this are the normative implications for urban environments and for experiences in urban environments.

Shortly after introducing this notion of the grounding of cultural values in the appreciation of the value of wilderness, Rolston clearly states that a full human life cannot be achieved in the city alone. Without being specific about what he is referring to, Rolston offers an account of the prevalence of environmental concern among urban dwellers (as opposed to rural people) to the "depravation" that they feel in the city. Whatever this depravation is, it causes them to look outside of the city and become concerned with the wild. Accordingly, the urban dweller who does not look outside the city for sources of value, or what Rolston terms a "mere" urban person, is "one-dimensional; three-dimensional persons will know how appropriately to respect urban, rural and wild environments."[20]

Now certainly, many will find this claim uncontroversial. After all, the suggestion appears to simply be that a full life is lived not in one kind of environment but in many, and Rolston is giving urban environments their due as the source of a uniquely human form of value which still has some positive content. Who then would want to object to the claim that, all things considered, a life is better where one can appreciate the three different environments that Rolston identifies? But Rolston is not claiming that all three environments are equal. It is not just that a human is comparatively worse off if they do not "respect," the wild, but that a human life is incomplete, that a human life is not wholly human, without the self-conscious knowledge of this respect.

To realize the gravity of this point it is important to note that, for Rolston, we humans do not live in the wild.[21] The wild is not our home, for presumably if it was then it would not be the wild. So, it isn't as if we have a choice to become better persons by living in the wild and then round out that experience by coming to appreciate our experience of the urban as a source of equal value. This makes it clear that Rolston's is not simply an appeal to the plural value of various environments. We can only live in the urban or the rural and still maintain the division of the three kinds of environments that Rolston identifies. If the urban, rural, and wild were equal then respecting only one of them as a resident would be equally as bad as only respecting any other one of them as a resident. But humans cannot live in the wild so wherever we live, in the rural or urban environment, it is the wild that completes us because the wild is the home of intrinsically valuable nature, or, in a sense, the home of our home as an evolved product of nature.

Further, note that Rolston is a bit more careful in his later work not to use the term "experience" in his description of the kind of interaction we should have with the wild. The term used is "respect." We must respect the wild. But what does this respect entail? In Rolston's earlier work (and there is no explicit repudiation of this point that we can find) a full human life is only possible through the actual experience of the wild: "Society is crucial for one aspect of persons, wilderness for another. Never to plunge into wilderness, never to expose oneself to it, is never to know either forest or self."[22] So, it is not just that one must respect or understand the importance of the wild, the rural, and the urban in relation to each other in order to be a fully three-dimensional human; one must actually experience the wild. The claim is then not simply that a richer life is lived, all things considered, by respecting the three environments of Earth, but that missing experience in one realm deprives one of a full life. An analogous claim might be that a full life cannot be lived without experiencing a high point of urban culture, such as hearing a live performance of Chopin's Impromptus or of watching a performance of Hamlet.[23] If Rolston were willing to say comparable things about urban experiences then perhaps those of us bothered by this suggestion could be partly pacified. But for every suggestion that the place of humans is in the *polis*, there are other suggestions that the city is a source of disvalue, specifically in terms of understanding one's self. "Lostness plagues the urban, mobile world," says Rolston.[24] And elsewhere:

> Big-city life in a high rise apartment—to say nothing of the slums—or a day's work in a windowless, air-conditioned factory represents synthetic life filled with plastic everything from teeth to trees. Such life is foreign to our native, earthen element. We have lost touch with natural reality; life is, alas, artificial.[25]

All life, apparently in the city, is not natural, nor a part of nature. It is something wholly different. Parks, trees, vestiges of streams, let alone buildings and cultural landscapes, are not a part of nature. They are anthropogenically derived and should only be anthropocentrically valued. Humans too, in some sense, are unnatural in the city, or at least one-dimensional and un-rooted, unless they experience the wild, the source of all value.

Anthropocentrism then is not simply a moral predicament as we suggested in the first part of this chapter, a hurdle to an environmental ethic which seeks to find a legitimate basis for the human-independent value of nature, but the bulwark of an inferior anthropogenically produced landscape. But why a rejection of anthropocentrism in ethics must lead to a denigration of the city as at best a second-class environment, is not clear at all. This is not to say that preservation of those statutorily designated wild areas is unimportant, nor are issues involving species preservation and biodiversity loss unimportant. It is only to say that for anyone sharing our position they are put in the unfortunate position of having to justify any other concerns.

In addition to the conceptual nature-culture value dualism that Rolston assumes in his environmental ethic, we find another dualism here which is potentially more damning: a geographical dualism between wilderness and urban environments which represents a bifurcation of two realms of existence, one containing "nature," however Rolston conceives of nature, and one not containing nature by definition. While simply pointing out that a theoretical position contains a dualism does not in itself justify its rejection, here the upshot of this dualism is distressing: natural values do not exist in urban environments because urban environments do not contain nature. There is another consequence however that is even more disturbing: what many people traditionally would call "environmental problems," understood as problems which concern the natural world in particular, do not exist in urban environments. As such, the domain of environmental ethics should be restricted to understanding the value of wilderness, and its environs, rather than topics such as the particular sorts of environmental hazards that the environmental justice movement has pointed to in cities, and the potential role that densely populated urban environments could play in an overall scheme of environmental sustainability. Such topics are clearly on the agenda of most major environmental organizations and have been subjects of theoretical and practical investigation by environmental professionals for decades.

Nonetheless, if Rolston's work is to be a guide, one would be left to believe that there is little contribution that environmental ethicists can make to these urban discussions and movements. For our field, the underlying dogma continues to be the following: since the primary of value of nature is non-urban, the primary focus ought also to be non-urban.

A PLACE FOR URBAN NATURAL VALUE FOR ROLSTON

Still, there is hope. As we suggested in the second section of this chapter, it is not the case that Rolston shuts the door completely on urban environments. Rolston's value schematic, for example, does include recognition of the importance of advancing an environmental ethic that works to balance cultural and hence urban values with natural values.[26] The problem is that Rolston does not clearly or completely investigate those potential values. He seems to recognize this himself and has even suggested that much more sustained theoretical work has to be done if we are to move toward a more harmonious balance of urban and natural environments.[27] To this end he outlines three very general principles of sustainability, what he calls "principles of culture, that are meant to be guides for the development of "sustainable cultures that fit in with the continuing ecological processes."[28] But what is problematic about this formulation is similar to the scant attention he pays to urban environments. It is incomplete and too limited in depth and scope, at least in terms of its ability to contribute to a more comprehensive environmental ethic and to a balanced appreciation of the relationship that all environments do and can have to each other.[29] This is not meant to suggest that Rolston should fully enter into the debates over sustainability. There are already, without a doubt, plenty such approaches. What we want to suggest instead is that there are resources within Rolston's original account on which to build a more complete treatment of urban environments.

Return now to the very basic question: What can and what cannot fit into the various categories of natural value Rolston outlines? More specifically, what categories of value for Rolston could potentially match up with urban environments? Consider first the anthropocentric and instrumental varieties of natural value identified by Rolston. Rolston has no problem recognizing the various types of anthropocentric and instrumental values that emerge in urban environments because all that is required of this type of value is the instrumental usage of spontaneous nature. This likely means that Rolston would view walks in urban parks, canoe trips down an urban river, and urban gardening as activities laden with various types of instrumental and anthropocentric value. We wonder, though, whether this is where the story has to end. Do spontaneous natural properties exist in urban environments independent of human interference? If so, are these properties a potential source of spontaneous natural values? Even though we are skeptical about Rolston's overall theory of value, and find anthropocentric values sufficient to sustain both a robust environmental ethic and one particular to urban environments, we want to consider whether Rolston's value schematic has room for more types of urban natural value than has thus far have been acknowledged. For this to be the case, all one would have to do is show that there are times when the spontaneity of urban nature is not necessarily "mixed" with human intentionality in urban environments. So as to not overlook this very real possibility, we need to more carefully canvas the urban terrain than Rolston has done so far. The most obvious place to look for these spontaneous natural properties would be the actual urban landscape, or more appropriately the non-human natural features of urban landscapes including non-human animals, waterways, ridges, ravines, rocks, air, and the soil.

Begin, for the sake of argument, with the many non-human animals that live in urban environments. For a traditional environmental ethicist such as Rolston, perhaps the immediate reaction to urban non-humans is to think of them as valuable, but perhaps as less valuable, all else being equal, than their more "wild" cousins. This move is similar to arguments that have been made about the status and value of captive non-human animals wherein the captive state of the non-humans is taken to change something about their essential nature (their spontaneous properties) to such an extent that they no longer can be considered "wild" or "authentic" examples of their species.[30]

While this type of argument is a compelling way to understand the status and value of captive non-human animals, we think it inaccurately describes the potential status and value of non-human animals in urban environments. Diane Michelfelder offers one framework for understanding the status and value of non-human animals in urban environments[31] that overcomes some of these issues.

It is probably not unreasonable to suggest that the dominant way most people understand the urban environment is as an environment for human beings, indeed as the primary environment for more than half of the global human population. This is, to a certain extent, to be expected. After all, human beings do much to craft, shape, and color urban environments with signs of human intentionality. Given this brute fact, and given human hubris, another tendency might follow—to view urban non-human animals as fully subsumed within a humanly made world. Michelfelder argues that this picture unnecessarily omits the very real possibility that urban non-human animals are living with urban human animals in a more integrated community. The idea that the urban environment is only for human beings overlooks this integration and arbitrarily diminishes the non-human urban animals to the status of non-native invaders, visitors, nuisances, pests, exotics, and somehow always "out-of-place."[32] If we think of non-human animals as neighbors in community, *pace* Michelfelder, these designations only work in some cases, not all.

The cases where such a diminishment of the status of urban animals is inappropriate are those where a non-human has, through opportunistic and adaptive behavior, carved out a particular, successful, urban niche. Urban raptors that nest, reproduce, and rear young in urban settings generation after generation hardly seem to deserve titles such as non-native invaders, visitors, nuisances, pests, exotics, etc. They are, after all, doing what raptors do in the wild. It is just that in this case their "wild" is an urban wild shared by human begins. Thus, while they may be urban, and while they may take advantage of human structures and an environment generated out of human intentionality, they also seem to be able to pass a reasonable litmus test required for some X to be spontaneously wild. Or, to use Michelfelder's preferred terminology, this "urban wildlife" seems to be a strong candidate for an urban X that is spontaneously valuable.

Now consider the urban soil. It obviously reflects a certain amount of human intentionality. This intentionality has helped to degrade, compact, and simplify that soil to a certain depth. Presumably though, at some depth in urban environments, there remains urban soil not "mixed" with human intentions. At these levels, however deep they might be, even Rolston would have to admit that the spontaneousness of the soil is valuable, indeed it is spontaneously valuable in Rolston's sense of intrinsic value because it is untouched by human beings. On the urban soil issue, then, it seems only appropriate that some spontaneous value be admitted.

Now consider urban waterways. They present a similar challenge and tell a related story. For certainly, they too reflect some human intentionality. Anthropogenic pollution and overuse by humans shape many urban waterways. Though still valuable, they are primarily instrumentally valuable insofar as they contribute to the human project that is urbanism. But here again this claim only goes so far. How "urban" is the water? If we recognize that global hydrological circulation literally guarantees a sharing of water worldwide, we must also recognize that the same water that fills a pristine wilderness lake at 14,000 feet will eventually show up, in some form, in the pond in an urban park. Since Rolston would be inclined to say that the "wild" water at 14,000 feet is spontaneously valuable, he would also likely want to say that the water in the urban park is also, somehow, spontaneously valuable. To say the opposite would be to undercut the claim that would have wilderness water be spontaneously valuable in the first place. The reason is that presumably a denial of the intrinsic spontaneous value of urban water would be fueled by the conviction that the water, corrupted by human intentionality, no longer passes the litmus test of spontaneous natural value. If this is the explanation for why urban water does not have spontaneous value, it also, by virtue of the fact that urban water and wilderness water are two sides of the same coin, would also eventually become the explanation for why wilderness water does not have spontaneous value. Rather than deny the spontaneous and intrinsic value of a pristine untouched wilderness lake at 14,000 feet, one can simply say that the qualities that make the water intrinsically valuable at 14,000 feet also exist in urban waterways, even if on different points in the spectrum of spontaneity. While the case for urban water is perhaps not as convincing as the case for urban soil, it would seem that Rolston should be disinclined in both cases to deny these instances of urban spontaneous natural value.

As if these cases were not enough to at least have us entertain the idea of intrinsic and spontaneous value existing in urban environments, the geological formations present in urban environments should be enough to convince even the most skeptical observers. Given shape by millennia of glacial activity, these formations often stand as the defining characteristics of urban formations. They do this regardless of and independent of human intentionality. Clearly there is little, if anything, human beings can do to substantively impact these massive geological formations. When humans build, whether they want to or not, they have to build in cooperation with such formations. It is no surprise that the highest buildings in Manhattan are in midtown and south in the financial district. It is there where the bedrock comes closest to the surface supporting the construction of taller structures. It is no surprise that the pressure of Lake Pontchartrain on New Orleans was too much for the levies to hold, and that the constant dismissal of this fact was a tragedy waiting to happen. Not only are these formations spontaneous, they are, on Rolston's scheme, spontaneously valuable. They are, in fact, a testament to how powerful of an idea spontaneity can be when attached to value since here even the presence of human intentionality does little to unsettle this value.

With these examples of spontaneous urban natural value on the table, one might wonder why this recognition is important. It is important because of the primary role spontaneous natural value plays in Rolston's view of the natural world, a view that has guided thinking in environmental ethics for three decades. Without a doubt, spontaneous natural value and wilderness areas represent two of the basic building blocks of Rolston's account of intrinsic value. What we have demonstrated here suggests that urban environments also become a basic building block of intrinsic value in his system because they too contain spontaneous natural value.

With this possibility on the table, we can rethink the value of urban environments even within one of the most conservative assessments of natural intrinsic value. As a result of this consideration other varieties of intrinsic value might need to be considered in urban environments as well. For example, when human beings mix their activities with the spontaneous properties of urban environments, do we not witness what Rolston calls systemic value? To deny that humans can initiate this mixing denies that humans are natural beings, something we are not willing to do. To admit to this type of systemic value is not to overlook the fact that this mixing often leads to negative ecological consequences. Without a doubt, this mixing historically has led to many instances of urban environmental degradation. There also, however, have been instances where this mixing has led to positive ecological consequences or at least more or less benign consequences (e.g., arboreta and urban gardens).[33] Both positive and negative consequences suggest that this systemic value of human beings in urban environments is a topic for future theoretical and practical work in environmental ethics. Among other things, such work should focus on understanding how humans should combine their intentionality with the spontaneous properties of urban environments.

What does this all mean for those interested in understanding the importance of Rolston's work to the field of environmental ethics in relation to his published comments on urban environments? First, it means that Rolston's prior insistence on a necessary disvalue for urban environments, and their relationship to human moral growth and development, is misguided. From the perspective of his view, to deny the spontaneity of basic properties underlying urban environments is to adopt a view in which urban water always stays in urban areas, urban soil is impacted by human intentionality to the earth's core, and that humans somehow substantively change the basic geophysical structure of urban geological formations. Second, it means that if the category of natural Xs to which intrinsic value is presumed to apply for Rolston is expanded to include urban environments, the tendency of environmental ethicists to focus their attention almost exclusively on wilderness will necessarily have to be questioned. Third, the fact that urban environments can be generators of systemic value because of, not in spite of, human beings means that we may have to reconsider the usefulness of binary nature-culture divisions. All three of these conclusions may represent important steps for environmental ethicists to take as a more serious consideration of urban environments proceeds. This consideration and its associated activity in the field of environmental ethics might enable the field in general more effectively to join the conversation already well underway in other disciplines about which paths ought to be investigated toward forming more sustainable human organization and settlement patterns.

With all of that said, what we would want in the end is not only a theory of value that can give reasons to hold dear and protect the green spaces of urban environments, but one that could say something substantial about the brown spaces, about the built environment, as well. Finding such a view will most likely go well beyond Rolston's value framework. Nonetheless, we hope that we have shown that it could be a starting point for this discussion. We also hope that we have shown that his value schematic—one which is very far in foundations, substance, and scope from our own—does not lead to a necessary denigration of the urban environment. Even so, it is likely that a more robust theory of urban and natural environments may require that we abandon nonanthropocentrism and searches for intrinsic value in nature for practical as well as philosophical reasons. But we will leave aside such considerations, at least, for now.

NOTES

[1] See Andrew Light, "The Urban Blind Spot in Environmental Ethics," *Environmental Politics*, Vol. 10, No. 1, 2001, pp. 7–35, from which part of this essay is adapted, "Urban Ecological Citizenship," *Journal of Social Philosophy*, Vol. 34, No. 1, 2003, pp. 44–63, and, with Aurora Wallace, "Not Out of the Woods: Preserving the Human in Environmental Architecture," *Environmental Values*, Vol. 14, No. 1, 2005, pp. 3–20. Also see James W. Sheppard, *The Paradox of Urban Environmentalism*, forthcoming 2006.

[2] Alastair Gunn, "Rethinking Communities: Environmental Ethics in an Urbanized World," *Environmental Ethics*, Vol. 20, 1998, p. 355. The few other papers on urban issues in environmental philosophy include our own already mentioned works, as well as Dale Jamieson, "The City Around Us," Tom Regan, ed., *Earthbound* (Philadelphia, PA: Temple University Press, 1984); Avner de-Shalit, "Urban Preservation and the Judgment of King Solomon," *Journal of Applied Philosophy* 2, No. 1 (1994): 3–13; Bill E. Lawson, "Living for the City: Urban United States and Environmental Justice," Laura Westra and Peter Wenz, eds., *Faces of Environmental Racism: Confronting Issues of Global Justice* (Lanham, MD: Rowman and Littlefield, 1995); Roger J. H. King, "Environmental Ethics and the Built Environment," *Environmental Ethics* 22, No. 2 (2000): 115–131; Sheppard, "The Nectar is in the Journey: Pragmatism, Progress, and the Promise of Incrementalism," *Philosophy and Geography* 6, No. 2 (2003): 167–187; and Robert Kirkman, "Reasons to Dwell on (if Not Necessarily in) the Suburbs," *Environmental Ethics* 26, No. 1 (2004): 77–95.

[3] It is not necessary here to offer a full list of these anti-urban accounts. It is enough simply to point out that many self-styled radical environmentalists, ranging from Dave Foreman and Edward Abbey to Murray Bookchin, have been some of the most vitriolic voices against cultural landscapes and existent urban formations and that many other more mainstream environmental ethicists have nudged up to and even crossed the border of outright misanthropy. For some examples see Andrew Light, "Contemporary Environmental Ethics: From Metaethics to Public Philosophy," *Metaphilosophy*, Vol. 33, No. 4, 2002, pp. 426–449.

[4] While reasonable people can disagree about the exact start of environmental philosophy as a recognizable philosophical endeavor, we take as a watershed year 1973 when three critical papers in environmental philosophy were first published: Richard Sylvan (then Routley), "Is There a Need for a New, an Environmental Ethic?" *Proceedings of the World Congress of Philosophy* 1973, pp. 205–210; Arne Naess, "The Shallow and the Deep: Long-Range Ecology Movements," *Inquiry* Vol. 16, 1973, pp. 95–100; and Peter Singer, "Animal Liberation," *New York Review of Books*, April 5, 1973, though the later has become more influential in the literature on animal welfare as opposed to environmental ethics proper.

[5] See Gary Varner, *In Nature's Interests?* (New York: Oxford University Press, 1998).

[6] Sylvan, "Is There a Need for a New, an Environmental Ethic?" p. 210.

[7] Paul Taylor, "Are Humans Superior to Animals and Plants?" *Environmental Ethics* Vol. 6, 1984, p. 151, n. 5, emphasis added. While quite interesting, it is striking that views like Taylor's and Sylvan's are false on broadly speaking both realist and anti-realist metaphysical grounds about moral claims. On realist grounds, one would expect that if there are conceptions of the value of nature that can stand outside of culture and determine obligations and duties regardless of cultural predilections, then there is no *a priori* reason why there could not be a foundation for human obligations to nature which did not depend on the attribution of nonanthropocentric value. And on anti-realist grounds, if there is no foundation for human conceptions of the value of other humans then there cannot be a *sui generis* sense of the value of nature that somehow transcends culture. Perhaps the way out of this dilemma is to say that for Sylvan, Taylor and others, a nonanthropocentric conception of nature is by definition outside of culture since it is a view independent of human prescriptions of value, culturally based or otherwise. More charitably then, what nonanthropocentrists want is a description of natural value that is culturally neutral, rather than resistant to cultural relativism.

[8] To be fair, this schematic represents our attempt to organize Rolston's many writings on the subject as well as his many ways of organizing natural values into one framework. We are sure that many such schematics could be offered, not all of which would cohere with our interpretation.

[9] For more on the connection between spontaneous properties and intrinsic natural value, see Rolston, "Values in Nature and the Nature of Value," Robin Attfield and Andrew Belsey, eds., *Philosophy and the Natural Environment* (Cambridge: Cambridge University Press, 1994), pp. 13–30.

[10] Holmes Rolston III, *Conserving Natural Value* (New York: Columbia University Press, 1994), p. 71.

[11] Ibid., pp. 173–174.

[12] Rolston, *Environmental Ethics: Duties to and Values in the Natural World* (Philadelphia: Temple University Press, 1988), p. 198.

[13] According to Rolston, these secondary instrumental values include the many ways nature is valuable to human beings such as its life-support value, economic value, recreational value, scientific value, aesthetic value, genetic-diversity value, historical value, cultural symbolization value, character-building value, diversity-unity values, stability and spontaneity values, dialectical value, life value, and religious value. For this list as well as Rolston's complete explanations, see Rolston, *Environmental Ethics*, pp. 3–27.

[14] *Homo sapiens* most likely emerged about 90,000 years ago. But it was 35,000–40,000 years ago that *Homo sapiens sapiens*, or modern human beings, emerged and began to spread across the planet. It is *Homo sapiens sapiens* presumably that have the capacity to make ascriptions of value.

[15] Earlier in his career Rolston was apt to consider species and ecosystems to be of primary significance. However, the later Rolston has entertained much more robust sets of natural value foci, with the extremes of this possible consideration going to the genome on one end and Earth on the other. Rolston entertains these questions in "Values in Nature and the Value of Nature," op. cit.

[16] Rolston, *Philosophy Gone Wild* (Buffalo, NY: Prometheus Books, 1989), pp. 134–135.

[17] Rolston, "The Wilderness Idea Reaffirmed," in *The Great New Wilderness Debate*, eds. J. Baird Callicott and Michael Nelson (Athens, GA: The University of Georgia Press, 1998), p. 371.

[18] Rolston, *Conserving Natural Value*, pp. 12–13.

[19] Ibid., p. 15.

[20] Ibid.

[21] Ibid., p. 14.

[22] Rolston, *Philosophy Gone Wild*, p. 230.

[23] We do not mean by these examples to suggest that only elements of high culture mark the value of the urban. If there is value to be found on a mountain path in the most pristine wilderness area then we would claim a certain kind of value on the most worn, disheveled, street in the most highly populated metropolis rich in the footsteps and textures of the thousands who have walked there before us.

[24] Ibid., p. 224.

[25] Rolston, *Environmental Ethics*, p. 36.

[26] Rolston, *Conserving Natural Value*.

[27] Rolston writes of this goal in *Environmental Ethics: Duties to and Values in the Natural World*, 330–335.

[28] The first principle of culture is that culture ought to rebuild wild nature, the second is that culture ought to be sustainable in relation to the ecological processes that support it, and the third is that culture ought to respect nature. Rolston, *Conserving Natural Value*, p. 87.

[29] For example, Rolston's specific discussion of sustainability is limited to six pages. Ibid., pp. 83–88. To be fair though, an argument could be made that the theme of sustainability does run throughout *Conserving Natural Value*. If this is the case however, it does so only implicitly rather than being the explicit focus of his work, as is certainly true now of the work of Bryan Norton. See Norton, *Sustainability: A Philosophy of Adaptive Ecosystem Management* (Chicago: University of Chicago Press, 2005).

[30] This argument is made most forcefully by Dale Jamieson in "Against Zoos" in Peter Singer, Ed., *In Defense of Animals* (New York: Blackwell, 1985). The example Jamieson uses is that of captive Mountain Gorillas. He argues that such gorillas are really what he calls lower-case gorillas in the sense that human intentions have corrupted their essential "gorilla-ness". As a result of their captivity, they are hence not upper-case gorillas.

[31] Diane P. Michelfelder, "Valuing Wildlife Populations in Urban Environments" *Journal of Social Philosophy* Volume XXXIV, Number 1 (Spring 2003): 79–90.

[32] Ibid., p. 85.

[33] See Andrew Light, "Elegy for a Garden: Thoughts on an Urban Environmental Ethic," *Philosophical Writings* Volume 14 (2000): 41–47.

HOLMES ROLSTON, III

14. LIVING ON EARTH: DIALOGUE AND DIALECTIC WITH MY CRITICS

I sometimes start with the pronoun "I," and, in my writing conversations, switch to the pronoun "we," mixing myself up with my readers, interacting with them, and frequently trying to persuade as well. Brenda Hausauer puzzles over this, and figures me out (Hausauer, chapter 9, pp. 156–158). When the readers are these sixteen thoughtful critics focused on my work, as in this volume, there is a quite a challenge to see how far "I" can get toward a "we"—whether by dialogue, complementarity, dialectic, rebuttal, repentance, regeneration, reformation, or synthesis.

I start with "I" to get readers to see where I am coming from, and then switch pronouns to confront readers with whether they wish to travel with me—maybe to get them moving along a path before they know whether they want to travel there. My justification for this seduction is that searchers have got to explore in order to know. I am doing that myself. "Rolston has pointed out that although the questions in his texts are often rhetorical, they once were autobiographical questions that were important to him in the development of his ideas" (Hausauer, chapter 9, p. 149). We need the context of discovery before we travel the context of justification (Hausauer, chapter 9, p. 153). This is a "Socratic methodology" (Hausauer, chapter 9, p. 164). "I" consider my interactions here as Socratic dialogue. Let's see what "we" "can learn from each other" (Davion, chapter 10, p. 167, cf. Lemons, chapter 12, pp. 203–205).

"I'm a lover of learning, and trees and country places won't teach me anything, whereas people in the city do" (*Phaedrus*, 230d). Socrates uses the personal pronoun too, but I have discovered that I don't want to follow. I know better than Socrates, I sometimes like to brag. And then I look to see whether I can get readers to come along. True, "the unexamined life is not worth living"; and people in the city can help. But "life in an unexamined world is not worth living either. We miss too much of value" (Rolston, 1997, p. 63). My critics here are variously willing and unwilling to follow; they may explore new paths. There is excitement in the woods; we will need emotion along with reason—as Mark Wynn so perceptively realizes.

A philosopher needs a *worldview*. An environmentalist needs to be a holist; an ethicist needs to be inclusive. Maybe this adventure is "master narrative" (Preston, chapter 3, p. 33), at least it is big picture. The "storied achievement" that I celebrate (Carlson, chapter 7, p. 106) is, if we must put it this way, a quite grand narrative, from big bang to people. Any philosopher who denies that these are remarkable events is simply insensitive to the story. Contra Socrates, none of us can be wise (*Homo sapiens*) without figuring out what nature has to teach us. Notice that "I" have switched to "we," because here "we" have consensus—well, maybe not yet with Christopher Preston—on how "real" this story is.

C.J. Preston and W. Ouderkirk (eds.), Nature, Value, Duty, 237–268.

WILD NATURE AND HUMAN CULTURE: DISCRIMINATING THE DIFFERENCES

Venturing into the woods, we are taking ourselves as we go, so we need some attention to *who* we are as well as *where* we are. Humans have two environments: *nature* and *culture*. Socrates is quite right that people in Athens have much to teach that trees can't. *Cogito, ergo sum solus* (*PGW*, p. 226, cited Hausauer, chapter 9, p. 148).[1] After a week alone in the woods at Lake Solitude, too much thinking was setting me apart from the scene. But this lonely "I" was a trope for "we" humans.

Distinctions versus dualism

I do have, if you like, a "dualism" of nature and culture, if that means, as Victoria Davion suggests "a kind of 'super dichotomy' in which difference is transformed into radical difference" (Davion, chapter 10, p. 167). In Keekok Lee's vocabulary, humans are "articulate"; animals are "mute." J. Craig Venter and over 200 co-authors, reporting on the Celera Genomics version of the human genome project, conclude that the genetic modifications in the human lineage producing cortical expansion and giving rise to language "culminated in a massive singularity that by even the simplest of criteria made humans more complex in a behavioral sense" (Venter et al., 2001, pp. 1347–1348). I take it that "massive singularity" is their term for the "transforming radical difference" disliked by non-dualists.

Our "mental incandescence" (*PGW*, p. 226; Hausauer, chapter 9, p. 148) enables us to rise to transcending freedom and overview. We humans are the most sophisticated of known natural products. The connecting fibers in a human brain, extended, would wrap around the Earth forty times. In our three-pound brain is more operational organization than in the whole of the Andromeda galaxy. Some trans-genetic threshold seems to have been crossed. The human brain is of such complexity that descriptive numbers are astronomical and difficult to fathom. A typical estimate is 10^{12} neurons, each with several thousand synapses. This network makes possible virtually endless mental activity. The result of such combinatorial explosion is that the human brain is capable of forming more possible thoughts than there are atoms in the universe (Rolston 2005a).

Davion, following Val Plumwood worries about my "fault-line dualism" (Davion, chapter 10, p. 167). Plumwood complains that I "cannot recognize the continuum of nature and culture," since my dualism "blocks recognition of the embeddedness of culture in nature." "This nature/culture dualism distorts the way we can think about land, obliging us to view it as either pure nature or as a cultural product, not nature at all . . . . Recovering the lost ground of continuity that dualistic conception has hidden from us allows . . . recognizing nature in what has been seen as pure culture and culture in what has been seen as pure nature." Andrew Light and James Sheppard want to see nature in their cities. In the wild, says Plumwood, we ought to recognize "the *presence* of the Other," reaching "a non-oppositional account of the relationship between humans and the wild Other." She wants to eliminate "the dualism," but not "the distinction." I have wondered if my mind is subtle enough to catch these distinctions (Rolston, 1999b, with Plumwood citations, p. 154). If we can get across this linguistic fault line, I think "we" (Davion, Plumwood, and I) are together in spirit. If we can't, my trouble must be "bewitchment by ossified language" (Preston, chapter 3, p. 33, quoting Bryan Norton).

Davion does praise my resistance to genetic reductionism (Davion, chapter 10, p. 175ff). But I do so by insisting on distinctive cultural determinants, and just this resistance separates humans from nature in (dualist? distinctive?) ways that may worry ecofeminists. Against the naturalists, I am reluctant to fit the richness of human lives in their cumulative transmissible cultures especially their capacities for doing science, for being ethical, for being religious– "into that [naturalistic] explanatory box" (*GGG*, p. 252, Davion, chapter 10, p. 177). That problem will return below with Light and Sheppard wanting to think of cities as natural.

Mixing nature and culture in national parks

John Lemons spirals around another side of this nature-culture issue. His main claim is that "let nature take its course" in any wilderness Other is an ideal that has become increasingly difficult to make operational in U.S. national parks, upset by internal and external anthropogenic causes. "No national park is large or isolated enough to avoid the impacts of ecological change and political decisions beyond its borders" (Lemons, chapter 12, p. 211). With that I do not disagree. At the same time, he and I and the National Park Service agree that, so far as is possible, within parks, spontaneous nature is an ideal.

Lemons is right that I have not addressed the way culture spills over into the parks to spoil nature, with threats often beyond the control of park managers (global warming, exotic species invasions, acid rain). As Lemons recognizes, he "is probably asking too much" if he expects that philosophers with their ethics "can yield practical solutions to complicated environmental problems" (Lemons, chapter 12, p. 212). This is "unrealistic" (Lemons, chapter 12, p. 217).

Philosophers can, however, be expected to give orienting directions. Thanks to Preston for defending the usefulness of my work, against the pragmatists' charge that I meditate on intrinsic value while the forests burn (Preston, chapter 3, p. 30ff). For example, I tried to give Yellowstone park managers some help when another philosopher (Alston Chase) charged that they were "playing God in Yellowstone." Half a dozen park officials, including the superintendent, personally thanked me for the conceptual analysis why their "preservationist mission" was not playing God (Rolston 1990, cf. Lemons, chapter 12, pp. 214–215).

I also gave the plenary Aubrey L. Haines Distinguished Lecture at the Fifth Biennial Scientific Conference at Yellowstone in October 1999, addressing at organizers' request how "Yellowstone National Park's mission and policy can be clarified by analysis of the natural and unnatural," with particular reference to invasive species. My argument, in sum:

> "Wild nature" successfully denotes, outside culture, an evolutionary and ecological natural history, which remains present on the Yellowstone landscape, jeopardized by numerous human influences, including the

> invasions of exotic species. . . . Natural processes can be preserved today, because of, rather than in spite of, park management. Over much of the North American landscape nature is managed and at an end. Yellowstone provides an opportunity to encounter and to conserve "untrammeled" nature as an end in itself, past, present, and future (Rolston, 2001, p. 267).

I was trying to help managers keep in focus the difference between nature and culture, and how a highly visited (and sometimes compromised) park can both conserve wild nature and educate Americans about who we are and where we are, and what we ought to do. (There's that inclusive "we" again. Do you want to follow?)

Down on the farm, off on the hunt

Clare Palmer thinks I am not careful enough delineating what humans in their human nature/culture may share with the animals in their nature. Both Davion and Palmer wonder whether some of our human activities might be hybrids of nature and culture: hunting, eating, mating, marrying, farming, breeding, keeping pets—maybe, with Lemons, visiting national parks. Palmer's analytic precision is welcome. She has thought more than I about the "complicated location of domesticed animals" (Palmer, chapter 11, p. 183), and brings me up short with her eight different contexts in which animals may be related to humans (Palmer, chapter 11, p. 196).

Davion and Palmer both highlight the twilight zones, and I agree this is a troublesome boundary. Here they both may be right that I have not given enough attention to developing a spectrum on which horses and cows, dogs and cats can be placed. I make them out to be living artifacts, compromised from their once-wild integrity and awkward hybrids of nature and culture. Every time I see a wolf in Yellowstone, I get goose pimples—yes, even one that I know is descended from reintroduced wolves. When I see a French poodle with painted toenails on the streets of Manhattan, I do not rejoice for the nature in the city. I pity it—preconceptual "yuk" factor, Mark Wynn would say. A bird dog on the hunt is much closer to the wolf ancestor, with some residual integrity.

As for domestic animals, I just try to do them no additional harm. Nor do I deny that factory farming is inhumane. I think my rural background has biased my view—or maybe it has given me clearer insight. As a child, it was my duty to kill the Sunday chicken. Chickens scratching around the barnyard dodging the mules still seem to me not too unlike those I much later saw wild in Nepal looking for insects at the feet of elephants. Pigs had the run of sixty acres. There were few horses, loved by their owners, and many mules which seemed degraded, sterile and stupid, bred for muscle bulk to pull and plow.

The few milk cows and many steers also seemed degraded. "A cow is a meat factory, pure and simple" (*EE*, p. 83, Davion, chapter 10, p. 168) is not a view that I meant to recommend. It is rather a description of the way modern agriculture regards cows, and my point was that it is better to be a gazelle, under constant flight for its life, than a contented cow, overfed and fat, oblivious to its upcoming death. Yes, I have "a persistent sense that the domesticated and rural, tainted by human contact, are inferior contexts to the wild, as are the animals within them" (Palmer, chapter 11, p. 193).

Degrading them so, one should be as "humane" as one can; perhaps that is a better rule than my homologous principle. Robin Attfield notices that animals often live well in the wild, even flourish, up to the event of their sudden death by a predator (cited Sideris, chapter 6, p. 92). Perhaps that can apply here with the farmer and his contented cows. But if we try to keep all eight of Palmer's human-animal contexts in focus the dialogue can become scholastic and casuistic. She concludes: "Domestication can be seen as a process such that, in creating relationships that close down domestic animals' abilities to live independent lives, creates special human responsibilities to provide for them" (Palmer, chapter 11, p. 199). Amen.

Davion worries about killing and eating these bred animals, Palmer too is a vegetarian. A main problem with that concern about causing domesticated food animals pain is that all these food animals would cease to exist if they did not live to be eaten, or produce milk and eggs that we do eat. Only the pets would remain, dogs and horses. Granddaddy would have needed the mules, draft animals; but today only the wilderness outfitters would be keeping a few. The cats too would be gone, if the farm has to be vegetarian.

Having distinguished culture from nature, human genius from animal integrity, I turn around and place human eating more in the realm of nature than of culture. "Wild animals prey on other animals in ecosystems, so humans prey on animals in agriculture" (Palmer, chapter 11, p. 192). That puts it with a raw bite, but it does get my central idea. Humans evolved as omnivores; they have captured (domesticated) animals that they breed and eat. Davion finds this "fraught with difficulties" (Davion, chapter 10, p. 180). Davion is but the latest critic to remind me that eating food has many cultural dimensions—what sorts of food we prefer, how it is cooked and served, what is polite at the table. Maybe Light and Sheppard should invite me more often to "dine" (= eat) at uptown restaurants in Manhattan. Nevertheless, our capacities to eat have a millennia-long evolutionary history. Biologically, eating is inescapable—short of intravenous feeding. Our cultural mores about eating are recent and transient.

I readily concede that humans can, within these cultures, choose to become vegetarian, perhaps out of concern for animal suffering. After an extended trip to India I could not deny this. But the question remains: Does a sensitive moral awareness today require us to cease eating animals? "Environmental ethics is . . . the most altruistic, global, generous, comprehensive ethic of all, demanding the most expansive capacity to see others" (*GGG*, p. 288). Why not then cease eating animals? I confess that this question stings.

I once hunted. I no longer hunt. I do reject "nothing but" sport hunting, as with the bear hunts. I reject trophy hunts. But I know too many hunters to reject "sport hunting" where it remains in the context of eating what one kills, participating in food chains. I give it a "best case" as "not *sport*; it is a *sacrament* of the fundamental, mandatory, seeking and taking possession of value that characterizes an ecosystem and from which no culture has ever escaped" (*EE*, p. 91; Davion, chapter 10, p. 170). Life feeds on life; hunters do know how the world is made in ways in which vegetarians who buy all their food at Safeway do not. In the wild, of course, females hunt quite as much as males. One of the most powerful experiences I recall is watching three lionesses stalk and kill a zebra on the floor of Ngorongoro Crater in Tanzania.

Here I have to join Lisa Sideris with her worries about the peaceable kingdom that the ecofeminist theologians desire. There is something of the Bambi syndrome here. Marti Kheel, *Reweaving the World*, writes: "Our sense of oneness with nature must be connected with loving actions" (cited Davion, chapter 10, p. 172). Davion dislikes hunters who get "involved in the bloodletting" to find more "identification" with natural processes (EE, p. 92; Davion, chapter 10, p. 172). But the Darwinian world is woven as a bloodstained world. "The unease with which the good hunter inflicts death is . . . an unease about the dialectic of death with life. The authentic hunter knows suffering as a sacrament of the way the world is made" (EE, pp. 92–93).

Ecofeminists and ecotheologians need to face the fact that without an evolutionary history of predators and prey, without human hunting (and gathering too), they would not have the hands, arms, legs, ears, and eyes that they do. We do now live in culture, not nature; perhaps we can escape these origins (I have already said that I no longer hunt, I do live in a city). But we do have these roots, and animals remain in the orders of nature out of which we have come. We need a different ethic for relating to ourselves than the ethic for relating to these wild others. Maybe that makes me guilty of discrimination; maybe I am just discriminating. Maybe I will soon cease eating meat. But if I did I might be losing some of my expansive capacity to see all those others, such as the wolves and the cats. If I did, I might have a less global, less comprehensive view of how the cruciform world is made.

Failure to be discriminating about the differences between nature and culture leads us astray. The "peaceable kingdom" model can be quite attractive for the ecofeminists, as their hoped for consummation of both nature and culture. But joining nature and culture indiscriminately transfers into nature what may be appropriate only in culture, as Sideris documents. Sallie McFague claims that physical health is a "basic right" in natural communities, just as it is in human communities (cited Sideris, chapter 6, p. 82). If so, eating and being eaten would vanish, trophic pyramids would collapse, and life vanish. McFague envisions a program of "healing the wounds of nature and feeding its starving creatures" (cited Sideris, chapter 6, p. 83). One might feed the herbivores by sacrificing only plants, but the predators need meat, especially the felines. McFague doesn't seem to have read the Psalms, where the lions roar and receive their prey from God (Psalm 104; Sideris, chapter 6, p. 90).

Nature in town: Paradox and complement in the political animal

Andrew Light and James Sheppard want an urban environmental ethics. Their concern is both theoretical and practical, and I welcome both. I can perhaps excuse my omission by recalling that I was reared in the Shenandoah Valley, first a country boy. I spent summers on granddaddy's farm in Alabama. But when I did get to town and read the philosophers, I have already said that I was quarrelling with Socrates. Now my problem, rather, is Artistotle, and this time because I think he is right. Man is by nature a "political animal" (Aristotle, *Politics*, I, 2.1253; Sheppard-Light, chapter 13, p. 227, citing *CNV*, also in *PGW*, p. 41). People are animals who build themselves a *polis*, a city, and love to socialize within it.

"Man is the animal for whom it is natural to be artificial" (Garvin, 1953, p. 378, *PGW*, p. 41). *Homo sapiens* is "the natural alien" (Evernden, 1993). The really natural thing for humans to do (our genetic disposition) is to build a culture differentiating (alienating) ourselves from nature. There is paradox here. An urban environment, so far as it is urban, is not natural; it is artifactual. With the technologies of our cultures, we adapt nature to ourselves; rather than, as in the wild animals, having bodies naturally selected to fit niches. We do not have hooves, fur, horns, but shoes on our feet, clothes, living in houses with furnaces, air-conditioning, and police protecting us from thieves.

Most of the inhabited world is a Hegelian landscape: nature is the thesis; culture is the antithesis; and what results is a synthesis of nature and culture. By contrast, I advocate an elliptical landscape (*EE*, pp. 329–335). Recall from geometry how an ellipse is generated under the constraints of two foci—not under the radius of one focal center, as with a circle. The glory of the Rocky Mountain landscape is its twin foci, nature and culture, that permits binocular vision. We see with more depth than those on the anthropocentric landscapes. That is why Denver and Salt Lake City are better placed than New York; the skyscrapers on the Western landscape are imposing if you look around town, but when you turn to the mountains also on the horizon, they get shrunk. We have "purple mountain majesties above fruited plains." That's what makes "America, the Beautiful." Environmental ethics needs to place New Yorkers on their coastal landscape and to give them wider horizons than Manhattan. They need to know where on Earth Manhattan is. Failing this sense of place, we are too much uptown and not enough down to Earth.

Light and Sheppard fear that my environmental ethic "entails a rejection of anthropocentric forms of value" and "anthropogenically created landscapes" (Sheppard-Light, chapter 13, p. 222). They worry whether "the direct and indirect implications of Rolston's theory of natural value unintentionally or intentionally condemn urban environments" (Sheppard-Light, chapter 13, p. 223). But I do not reject human-created value, as Light and Shepard recognize. I think I celebrate it:

> Over great stretches of Earth, wild nature has been already or likely will be diminished in favor of civilization. In some sense, that ought to be so. This ending may be always, in its own way, a sad thing; but it is an inevitable thing, and the culture that replaces nature can have compensating values. It would be sadder still, if culture had never appeared to grace the Earth, or if cultures had remained so modest that they had never substantially modified the landscape. We do not always lament our presence, even though we do want some lands where humans only visit. Humans too belong on the planet; and the epoch of evolutionary nature, and even of ecological nature is over. That is what is right about the view that with the arrival of humans pristine nature vanishes. Nature does not vanish equally and everywhere,

> but there has been loosed on the planet such a power that wild nature will never again be the dominant determinant of what takes place on the inhabited landscapes (Rolston, 1999c, p 151).

So I do think, in a way, that humans are the masters in this master narrative. They have dominion on Earth.

I have claimed from the start, as Sheppard and Light realize, that human well-being requires three environments, the urban, the rural, and the wild (Sheppard and Light, chapter 13, p. 228; *EE*, chapter 1, p. 40; *PGW*, pp. 40–46). I require Aristotle's *polis*, Socrates' Athens, to be human. So where is the "anti-urban bias" in my early work by which I "exclude" value in the "urban environment" (Sheppard-Light, chapter 13, p. 226)? Is this a "denigration of the city as at best a second-class environment" (Sheppard-Light, chapter 13, p. 229)? Where is my "misguided" "insistence on a necessary disvalue for urban environments" (Sheppard and Light, chapter 13, p. 233)? I do think that the all-city-nothing-but-the-city person is a one-dimensional person; the full human life needs experiences of the urban, rural, and wild—and I stand by that claim.

Sheppard and Light complain that Rolston holds that "natural values do not exist in urban environments because urban environments do not contain nature" (Sheppard and Light, chapter 13, p. 229). That is right. So far as an environment is urban, built, artifacted, it does not contain spontaneous natural value; what was once wild there has been re-sourced, made over into an urban, built landscape. Of course nature remains in, with, and under such re-building; there is soil beneath, wind and sky overhead, pariah species such as pigeons and sparrows in the streets, and weeds in the pavement cracks. There are some trees and roses in the city park, gardened nature, if you like, domesticated nature, as are the dogs and cats. Don't pretend that the urban environment is rural, or wild. Don't pretend that it is natural. Celebrate it, keep it human, humane. And don't suppose that I think it worthless. I just don't think it is natural.

"Since the primary value of nature is non-urban, the primary focus [of environmental ethics] ought also to be non-urban" (Shepard-Light, chapter 13, p. 230). That is a dogma I need to abandon. Well, I am not abandoning it. I would indeed like to re-focus urban peoples so that they see their cities in a more-than-urban context. I want such cities to be built in harmony with their surrounding landscapes. I want urban people to be concerned about their sustainability. On any national or global agenda, that clearly is a major concern of environmental ethics. I'd like for the city people to get three-dimensional.

At this point I concede that, over the decades of my career, the concerns in environmental ethics have become more inclusive and comprehensive, going from Leopold's love of the sand counties in Wisconsin to fears of global warming. I started out saving the whooping cranes and found out that this required a more equitable distribution of wealth in global capitalism from the wealthy in those towering skyscraper offices to the poor in Africa or Asia being exploited by the rich. Yes, I have advocated saving endangered species, but equally I have advocated justice in environmental business and sustainable development (Rolston, 1995a, 2002a).

I'm not sure I know what an "urban ecosystem" is (Sheppard and Light, chapter 13, p. 226). I have regularly said that culture remains tethered to nature. If that is what is meant, I agree entirely. Manhattan needs a water supply, air to breathe, food to eat. But I do not think that Manhattan Island as an urban metropolis is an ecosystem. I am unable to apply any of the leading concepts in ecology (succession, niches, tropic pyramids, energy flow, nutrient cycling, keystone species, natural selection) to the way people go about their business on the island. The average bite of food eaten there has traveled some 1200 miles; the price of real estate (or meals in those uptown restaurants) does not reflect any ecological principles; nor does the behavior of sales persons in stores. The "urban environment," like the "business environment" or the "political environment," is not an "urban ecosystem." Meanwhile, New Yorkers need to know they are earthlings.

If Sheppard and Light have an improved environmental ethics that can get New Yorkers into their natural world, go for it. Maybe the place to start is with nature lingering in the gardens and vacant lots, with peregrine falcons on skyscraper ledges, or boating up and down the Hudson River (the waterways of Sheppard and Light, chapter 13, p. 232). Maybe the place to begin is with the soil, as Sheppard and Light suggest (chapter 13, p. 232). Pause by the excavation for the new bank building and cheer when the bulldozers hit bedrock, the geological basis of Manhattan (Sheppard and Light chapter 13, p. 233)! Urbanites in coastal cities experience wild nature with their storms and at the sea; but of course neither storm nor sea is urban, any more than bedrock.

Sheppard and Light need to take Chicago Wilderness as their model, a consortium of more than 170 public and private organizations to protect, restore, study, and manage the natural ecosystems of the Chicago region. This has surprisingly revealed how many fragments of nature survive in that metropolitan area, totaling over 220,000 acres and how much nine-million regional Chicagoans can be brought to care about their migrant birds, their wetlands, about Lake Michigan, sand dunes, and prairies. If you want to find out what frogs are calling in what ponds, or what wildflowers are in bloom in one of the prairie parks, you check their website (http://www.chicagowilderness.com).

One such fragment is Montrose Point, jutting out into Lake Michigan. An earlier generation desired there an elaborately fashioned "landscape park" designed to intensify aesthetic experience, "landscape art." Many Chicagoans now want there a more naturally functioning environment. Seasonal bird migrants discovered the left-over honeysuckle hedge, planted by the Army to cover an eyesore, turning it into the "Magic Hedge," an icon attracting Chicagoans, as well as birds, to an underlying nature lingering in, with, and under the dominant culture. Watching what happens on Montrose Point is not to lament that nature is gone, but to rejoice that nature is coming back. That the landfill Point did not even exist prior to the twentieth century does not mean that the nature reasserting itself there is not authentic. This is, if you like, "urban wildness." The message coming through is of the deeper mystery, the nature seemingly at the margins but actually in the depths of human life. The ground under our feet makes us wonder about the ground of our being. In, with, and under culture, there is always this once and future nature.

NATURAL FACT AND INTRINSIC VALUE: FROM *IS* TO *OUGHT*

Material fact: Geophysical nature

Geophysical and biological nature is out there surrounding us, fact of the matter. The main fact is: matter. Lee is insistent about this. Light and Sheppard found bedrock beneath Manhattan. On Earth there are humans, with myriads of other species, receiving energy from the sun, orbited also by other planets. Maybe we humans are a "massive singularity" (Venter) in our cumulative transmissible cultures, in which we have figured out (scientifically) how much is out there, but this leaves us with a radical question (philosophically) about how to value what we find. Start, Lee says, with the material stuff of Earth (and planets).

Lee is correct that I have paid "little explicit attention . . . to the domain of abiotic nature *per se*" (Lee, chapter 2, p. 18). I have not been challenged enough by "the most radical type of environmental philosophy" (Lee, chapter 2, p. 18). She pushes me further in the direction of the challenges I felt in icy Antarctica. There did seem to be something there to wonder at, even with no living things in sight. This was at least aesthetic (where I need to consult Eugene Hargrove and Allen Carlson). I did celebrate several dozen nations, in the *Madrid Protocol* [1991] to the *Antarctic Treaty*, seeking to protect "the intrinsic value of Antarctica," even though I remain unsure what this means (Rolston, 2002b, p. 115).

In my account of "projective nature," I start with "trajectories" (using a term also used by Lee, chapter 2, p. 19):

> A "mere thing" can, however, be something to be respected, the project of projective nature. Crystals, volcanoes, geysers, headlands, rivers, springs, moons, cirques, paternoster lakes, buttes, mesas, canyons—these also are among the natural kinds. They do not have organic integrity or individuality. . . . They do not have wills or interests, but rather headings, trajectories, traits, successions, beginnings, endings, cycles, which give them a tectonic integrity . . . [and] which ultimately must be valued and of value. . . . There is value wherever there is positive creativity (*EE*, pp. 198–199).

For a striking example, I am originally indebted to Hargrove, the angel hair—delicately spun gypsum crystals—in Mammoth Cave, in a section named Turner Avenue. To prevent damage by visitors, this part of the cave is closed, to preserve (in my terms) a "project" of "systemic nature." Angel hair counts morally in the sense that natural value here lays a claim on human behavior (*EE*, p. 199).

Vital facts: Biological nature

Nevertheless what is most "radical" on Earth is its evolution of life. If "radical" means going down to the "roots," yes we need to do that. But "roots" are alive; there are none in bare rocks on the moon. My real concern is "fruits" as much as "roots." Endings are as important as beginnings. "Life . . . is the principal mystery that comes out of nature, and our emotions are stirred proportionately" (*PGW*, p. 255, Wynn, chapter 4, p. 57). I am much more moved by biological than geological nature. We need to value the startling fact of the storied history of the matter on Earth, its uphill (negentropic) climb, until, to use Lee's terms, the "mute" becomes "articulate."

So now I am setting out to value cybernetic nature, producing storied natural history, until it reaches an apex in *Homo sapiens*, cognitively empowered with reason and emotion, who can and ought value what has been going on (as well as to build cities). I am tracking the climb from nature to spirit. That is going to be uphill from here on, but the view from the top, I promise, will be rewarding. Wynn is following, so are Sideris, Preston, Palmer, Davion, Hausauer, and Lee, silently, wondering when to get articulate. Since we are in Rocky Mountain National Park, Lemons is way up front, trying to see spillover damage from outside the park. Ned Hettinger is also up front, wondering what kind of "spirit" might be up top. Katie McShane has her doubts. Maybe I can convince her if we find a pasqueflower along the way. Light and Sheppard first said they would spend the day at Denver Botanic Gardens, but when I told them you could see Denver from the top, the two joined us enthusiastically; they want to see a metropolis, fruited plains, and mountain majesties all in the same vista.

I forgot, I almost lost Hargrove, who is wondering whether the beauty in the flower is objective or subjective (Hargrove, chapter 8, p. 125ff). He continues to follow when Lee speaks up and assures him that the flower enacts "mute" value which he can "articulate." The flower is valuable "in itself," and he can, if he wishes, also value the flower "for itself" (Lee, chapter 2, p. 22). But he then pauses to wonder what this means, and starts up again when I explain to him that the persistence of life, present in the pasqueflower, is fact of the matter, although the experience of this as aesthetically stimulating comes only with the sensitive beholder. Lee gets up off the rock on which she has been seated, decides to become articulate herself, praises its "independent value" and resumes the climb, leaving the rock to its own value all "by itself" (Lee, chapter 2, p. 24).

Valuing aesthetic nature

Hargrove seems, at least at first, to read me 180 degrees opposite from Carlson, and also to think that I can do 180 degree turns myself (Hargrove, chapter 8, p. 133). Carlson finds my aesthetics science-based and objective in nature, something like fact of the matter. Hargrove is disappointed with my subjectivity, more like value projected onto nature. That leaves me with the worry that I am the one confused and that they are tracking my confusion. Justifying myself a bit, however, I find each author tracking a different dimension of my environmental aesthetics.

As Hausauer says of my "musings-out-loud": "It's not always crystal-clear whether Rolston is saying what he believes, what he thinks most people believe, what he thinks most people ought to believe," when he is "presenting and refuting alternate points of view" (Hausauer, chapter 9, pp. 158–159). Hargrove mistakes my exploration of whether "Let the flowers live!" might be interpreted instrumentally as if it were my own view (Hargrove, chapter 8, p. 133). That passage rather concludes that "the value . . . we will claim, is objectively intrinsic to the nonsentient life and not merely instrumental" (*EE*, p. 112, Hargrove, chapter 8, p. 133). Later Hargrove realizes that "seemingly conflicting statements may actually be statements explicating a view Rolston does not actually hold together with statements criticizing that view" (Hargrove, chapter 8, p. 140). Sometimes, I suppose, this does look like a "magic show, brought off with smoke and mirrors" (Preston, chapter 3, p. 33, quoting J. Baird Callicott).

Nevertheless Hargrove follows me well along one main track. The aesthetic experiences in nature with which most people start out, colorful and picturesque scenic beauty, however worthwhile and treasured, are epiphenomenal and human-generated. Such aesthetic experiences in nature are "rainbow-like." The fall leaf colors are an epiphenomenal beauty, as was the beauty in the memorable southwest Virginia sunset (Carlson, chapter 7, p. 104). Here I do not differ from J. Baird Callicott with his projected, quasi-intrinsic, truncated intrinsic value.

Now Hargrove reads me quite correctly. "Let the flowers live!" is a richer appreciation of their aesthetic properties than "Please leave the flowers for others to enjoy" (*EE*, p. 100, Hargrove, chapter 8, p. 132). The former appreciates plants themselves as having a good of their own to be respected. But where is the 180 degree turn when I claim that in treating the flowers one must argue a greater good if the flowers are picked? Ordinarily, I pick no such bouquets, but I can imagine a scenario in which I might. Yes, I did pick a bluebonnet, one among hundreds in his yard, to key it, and I would do so again. But I no longer cut Christmas trees in the wild (*EE*, p. 122).

The problem is that aesthetics is the wrong place to start when tracking facts and values in nature. Or if one starts there, one needs to follow me all the way, as Carlson does.

My fallacy of misplaced wonder "is to value the *experience* of wonder, rather than the *objects* of wonder" (*EE*, p. 131, Hargrove, chapter 8, p. 127), and Hargrove in his paraphrase wonders if I am not myself committing a parallel fallacy of misplaced beauty, locating beauty in human experience and ignoring objective beauty in nature. But what if we are aesthetically moved by objective natural history, marveling at the life struggle—as so well emphasized by Carlson, Wynn, and Sideris? This is a miracle, something to be wondered at—though Callicott would get troubled if I called it a "magic show." I wonder what he thinks of the name "Magic Hedge" in the Chicago Wilderness, whether the Chicagoans have misplaced wonder when they delight in the migrating birds in their colorful breeding plumage on the once-ugly trash dirt heap.

But to keep my emotions under control, I argue with an analogy to mathematics. The mathematical properties are there in nature, but the discipline of mathematics is human creation in interaction with what is there in nature. Similarly the aesthetic properties are out there; the aesthetic experience arises relationally (*EE*, pp. 235–237). My commentators neglect exploring this. Perhaps this is because the mathematical dimensions of nature are often also aesthetic in physics, astronomy, chemistry, crystallography—in that formed integrity in the abiotic world on which Lee insists. But the struggle for life, characteristic of Darwinian evolutionary biology, is not evidently mathematical.

Nor is it evidently aesthetic, not at least at first. That's the problem of the elk carcass with maggots—ugly fact of the matter (Carlson, chapter 7, p. 111; Hettinger, chapter 5, p. 63). I once complained: "All is a slaughterhouse, with life a miasma rising over the stench" (*PGW*, p. 129). But look more deeply. Leigh Van Valen, in theoretical biology, remarks that evolutionary history "resembles in some ways a symphony, although its orchestration is internal and caused largely by the interactions of many melodic strands" (Van Valen, 1991). I began to hear the music, but it was music in a minor key.

Valuing genetic nature

To continue that metaphor, McShane can't seem to hear any symphony, or even melody. She doubts that the genes are that well orchestrated internally. At the beginning, McShane keeps chasing better ideas and arguing away her gains. Careful criticism is both good philosophy and good science. She would probably add that such talk of music in natural history needs to be supported with hard argument. Could the biological (genetic) fact of the matter also be a good (value) that counts morally? Maybe because this genetic activity is good for individuals? No, because individuals can be maladapted. Well, maybe it is good for the species. Well, no, a species might be doing its genetic thing and still go extinct. She can't find an argument that gets her from genetic *is* to moral *ought*.

But start at the end. McShane concludes: "There are many parts of the natural world, one might argue, that can be benefited or harmed independently of the existence of humans. . . . Instead of focusing on which states are sought by the genetic set of an organism, one might focus on which states would be good for the organism, . . . on which states the organism would be better off seeking." Then "one can make the case that the well-being of other things makes a moral claim on us" (McShane, chapter 1, p. 13). She is quite right that the well-being of other organisms makes a moral claim on us. But I also have to wonder if she can't hear me or the music because she is deaf to what the geneticists have been shouting: that genetic sets are naturally selected to keep organisms well adapted, flourishing in their niches.

"The achievement of its genetic goals may not *be* good for the organism" (McShane, chapter 1, p. 7). McShane seems to think that we human philosophers should figure out what's good for the coyotes because the coyote genes often do not know. Then human geneticists should do genetic modification to get such mistaken coyote genes better tracking their coyote welfare. But of course the problem is that, in an uncertain world, we wouldn't know either. Oh, we might, say, re-design the coyotes to dislike lambs, or to resist canine distemper, because humans can anticipate the near future better than coyotes, when we are causing most of the disruptions. But suppose we had been monitoring those coyote genes for the last five million years, spotting and fixing the inept ones.

"Attainment of an organism's genetic goals is good for that organism, i.e. it increases that organism's well being." Exactly. But no, there is more: "It is only when an organism is in an environment to which it is well-adapted that the ways its genetic set directs it to operate will increase or at least sustain its well-being" (McShane, chapter 1, pp. 6–7). This is what I call "situated environmental fitness" (*EE*, p. 102). "The organisms we come across tend to be fairly well adapted to the environments in which we find them" (McShane, chapter 1, p. 7). If that were not true, of course, Darwinism would be false. (Make some allowance here for chaotic upsets punctuating this equilibrium.)

McShane then worries about all the maladaptation, from humans liking sweets to climate change that organisms cannot track. If we are wondering whether genes are valuable for defending life, cases drawn from cultural upsets are not relevant. Natural selection is no longer the principal determinant of events in Manhattan. Nor is the analogy relevant that she draws with Nazi preferences. "Hitler's trajectory is very different from that of the oak in the forest or the lion in the savannah" (Lee, chapter 2, p. 19). Humans choosing their preferences are moral agents; genetically generated plants and animals are not, Mother Teresa versus the trilliums. In Lee's terms: Humans deliberate; wild animals and plants are mute. "As valuers we are capable of getting it wrong—in some cases quite drastically wrong" (McShane, chapter 1, p. 11). In natural history, those who got it quite wrong are extinct, replaced by those who got it right.

Get into the wild and get the big picture. The phenomenon of maladaptation is contained in the larger context of genetic exploring of new possibility space in a changing world. In a sometimes uncertain world, "maladaptation" is of great value (Jansen and Stumpf, 2005). McShane first seems to know this: "It might be bad for the individual organisms that have these maladapted genetic instructions, but it is ultimately good for the species because it is what allows advantageous traits to become more frequent and disadvantageous traits less frequent in the population" (McShane, chapter 1, p. 7). But then she recalls all those turnover extinctions in natural history and starts worrying that all those genes got it wrong.

True, the genes in trilobites went on doing their thing, defending trilobite life, and the trilobites went extinct. But that's only half the truth. The other half is that any such trilobite genes did not do well enough what good genes need to do: that is, explore alternative possibilities of defending trilobite life. Good genes need to be "searching genes" (*GGG*, pp. 23–26). Genes are tested for their evolvability. Genes are tested for their capacity to send more of survival value forward to the next generation. Survival of the fittest is "survival of the senders" (*GGG*, p. 64). "These elements of trial and error are incorporated into a larger generative process" (*GGG*, p. 28, cited Hausauer, chapter 9, p. 159), "It is the maladaptation that permits adaptation, the imperfection that drives the world toward perfection" (*GGG*, p. 30; cf. *EE*, p. 148). In the big picture, this "genetic load" (*EE*, p. 147; *S&R*, p. 142) re-configures into a cybernetic nature with a "trajectory" that generates what is most valuable of all: Earth's storied natural history. That is where genetic natural history has been "headed" (non-deliberately); and that is where we are headed, searching whether we can follow and evaluate what is going on.

HOMO SAPIENS: REASON AND EMOTION

Humans use words, lots of them, as the contributors to the conversation in this anthology prove so well, myself included. Ideas pass from mind to mind; and, we hope, these ideas connect in some way with the world—to help us figure out who and where we are on this Earth. These ideas cumulatively over millennia have lately come to wrap around the world. They fill big libraries in thousands of big cities. Hausauer, Preston, and Wynn all address the encounter of words with the world in my thought (epistemology, to use the technical term)—how my words are both arguing about (reason) and lighting up feelings about (emotions) that natural world, and just how good (real? true?) this encounter is.

I will continue to defend a "massive singularity" (Venter et al.), a "transforming radical difference" (Davion, chapter 10, p. 167), a "mental incandescence" (*PGW*, p. 226) producing a wisdom distinctive to *Homo sapiens*, as we have named ourselves. We are the only species wise enough to figure out who and where we are (and, *pace* Light and Sheppard, to build cities). After that, I move on to welcome Wynn's contribution discovering wisdom also in my emotional responses to nature. This will lead us forward, toward a deeper appreciation of the life struggle in which we are engaged. "Our encounter with nature is as passionate as it is cognitive" (*PGW*, p. 248, Wynn, chapter 4, p. 59). As Wynn realizes, "feelings themselves give us access to the nature of things" (Wynn, chapter 4, p. 46), a claim that Davion and her feminists are also likely to endorse.

Reasoning about real nature

But first the powers of reason to know the world. I don't know that I have claimed "human ability to get out of our skins" (Preston, chapter 3, p. 34). I do claim that the skin is a semi-permeable membrane (so to speak) and that we can see what is outside of our skins, with our perceptual and conceptual apparatus. I regularly claim that humans can transcend their sector, unlike any other creature, and take up the viewpoint of others. That puts me in "double trouble" (Preston, chapter 3, p. 37) because I think I can not only see lions but take up their point of view and see what they value. I want to "transfer *tout court* into the philosophical relalm the kind of commitment to the actual existence . . . that we might employ on a backpacking trip *tout court* into the philosophical realm." When Rolston does that, claims Preston, he "impales himself on the realist horn" (Preston, chapter 3, pp. 36–37)

"The realist, believes Fine, simply deludes him or herself when claiming to be able to 'stand outside the arena watching the ongoing game, and then . . . to judge (from this external point of view) what the point is' " (Preston, chapter 3, p. 38, quoting Arthur Fine). When I watched the three lionesses stalk that zebra, watching the ongoing game, I did not have any trouble at all seeing what the point of the stalk was. One lioness guarded the kill, the other two went back and led the cubs to the kill. I don't think Fine would have any trouble making the transfer of events from field experience to the philosophical realm and see the point. If it comes to that, I want to add a "desk-thumping, foot-stamping shout of 'Really!' " to this claim. I wouldn't shout of course until after the kill, lest I alert the zebra and spoil the stalk. (I wonder what Davion, Palmer, and the vegetarians would have done.) If I have to use Fine's terms, lion "existence" is as important as lion "essence" (Preston, chapter 3, p. 38).

We are not going to have realism "and not anti-realism either." "Undercut both!" (Preston, chapter 3, p. 38, notes 35, 36). Notice, for example, how anti-realist Richard Rorty undercuts himself. "The world is out there," but "truth is not out there." Truth arises when we humans use our language to create claims about the world out there, with pragmatic claims permitted but claims of correspondence forbidden and even ridiculed. "The world is out there but descriptions of the world are not" (Rorty, Preston, chapter 3, p. 39). But if "the world is out there" is not some true statement made in my mind with successful reference to an outside world, Rorty has undercut all capacity to understand that sentence, either his own capacity or mine. If Rorty finds himself unable to claim (descriptively) that the lion out there killed a zebra out there, a statement corresponding to what happened, that is "a very strange game" (Preston, chapter 3, p. 39). We may want to undercut the anti-realism, but that we simultaneously want to undercut all the realism is not so clear.

When I say that these epistemological misgivings do not really upset our ontological commitments I mean that all those discussions about whether the concept of "lion" is socially constructed do not in the slightest undercut my convictions that the lions are out there. Fuss around all evening about what is a "canid" (family? genus? species?). I did see wolf, coyote, and fox all in one great Yellowstone day; nor do I have any doubts that these are related natural kinds (Rolston, 2005b, p. 147). I don't have any doubt that all these animals were hunting because they valued their own lives. I don't know about foundations, but keep this discussion on the ground. True, we humans make up our categories as we know the world ("canids," "lions"); that is epistemology. But it is also true that the world made up these natural kinds once upon a time. That is ontology. In claiming this is Rolston so "frightened of relaxing his grip on a picture of knowledge as correspondence with the real" (Preston, chapter 3, p. 39)?

Preston recommends "a dignified surrender of realism" (Preston, chapter 3, p. 37). For help he turns to Arthur Fine, who notices that realists and anti-realists behave similarly in everyday life regarding tables and chairs because to do so helps both to secure their ends. Settling the metaphysical issues is not important in practice. So Fine wants us to drop all these hopeless "realist and anti-realist metaphysical additions" (Preston, chapter 3, p. 40) and adopt the "natural ontological attitude." We take up an ontological attitude with a great dislike for metaphysics! Metaphysically "less can be more" (Preston, chapter 3, p. 38). "California natural." "No additives please" (Preston, chapter 3, p. 40). That seems to be a quirky way of being a metaphysical agnostic. Realist or non-realist? I don't know. Rest content with these skimpy ontological commitments—Fine's "minimalist solution" (Preston, chapter 3, p. 40). That might please McShane too; she has a hard time adding value to the genetic facts. But it's hard for an inclusive holist like myself to be a minimalist. It's not hard for me as an environmentalist to appreciate "less is more" for living on Earth, but in environmental philosophy the decision to see less, not more on Earth could be premeditated poverty.

Fine will probably reply that he and everyone else believes that lions (like tables and chairs) are out there, macroentities that we can observe (see, hear, taste, touch). It is the unobservables about which we should be minimalist. Some of these unobservables are micro. With a good light microscope you can see chromosomes, with an electron microscope molecular genes, but not yet the triplet sequences. You can't see photosynthesis. Cut off realist convictions there. Even at the macrolevels, one cannot observe energy flow, or gravity, or speciation, or species, or ecosystems, or wilderness, or the Pleistocene period, much less instrumental or intrinsic value.

So we can see that the lion killed the zebra, but we will have to be agnostic about whether the zebra eating grass was today capturing energy that the grass yesterday had captured in photosynthesis. We can behave "as if" the lions have in their genes DNA of value to them in maintaining their metabolism, fueled by energy in the zebra meat. Are the lionesses passing along anything of value to these cubs? Are they defending a species line? Well, it is "natural" to believe that, but metaphysically, we should keep skimpy beliefs about anything beyond observable, preferably with our eyes. "No additives please."

But straight on "Fine contends that the best reasons to believe in the existence of unobservable entities are reasons taken from the methods and practices of science itself" (Preston, chapter 3, p. 41). Even Rorty seems to think we know "the effects of causes" out there (Preston, chapter 3, p. 39). Legitimate these claims using science, but please don't posit any descriptive powers of science to know what is going on out there. So we get the phenomena all back, only we have very little conviction about it. "Once we have used all reasonable evidence to posit the existence of natural values we won't need to stamp the foot and insist on their reality" (Preston, chapter 3, p. 41). See the lions kill the zebra but don't be too sure they were hunting or valued the kill, or the cubs they fed. You can't see any values out there, so don't claim that there is any value at stake with those so-called canids: the foxes, coyotes, and wolves. "Values are not immediately revealed by the methods and practices of science" (Preston, chapter 3, p. 41). Nothing about what predators value? Such realism would be "extra-scientific" (Preston, chapter 3, p. 42). Strange game!

If I take any reasons from the methods and practices of science itself, the first fact I have to deal with is that 99% of these biological scientists are realists themselves. They will claim that value is intrinsic to the observation of the hunt, intrinsic to the biology of predator and prey, intrinsic to the emotional excitement of the stalk and kill, intrinsic to survival and adaptive fit. If I turn to policy makers, environmental managers, they want to couple ethics with science, as Lemons points out (chapter 12, p. 204). But suppose I tell them that neither they nor their scientist advisors have made any true statements descriptively corresponding to plants, animals, events in the natural world out there, or to anything of value present there? They should rather rest content with all they can get: minimalist agnosticism. Figure out how to use this nondescriptive operational functionality to get what you want done inside and outside the park. They don't get up and do it; they just listen to philosophers less, not more. That must be why Andrew Light recommends that philosophers keep these "arcane debates" in our "private conversations." Adopt "a much more straightforward and robust policy-oriented approach to environmental problems when talking with the public" (quoted in Preston, chapter 3, p. 30).

When I spoke to the Yellowstone park managers, mixed scientists and decision makers (Rolston, 2001, see above), I opposed ideas, such as those of David Graber (Lemons, chapter 12, p. 209), that parks might be "socially constructed" to provide a "mythology" for tourists by arranging "a landscape that is *managed* to reveal as few traces of the passage of other humans as possible" (Rolston 2001, p. 268, citing Graber). This gives the illusion of pristine nature, when the landscape has been in fact engineered to appear wild. I argued that Yellowstone is not a museum of pre-Columbian America. I gave a reasoned analysis of why invasive exotics were not "natural," why the quest for elemental nature in parks was not hopeless, how restoration biology could let nature return. My account and the policy that followed were based on ecosystem science and natural history in Yellowstone (invasives versus natives, changes since Pleistocene times, fire ecology, native American impacts at ecosystem levels). If none of these words were descriptive of the Yellowstone world, if I was only preaching a socially constructed European-American wilderness mythology, my listeners were wasting their time.

But, rather, they cheered my conclusion:

> Is it the case that we have lost any possibility of letting Yellowstone be natural? In an absolute sense this is true, since there is no square foot on which humans have not disturbed the predation pressures, nor any on which rain falls without detectable pollutants. But it does not follow that nature has absolutely ended, because it is not absolutely present. Answers come in degrees, with Times Square on one end of a spectrum and the Absaroka Wilderness on another. Events in Yellowstone can remain 99.44% natural on many a square foot, indeed on hundreds of square miles. We can restore nature. We can put the wolves back and clean up the air, and we have recently done both. Wildness can return. . . . Let Yellowstone teach . . . that nature is forever lingering around. . . . Let Yellowstone be the place that Americans can forever encounter once and future nature (Rolston, 2001, p. 272, p. 275).

Pace Lemons, this does not set standards for how much herbicide can be used to kill exotic species in the park, while not damaging the native species (which also would require description of unobservable microscopic parts per million). Rather my analysis, I hope, uses reason to clarify the "preservationist" mission of the parks, at one end of a spectrum of culture mixing with nature—as Lemons too desires (chapter 12, p. 213). I also intend that my words have a "motivational and inspirational" imperative (Lemons, chapter 12, p. 217). I'd like to think that on that podium, with elk in sight outside, I both was "getting the words right" and that these were "words gone wild" (Hausauer, chapter 9, p. 145, p. 143) about values intrinsic to the park. What the park managers wanted to hear was a persuasive account of how to justify to themselves and their public the presence and preservation of values they daily encounter on that Yellowstone landscape. That's no "arcane debate about value theory" (Light, quoted by Preston, chapter 3, p. 30). That's truth about the world.

No! No! How antiquated! Give up the "realist fantasy" that you know anything corresponding with what is really taking place in Yellowstone or on the Serengeti plains, either facts or values there. Get "better ontological grounds" (Preston, chapter 3, p. 43). Science gives us a natural ontology without realist metaphysics. Sorry, but my mind is not subtle enough to catch the difference. To use Lee's terms, this skimpy metaphysics leaves me "mute" and "inarticulate." I'm not a foot stomper but I do like to be robustly articulate, especially about Africa and Yellowstone.

Well, maybe we can be realists on local landscapes, but we should stay skimpy epistemological ex-realists when we extrapolate comprehensively to global knowledge of natural history. Be agnostic about the evolution of biodiversity and biocomplexity? If so I lose my grand narrative, any capacity to speak to my deeper, less particular and more universal concerns. I need to be a historian. I need for the sentence: "Life persists in the midst of its perpetual perishing" to correspond to reality—in Yellowstone, on the Serengeti, and around the Earth. I need an epistemology of the life-death-life-death-life-death lineage, which I need to be ontological fact of the matter. If that isn't epistemlogically accurate ("true!") about what's going on out there, there isn't any evaluating to do. I'll stamp my foot about that. So will Carlson, Wynn, Sideris, and Hettinger. Lee will get quite articulate.

Caring in and for nature

In fact, now I am getting emotional. "This argument seeks to be reasoned, but if it is mixed with emotion, that too is intentional" (*PGW*, p. 248). Philosophers like to reason, but they ought not to forget that "philosophy" has in its etymology a "love" for this wisdom that *Homo sapiens* seeks. I closed the first paper I ever wrote in the field with the hope for an ethic to which we "can only be drawn in love" (*PGW*, p. 27). The last chapter in my *Environmental Ethics* is essentially the (reasoned) argument that an ethic needs a "personal backing" (*EE*, p. 353).

> Our emotions defend the organic self, but they also stretch it out to integrate it into its place. . . . Human emotions fit us for defending the self, aloft and transcendent over nature, but they ought still the more to fit us to that natural environment which transcends us. These are emotions that we all live *by*, but they are emotions that some of us live *for* (*PGW*, p. 253, p. 255).

Wynn worries that "emotion" is not much indexed in my books. True, but the word "value" is on almost every page, which I intend, *pace* Preston and Fine, to refer to value out there in the world. Early on I note that our human valuing is always emotional. There is "inevitable human excitement" (*EE*, p. 31): "Value is not received as the conclusion of an argument, or by the indifferent observation of a causal series. A value or disvalue is recognized as whatever has got some bite to it. . . . In the case of bare knowing, assuming the correspondence theory of truth at least at some ranges, the knower has an internal *representation* of what is there, perhaps calmly so. Valuing requires more, an internal *excitation*. That brings emoting" (*EE*, p. 28).

> Being among the archetypes, a forest is about as near to ultimacy as we can come in the natural world—a vast scene of sprouting, budding, flowering, fruiting, passing away, passing life on. We get goose pimples with mountaintop experiences, hearing the wind in the pines, with solitude in a sequoia grove, watching the falling autumn leaves. We feel life's transient beauty sustained over chaos (Rolston, 2004, p. 296).

Why isn't that mixing reason with emotion, fact of the matter as verified in science, with emotion overlooking it all from the mountaintop or gazing up at the perennially persisting sequoias? "The language of aesthetics offers a deepening of the kind of understanding that is available in the language of science" because "the language of aesthetics is more directly tied to our particular, human sensibility" (Wynn, chapter 4, p. 56). Also, vice-versa, science expands our local human sensitivity (the stink of the rotting elk carcass passed en route) and deepens our understanding. This more inclusive understanding (the recycling web of life, the age of the sequoia, scarred by fire) enables us "to see beauty now where we could not see it before" (Carlson, chapter 7, p. 106, citing Rolston).

"One important contribution of the emotions to the economy of human life is to establish pre-reflectively the salience of certain features of our surroundings" (Wynn, chapter 4, p. 59). Here I am on a solo backpack, again in the alpine Rockies:

> I am five miles from the trailhead; I am quite on my own. The storm is coming up, the spruce are bending with the wind, supper is not cooked, and it is getting dark. . . . One has to take care against disorientation. But that is again to realize our limits, to sense vulnerable embodiment, and to risk engagement with the sublime. . . . Bodily participation in the forest, the competence demanded and enjoyed there amidst its opportunities and threats, the struggle for location in and against the primordial world—this engagement enriches the aesthetic experience (Rolston, 1998, p. 163).

Such an emotion is, I might have said, "a moment of truth" (*PGW*, p. 261, Wynn, chapter 4, p. 25). In Wynn's terms, "such responses may be called `primitive' in the sense that they operate independently of any discursive appreciation of the character of our surroundings" (Wynn, chapter 4, p. 47).

I get first the "gut feeling," but then I do say something to myself, and verbalize how long it is until dark and recall that there is a large boulder not far from my tent where I can cook out of the wind. At this point our "affects build on what we know by verbal means, but offer a deeper kind of intentionality than is available by those means" (Wynn, chapter 4, p. 47). There is deeper embodiment. "The difference also lies in the content, although it might be that this difference cannot be captured in words" (Wynn, chapter 4, p. 47, quoting Goldie). Such feelings have, as Wynn says, "intentional content in their own right" (Wynn, chapter 4, p. 48).

But there must be dialectic between reason and emotion, each checks and balances the other. In my own career, I could say of the kind of Appalachian experiences recounted in "Mystery and Majesty in Washington County": "Rather than simply generating rationalisations of themselves, they may help to constitute a research programme" (Wynn, chapter 4, p. 50). So my "naive emoting" led to a return to study philosophy of science (rather naively thinking this was philosophy of nature) and despite the fact that I already had a Ph.D. from Edinburgh.

I felt the crossfire of reason and emotion in Hawaii. I first despised at a crater's edge the superstitious offering of fruits and flowers to Pele, placating her to stop the lava flow. I knew better. But later, in the twilight, watching red lava pour down into the ocean, I knew the science but was overcome with the numinous. "Yet, in my scientific superiority, I too there experienced the sublime, a virtually religious experience, as lava out of the bowels of the Earth created new landscape at the edge of the sea" (Rolston 1995b, p. 374). I found the science "in some fashion deepened or re-channelled" (Wynn, chapter 4, p. 50). As once before, at the Precambrian contact, thinking how rocks evolved into spirits, I found myself "naively emoting." "My soul is hidden in this cleft of rock" (*PGW*, p. 234, p. 239, Wynn, chapter 4, p. 50).

"Rolston asks much of his readers; he asks readers to utilize more of their whole selves in thinking and feeling through philosophical ideas than many philosophers do." So concludes Hausauer (chapter 9, p. 164). With that in mind I might say to McShane that she should put on her pack, get winded on a climb, and, continuing with her argument now embodied, ask herself whether she values the genes that make cytochrome-c molecules. And if she values them herself, she should give me a reason why the squirrel, alarmed and scampering up a tree as she rounds a bend, does not also value his genetically-made cytochrome-c molecules. Climbing uphill has turned on some feedback loops, in hiker and in squirrel. Maybe neither she nor the squirrel knows anything about cytochrome-c molecules or how this goal-directed metabolism and behavior operates. But meanwhile the genetics is defending life, and if she thinks that a good thing for herself, mutatis mutandis, it is good for the squirrel. I'd like to lead her out of the context of justification and into the context of discovery (cf. Hausauer, chapter 9, p. 153). She is too good a philosopher not to know that they are equally important. Values in nature are as much discovered as are they justified.

STORIED NATURAL HISTORY: LIFE PERSISTING IN PERISHING

Let's climb higher, toward the evil problem, the after-Darwin problem (Carlson, chapter 7, p. 109). "We are born to die, but it is life rather than death which is the principal mystery that comes out of nature, and our emotions are stirred proportionately" (*PGW*, p. 255, Wynn, chapter 4, p. 57). In her conclusion, McShane asks: "What is it about being valued by a genetic set that makes something ethically valuable—i.e. the sort of thing that moral agents have a duty to protect, promote, endorse?" My short answer is: Life (McShane, chapter 1, p. 12).

Cybernetic nature

As a result of the cybernetic character of genes, their searching out and storing of information, life has been on "trajectory." In a mood like McShane's, Darwin feared that there is "no more design in the variability of organic beings and in the action of natural selection, than in the course which the wind blows" (cited Sideris, chapter 6, p. 79). Darwin fails to see that the course the wind blows is not a design-accumulating system, but that his natural "selection" is. Vital information accumulates into storied evolutionary achievement (*GGG*, p. 50ff, Carlson, chapter 7, p. 105).

> Something is learned across evolutionary history: how to make more diverse and more complex kinds. . . . On Earth there is no such learning with the passing of cold and warm fronts; they just come and

> go. . . . But the life story is different, because in biology, unlike physics, chemistry, geomorphology, or astronomy, something can be learned (*GGG*, p. 1).
>
> Though it can seem indifferent to life, this system in which all individuals are perpetually perishing is the system that creates life and exemplifies it in individuals of increasing number, kind, and quality over several billion years of evolutionary time. . . . The story seems some kind of steady, if statistical, heading of the system (*EE*, p. 186).

Karl Popper called Earth "a world of propensities," with possibilities, probabilities, headings, trajectories.

> This view of propensities allows us to see in a new light the processes that constitute our world: the world process. The world is no longer a causal machine—it can now be seen as a world of propensities. . . . Especially in the evolution of biochemistry, it is widely appreciated that every new compound creates new possibilities for further new compounds to synthesize: possibilities which previously did not exist. . . . Our world of propensities is inherently creative. . . . The variety of those [organisms] that have realized themselves is staggering. . . . In the end, we ourselves become possible" (Popper, 1990, pp 18–26; *GGG*, p. 353).

Natural history is like Wynn's dominant seventh chord, pushing further (Wynn, chapter 4, p. 48). Nature doesn't seem to prefer a "minimalist account" to use Fine's terms, "California natural"—without "additives" (quoted, Preston, chapter 3, p. 40). Nature seeks "additives." There is a genetically-impelled "projective nature" (*EE*, p. 192), with an "inner conative urge. The philosopher may have trouble supplying rational arguments why this life, now instantiated in me, should want to flow on, but he is unlikely to eradicate this natural passion" (*PGW*, p. 63). Preston knows that first in himself, and he can look around and "add" that this (really) is descriptive of objective natural history.

Caution, Preston might say. A minimalist view claims only that these things did happen, not that they were a heading of the system. True, biologists are spread across a spectrum whether the story is random, contingent, caused, unlikely, likely, determined, open (*GGG*, Chapter 1). Some of this is packed into the word "wild": spontaneity, openness, order mixed with disorder, autonomy and loss of control, chaos with strange attractors; these put adventure into the system. Those who put some arrow on evolutionary time may still doubt that Darwinian theory predicts the long-term historical innovations that have in fact occurred (Maynard Smith and Szathmáry, 1995, p. 3). But no biologist doubts that the story did happen with creative achievements that now ought to be conserved. Even Stephen Jay Gould, finding life quite contingent, also found it "wonderful life" (Gould, 1989).

Maybe the ecofeminists can add in their "caring." They first find this in themselves, and then go on to find how organisms, defending their lives perennially, care that life continue perennially. Organisms are pushed; they push to carry life on. Giving birth is a rite of passage; newborns are a moment of truth.

Seeking survival, learning of behavior, and crooked lines

But McShane still wants reasons: Why value the "seeking and avoiding behavior" that genes have learned? She prefers the minimalist ontology, without additives. But nature has added up these genetic subroutines into wonderful life. What is being sought is life and what is being avoided is death. What's going on is life persisting in the midst of its perpetual perishing. Get out a biology text. What McShane seems to overlook is that the concept of "survival value" underwrites every page of the text. Organisms that are good at finding such value survive; organisms that are not perish.

McShane grants that non-psychological organisms (plants) have evolved other ways to stay alive, and she recognizes that I have repeatedly developed this idea (McShane, chapter 1, pp. 4–5). But then we get the protest that, well, this is just "seeking and avoiding behavior" and why should that be of any value? To answer, again, we need the bigger picture. "The genes are the species writ small, the macroscopic species in microscopic code; the species is the genes writ large, the microscopic code in macroscopic species. A genome is a map of a life form . . . Its genes code the kind, representatively; and the organism, an expression of the kind, presents and re-presents the kind in the world" (*GGG*, p. 63). Where we find living things valuing their lives, have we reason to count this morally? The question is essentially: Ought we to respect this ongoing life?

Being the well-trained analytic philosopher she is, McShane keeps longing for tight (= straight) logic, and complains that I have insufficient reasons. My work doesn't meet her standard of what Hausauer calls the "straight argumentative norm" typical of philosophy journals (Hausauer, chapter 9, p. 143). My straight reply to McShane is that: "Earth is not so much a syllogism with premises and conclusions as it is a text to be interpreted. It is stories being told" (*EE*, p. 343, Hausauer, chapter 9, p. 162).

Nature isn't so logical; nature doesn't like straight lines. "The communities constructed are sometimes wayward; there are false starts, trials and errors, but there is much fitting together of rationality and positive value" (*EE*, p. 173). If I were to paraphrase William Blake (in Sideris' epigraph): "Reason demands straight roads, but crooked roads are the genius of genetic natural history." McShane needs to fit her maladaptations and my searching genes sending life forward into the context of "writing straight with crooked lines" (Sideris, chapter 6, p. 77, p. 89, quoting Rolston). Paradoxically, dialectically, the crooked roads are what make possible improvement and genius.

That is fact of the matter about the world past. I am also in the world present trying to get ideas passing from mind to mind, and if this happens by straight argument well and good. But if it happens by evocative triggering spiralling around some word subtleties, also well and good (cf. Hausauer's worries about repetition rather than reasons, chapter 9, pp. 147–148). If it requires some wandering around, exploring seemingly likely paths that prove unlikely, we are still finding our way with some success. Sometimes with head as well as feet, crooked paths are the shortest way to get to trail's end, especially if we are ziz-zagging up toward the top.

"Even if we can say in general that the continued existence of life on earth is a good thing, this won't be enough to let us conclude in any particular case that the achievement of an organism's genetic goals is a good thing" (McShane, chapter 1, p. 9). McShane has a hard time moving from particulars to universals—which her hard logic does not permit her to do. Life as a whole is good, but nothing follows about specific lives. I invite her to think of it collectively and distributively.

> "Ought species *x* to exist?" is a distributive increment in the collective question, "Ought life on Earth to exist?" . . . The answer to the species question is not always the same as the answer to the collective question, but since life on Earth is an aggregate of many species, the two are sufficiently related that the burden of proof lies with those who wish deliberately to extinguish a species and simultaneously to care for life on Earth (*EE*, p. 145).

Individuals in storied nature

But that introduces a further problem, again collective and distributive. The collective process seems to sacrifice individuals. "Whole organisms are ephemeral. The genes have more of an eye on the species (so to speak) than on the individual. The solitary organism, living in the present, is born to lose; all that can be transmitted from past to future is its kind" (*GGG*, p. 63). Now my critics worry about what happens to these ever-perishing individuals (Sideris, chapter 6, p. 78, p. 87). Darwin too was dismayed by how individuals, in their pains, are caught up, unrequited, in the selection process, even though the system does select against "counterproductive pain" and there is a "generally beneficent arrangement" (Darwin, cited Sideris, chapter 6, pp. 80–81).

I give this account: (I hope McShane is still listening.)

> Life is sustained in any individual in some relative proportion to its fitness for it. Thus, in their evolutionary development, there never were any really ill-fitted woodpecker species. All the preceding forms that enjoyed life did so in that extent to which they were impressively well formed, 99.999 percent hit, and .001 percent miss or mutational gamble, "blessed," we might say, with the cumulative tradition of a billion years, "cursed," if you will, by fractional mutations, random variations, sometimes detrimental, usually harmless. But it is just the "curse" they bear in which lies the possibility of there being woodpeckers better yet, or even continuing on, in the future (*S&R*, p. 136).

Robin Attfield makes a similar point, that "the overall system is a precondition for the flourishing of individuals," who may live well even in the midst of their struggles (cited Sideris, chapter 6, p. 92).

I celebrate individuals. But individuals always have the bigger story going on over their heads. "Nature evolves a flourishing community in some indifference to the pains and pleasures of individuals, even though pain and pleasure in the higher forms is a major evolutionary achievement" (*EE*, p. 108). I claim that the developing system escalates individuality (*EE*, p. 216, Fig. 6.6). The evolution of immune systems takes such individuality to extremes (Rolston, 1996).

> The community beauty, integrity, and constancy includes a persistent selecting for individuality. That is a strange, liberating "priority" or "heading" of the system: escalation of individuals in kind and complexity, in quantity and quality. . . . The goods and "rights" of individuals (their flourishing and freedom) belong in such a system; the ecosystem itself promotes them in its own way. . . . Individual welfare is both promoted by and subordinated to the generating communal forces (*EE*, p. 186).

"A disvalue to an individual may be a value in the system and will result in values carried to other individuals. . . . Excellence is not a matter of encapsulated being, but of fittedness into a pervasive whole" (*PGW*, 132–133, Carlson, chapter 7, p. 108).

CRUCIFORM NATURE

The story we have from Darwinian natural history echoes classical religious themes of death and regeneration. Hiking in the Rocky Mountains, I confronted this in the pasqueflower: life beset by storms, persisting through winter, and flowering again in spring at Easter. I confront a "cruciform creation," life dying and regenerated through death—nature as "*via dolorosa*." I find "an encouraging beauty" in life's perennial regeneration (*PGW*, pp. 256ff; Sideris, chapter 6, pp. 86–90; Wynn, chapter 4, p. 53 p. 57; Carlson, chapter 7, p. 107, note 18).

> Every life is chastened and christened, straitened and baptized in struggle. Everywhere there is vicarious suffering. The global Earth is a land of promise, and yet one that has to be died for. All world progress and directional history is ultimately brought under the shadow of a cross. The story is a passion play long before it reaches the Christ. Since the beginning, the myriad creatures have been giving up their lives as a ransom for many. In that sense, Jesus is not the exception to the natural order, but a chief exemplification of it (*S&R*, p. 291).

Both in the divine Logos once incarnate in Palestine and in the life incarnate on Earth for millennia before that: "Light shines in the darkness and the darkness has not overcome it" (John 1).

I am being drawn into celebrating "green pastures in the shadow of death, a table prepared in the midst of mine enemies" (Psalm 23). I know the evolutionary science, I know there is life-death-life-death, but to stop there is like ending with a dominant seventh (Wynn, chapter 4, p. 48). When I encounter the lovely blossoms breaking through the snow, I get stretched toward the flower's "exuberance in the fundamental, etymological sense of being more than expectedly luxuriant" (*PGW*, 257, Wynn, chapter 4, p. 53). That is a distributive token taken as type for the collective Earth, with its millions of species, continuing after a turnover of billions of species. I am pulled (what McShane dislikes) from the particular to the universal.

The science reinforces native range experience. All living things are caught up in the struggle for life; we humans are too, and the backpacker in the woods, the hunter, knows this in embodied participatory experience. But the science frames this into the comprehensive evolutionary and ecological scene. "Life in green pastures at the valley of the shadow of death"—one needs no biology text for that. But the biology text puts the poetry into prose. One needs the "on the ground immediacy" to survive, one needs the textbook science read into the field experience for the "deeper beauty that science can unfold" (Carlson, chapter 7, p. 111)

The central fact of the matter biologically is the survival of life over millennia, life-death-life-death-life-death; but such fact of the matter is *ipso facto* valuable, vital. For any beholder entering that scene today, life is a gift out of the primordial past. Yes, but does not Darwin say it is a nasty gift, nature red in tooth and claw? Look again: there is "storied achievement" in a "cruciform nature." This is a "new picture painted over the old, although some of the old picture shows through. Wildness seemed a great struggle, and so it is; but it is also a great flowing of opposites into each other. Wildness is a complex tapestry of value" (*PGW*, p. 131, Carlson, chapter 7, p. 107). This does not deny the dark clouds; it sees silver linings.

At such depths my aesthetic experience becomes metaphysical, theological, as Carlson recognizes. In this tapestry interweaving life and death, there is positive aesthetics because the struggle has been reframed in an ambience vital for creativity. The re-framing, going on in my mind, is equally a discovery of what is taking place on the landscape. When I say the scale runs from zero upward with no negative numbers, I am not referring to individuals but to landscapes. A dead elk is a negative. But on the landscape scale, one does not just see the rotting elk, one "sees into" the patterns and processes that have produced elk and kept them on the Rocky Mountain landscape since the Pleistocene and before, part and parcel of the "encouraging beauty" of the pasqueflower, fertilized by the rotting elk. When I hike in the forest, every footstep is on humus, the debris of death. The eating and being eaten is in trophic pyramids that make life interdependent and all higher life possible. The feed loops are feedback loops in a life support system that generates storied history. With such encounter I face the "sublime," taken both aesthetically and cosmologically (as the etymology of the word suggests) "to the edge of [my] limits."

Lee thinks that I avoid the most radical environmental philosophy, concern for the value of inanimate nature. Perhaps. I do argue that physicists and astronomers ought to be deeply impressed with the generative powers of their inanimate nature. But there is another way in which I spend more time confronting the most radical challenges. The inanimate world is a non-pathetic world. Rocks and ice do not suffer, so there is no cruciform nature there. "When Comet Shoemaker-Levy crashed into Jupiter in 1994 and upset the flow bands, I was not prompted to ask questions of good and evil. There does not seem to be anything evil out there in space. The place to look is here on Earth. . . . I am not asking whether this is the best possible world, but more modestly, whether this Earth is systemically prolific at increasing biodiversity and biocomplexity, and whether the evils here integrate well into those powers" (Rolston, 2003, p. 67). The biological re-generation moves me more profoundly than the physico-chemical generation.

Such a science-based aesthetic of "pathetic" genesis moves us more deeply and "we see beauty now where we could not see it before." Carlson goes right to the heart of the matter when he lifts that up as epigraph to his article. He is generously appreciative of my own struggle, personal and experiential, to make sense of the dimension of struggle objectively present in natural history, inescapable yet productive of the richness of biodiversity on wonderland Earth. One needs such aesthetic to be science-based, but the Darwinian account is opening up ultimate questions. Carlson's inquiry is continued when Sideris pushes down more deeply, reading me theologically, as do Wynn and Hettinger.

This is "a sometimes tragic view of life, but one in which tragedy is the shadow of prolific creativity. . . . A world without blood would be poorer, but a world without bloodshed would be poorer too, less rich in biodiversity and less divine" (Rolston, 2003, p. 85). Malcolm Budd, Yuriko Saito, and others complain that aesthetics is immediate experience, but that I require seeing across millennia to take the smell out of the rotting elk (Carlson, chapter 7, pp. 112–113). If one were to get the elk in camera viewfinder, one would not snap the picture; it's not a pretty sight. The experience of ugliness arises (Hargrove, chapter 8, p. 138). But this is short-sighted aesthetics, troubled with a framing problem. Put the camera away, and see the elk in larger view, with the sweep of the hills, the wind in the spruce, insects from the maggots pollinating the pasqueflowers, recall the cow elk with newborn calf seen alive and well earlier in the day. The ugliness now seems local in a more inclusive "encouraging beauty." The observable is the dead elk; the unobservable but more real is sublime life.

There is ready explanation for the local ugliness. Over their evolutionary history, humans with such repulsive "anti-aesthetic" experiences will avoid rotting carcasses, which can also carry diseases (cf. Hargrove, chapter 8, p.139). At this level there is "horror at what is gruesome, and disgust at what is foul. These properties characterise the way things look, sound, taste, and smell" (Wynn, chapter 4, pp. 46–47, quoting Deigh). We are disposed to both biophilia and biophobia, to use E. O. Wilson's terms. Or, in the more homely vocabulary of my Shenandoah Valley rearing, we react "at gut level." This is preconceptual and moves us to act appropriately almost instinctively.

Take the more inclusive view, however, and we see beauty where we could not see it before. One might not need millennia, perhaps a surrounding ecology will do. I was, minutes ago, bending over a pasqueflower and rejoicing in the regenerating life. Perhaps I did encounter something unobservable but objectively there in that pasqueflower. I invite my local aestheticians to re-illuminate the elk and see the recycling of life. Now I put McShane's troublesome maladaptations into "cruciform nature." "The natural world is godly—cruciform and deiform—Rolston claims, precisely because of the struggle and suffering, not in spite of it" (Sideris, chapter 6, p. 78). Darwinian nature is precursor to the Christian account of the abundant life. "We see life produced out of death, crucifixion and resurrection intertwined, new life rising out of old, 'life suffering through to something higher' " (Sideris, chapter 6, p. 89).

NATURE AND SPIRIT

Now Hettinger wants into this conversation because he fears, alas, that I "undermine" myself again (Hettinger, chapter 5, p. 63, p. 73), this time with a quite un-real "unobservable": God!! Or, since "real" is now a prohibited word and since God is an "antiquated" unobservable, maybe we should start with "spirit." "Spirit" too is unobservable (so is "nature" if it comes to that), but in my *Existenz*, torn over the elk, stunned by the pasqueflower, worried about death, I seem to embody some kind of "spirited nature." In me, in us, "mute" nature has become quite "articulate" (Lee). Just this searching for "spirit" paradoxically reveals human uniqueness, for no other animal is driven religiously. None of them build cathedrals in Sheppard's and Light's cities. None has this order of "spirited" or "spiritual" concern.

Agreed, Hettinger will say, but keep it natural; celebrate a nature than can evolve into spirit, perhaps uniquely so in humans. Distinguish culture from nature if you like, cathedral art in Manhattan from sunsets in Washington County, but no dualism allowed metaphysically. Culture too is natural, not supernatural. Keep it all in "that [naturalistic] explanatory box" (*GGG*, p. 252, Davion, chapter, 10 p. 177). Perhaps, but now the *is/ought* problem returns, more urgently. In nature I find values to be biologically-generated fact of the matter; in culture one needs value guidance that is not fact of the matter either in nature or culture. Religion is for regenerating people ethically, not for elk or coyotes, who are regenerated genetically. Remember that Hitler's trajectory is very different from that of the lion in the savannah (Lee, chapter 2, p. 19).

We humans must make religious choices confronting all of occurrent reality: culture as well as nature. The grand narrative, the storied history, produces persons, not just *Stegosaurus* but Mother Teresa. Acting as spirit, I need an account of evolutionary history, I also need to know whether to be just, charitable, whether and when to forgive sins. I need an ethic for Light and Sheppard in their cities. Hettinger needs a lot of *ought* out of his *is*. Here are "real" "additives" to get into the big picture.

When nature evolves spirit, I need some form of spiritual explanation. Hettinger would reply that he offers that: a spirited nature. So let's recall the story: With humans, nature gets more into spirit, and do we have adequate "naturalistic" explanation for that? Certainly not if "naturalistic" means "scientific" because there is no set of sciences that predicts or retrodicts the natural history from big bang to Einstein, from quarks to Jesus. No science predicts that trilobites will evolve into political animals that build cathedrals in their cities.

Posit lots of emergence, opening up of new possibility spaces; make Earth, with Popper, a world of propensities. But keep it immanent, Hettinger advises. Keep it deep in nature, but don't go for any transcendence. He knows, I suppose, that metaphysicians have long argued that if you go deep enough, immanence is transcendence. To go down to an energy pit beneath the quarks, out of which all bubbles up, radically transcends any form of reality that we know at our native ranges. A quantum fluctuation in a vacuum that explodes and suddenly inflates into a universe—this quite transcends common experience even more. You can get beyond by going beneath as readily as by going up and out.

"Naturalistic spirituality sees earth as a holy place and claims that if anything is sacred and worthy of reverence and devotion, it is this miraculous earthen community of life. . . . It treats the earth as our creator and the ground of our being" (Hettinger, chapter 5, p. 70). I make those claims myself, as Hettinger realizes: "If there is any holy ground, this is it" (*CNV*, p. 236, Hettinger, chapter 5, p. 71). E. O. Wilson, a secular humanist, ever insistent that he can find no divinity in, with, or under nature, still exclaims, with emphasis: "The flower in the crannied wall—it *is* a miracle" (Wilson, 1992, p. 345). "The biospheric membrane that covers the Earth, and you and me, . . . is the miracle we have been given" (Wilson, 2002, p. 21).

Daniel Dennett, as resolute a naturalist as one can find, still ends his survey of natural history: "The world is sacred." Apparently not even Darwin's "universal acid" can dissolve that claim, dissolve God though this acid can (Dennett, 1995, pp. 520–521). Even Stephen Jay Gould, after a career advocating contingency in natural history, closed his massive paleontology text, among the last words he wrote, calling call the Earthen drama "almost unspeakably holy" (Gould, 2002, p. 1342). Accept the holy; but don't add in transcendence. Maybe these words: "miracle," "sacred," and "holy" are just poetic rhetoric. But I suspect that these provocative words are there because even these secularists are tugged by a deeper undertow than they realize by their encounters with these archaic orders (Rolston, 2004, p. 297, Wynn's dominant seventh chord in subliminal ranges).

The question is how deep do you need your explanations to go? In a more comprehensive metaphysics, one has to get Earth created, and get the set up that grounds life. The information creating the community of life was not "on the ground," not "in the dirt (earth)," seeded into the rocks or atoms, but appeared somehow. It does not follow that when one finds holy ground (earth) that the ground is self-explanatory. The phenomenon of radical interest on Earth is the information explosion, Earth's own big bang. To the matter (taken off the ground) and energy (captured from solar input), one needs to add "information floating in from nowhere" (Hettinger, chapter 5, p. 68, citing Rolston), vast quantities of it. New cybernetic possibility space opens up. It does not matter whether you model this possibility space as coming from above, or below, or left or right, East or West, or zig-zaging in from behind (or beyond), we get something where we had nothing of that kind before.

Hettinger is content with a "soft naturalism," which I complain is a "mystic chant over an unintelligible universe" (*S&R*, pp. 253–258, Hettinger, chapter 5, p. 72). Maybe I should put this less pejoratively. Watching Earthrise from the moon astronaut Edgar Mitchell was deeply moved by the "sparkling blue and white jewel . . . rising gradually like a small pearl in a thick sea of black mystery" and continued that his view was "a glimpse of divinity" (cited Rolston, 2004, p. 297). Hettinger accepts the pearl in the black mystery, a sacred Earth, but he cannot glimpse any divinity, at least none in the mystery encompassing Earth. Stay with the wonderful Earth, minimalist account. Skip the transcendence. No additives please. Be spiritual, California style.

Hettinger thinks there is a danger that if people do add God they will worship God and skip creation. Go straight to the chief. Hettinger makes this out to be a zero-sum game. Any points we give to God have to be subtracted from small pearl Earth. If you praise God, by just that much you celebrate nature the less. Vice-versa, when nature scores many points, God loses points. Give nature all the points and God isn't in the game any more. I make the world out to be a gift of grace. God cares that there be others with their integrity and autonomy. But it is, paradoxically, the kind of gift that has to be died for.

If Earth is God's artifact, it can't be wild spontaneous nature (Hettinger, chapter 5, p. 73). But that is a far too mechanistic, civil-engineering concept of creation, as though natural history were all front-loaded in at the big bang, or inserted intermittently by intelligent design. The Genesis God says: "Let the earth bring forth of itself." If we need analogies, God is like coach, catalyst, prompter, orchestrator, father, educator (Hettinger, chapter 5, p. 73). God is like mother, giving birth in labor (the etymological root of "nature"). God is in, with, and under this persisting in perishing, struggling though to something higher. The bright pearl shining in black mystery is light shining in darkness that does not overcome it.

If God did it, it is not so wonderful as if nature did it all by itself? But that has to be traded off against a need for adequate explanation. There is certainly no ultimacy in the ultrastructures as now known. We have hit no "rock bottom" in physics, and have few signs that we ever will or can, or would know when we had. We are nowhere close to an account of nature that makes it self-explanatory, not at the Big Bang, not at the origin of Earth, not at the origin of life, not at the origin of human life, not at the origin of culture, not at the origin of science, ethics, and religion, not with nature evolving into spirit (*GGG*, chapters 3–6). All these things, when they happen, are a considerable surprise. The element of escalating serendipity needs more explanation than just to say: "Wonderful surprise!" In the soft naturalist account, there is too much "bootstrapping," nature lifting itself up and up and up by its own bootstraps.

> In that sense, we do not say that naturally inexplicable things never occur; they occur in every emergent increment that breaks previous records of attainment and power. They would come as a surprise to any science based on previously known nature. When they come, it may be possible partially to develop a science of the new phenomena; but they also come, above all, as developments in a story with increasingly rich historical dimensions (*S&R*, p. 299).

Yes, Hettinger replies, but why not revise one's concept of "nature"? Spell it, if you like with a capital N, and let Nature regularly be transcending itself. That happens when the political animal builds cities. But believing in God is incredible. Maybe it is a stretch, but my experience encountering nature is that the story is already incredible, progressively more so at every emergent level. The story is quite fantastic, except that it is true. Yes: (Really) True! I am getting emotional again, so let's be more reasonable:

> Both good induction and good historical explanation lead us to believe in surprises still to come and powers already at work greater than we know. For all the unifying theories of science, nature as a historical system has

> never yet proved simpler or less mysterious than we thought; the universe has always had more storied achievements taking place in it than we knew. To suspect the work of spiritual forces is not, in this view, to be naïve but rather to be realistic (*S&R*, p. 302).

This rational need for explanation compounds when we reckon with the dramatic cruciform creation: "Experiences of the power of survival, of new life rising out of the old, of the transformative character of suffering, of good resurrected out of evil, are even more forcefully those for which the theory of God has come to provide the most plausible hypothesis" (*S&R*, p. 135). Or, if that doesn't seem reasonable, then listen again, feel the pull of the dominant seventh chord in that minor key: "I succumb again to the naïve emoting from which I, so deeply disciplined in scientific criticism and positivist analysis, had thought myself to have escaped." "Need I apologize for my wonder? I think I shall rather boast of it" (*PGW*, pp. 239–240, Wynn, chapter 4, p. 46).

Hettinger wonders too, as do all my wonderful critics—we are still leaning toward a consummation. After a lifetime in biology, J.B.S. Haldane, writing the year I was born, found that the marks of biological nature are its "beauty," "tragedy," and "inexhaustible queerness" (Haldane, 1932, pp. 167–169). Across my lifetime, I find that this beauty approaches the sublime; the tragedy is perpetually redeemed with the renewal of life, and the inexhaustible queerness recomposes as the numinous.

NOTE

[1] In his textual references to his own major works, Rolston abbreviates them as follows: *PGW* = *Philosophy Gone Wild*, *EE* = *Environmental Ethics*, *GGG* = *Genes, Genesis and God*, *S&R* = *Science and Religion*, *CNV* = *Conserving Natural Value*—Eds.

REFERENCES

Dennett, Daniel C. 1995. *Darwin's Dangerous Idea*. New York: Simon and Schuster.

Evernden, Neil. 1993. *The Natural Alien: Humankind and Environment*. Toronto: University of Toronto Press.

Garvin, Lucius. 1953. *A Modern Introduction to Ethics*. Cambridge, MA: Houghton Mifflin.

Gould, Stephen Jay. 1989. *Wonderful Life: The Burgess Shale and the Nature of History*. New York: W. W. Norton.

Gould, Stephen Jay. 2002. *The Structure of Evolutionary Theory*. Cambridge, MA: Harvard University Press.

Haldane, J.B.S. 1932. *The Causes of Evolution*. Ithaca, NY: Cornell University Press.

Jansen, Vincent A. A. and Michael P. H. Stumpf. 2005. "Making Sense of Evolution in an Uncertain World," *Science* 309:2005–2007.

Maynard Smith, John and Eörs Szathmáry, 1995. *The Major Transitions in Evolution*. New York: W. H. Freeman.

Popper, Karl R. 1990. *A World of Propensities*. Bristol, England: Thoemmes.

Rolston, Holmes, III. 1987. *Science and Religion: A Critical Survey*. New York: Random House. Abbreviated *S&R*.

Rolston, Holmes, III. 1988. *Environmental Ethics*. Philadelphia: Temple University Press. Abbreviated *EE*.

Rolston, Holmes, III. 1989. *Philosophy Gone Wild*. Amherst, NY: Prometheus Books. Abbreviated *PGW*.

Rolston, Holmes, III. 1990. "Biology and Philosophy in Yellowstone." *Biology and Philosophy* 5:241–258.

Rolston, Holmes, III. 1994. *Conserving Natural Value*. New York: Columbia University Press. Abbreviated *CNV*.
Rolston, Holmes, III. 1995a. "Environmental Protection and an Equitable International Order: Ethics after the Earth Summit," *Business Ethics Quarterly* 5:735–752
Rolston, Holmes, III. 1995b. "Does Aesthetic Appreciation of Landscapes Need to be Science-Based?" *British Journal of Aesthetics* 35:374–386.
Rolston, Holmes, III. 1996. "Immunity in Natural History," *Perspectives in Biology and Medicine* 39:353–372.
Rolston, Holmes, III. 1997. "Nature for Real: Is Nature a Social Construct?" Pages 38–64 in T.D.J. Chappell, ed., *The Philosophy of the Environment*. Edinburgh: University of Edinburgh Press.
Rolston, Holmes, III. 1998. "Aesthetic Experience in Forests," *The Journal of Aesthetics and Art Criticism* 56:157–177.
Rolston, Holmes, III. 1999a. *Genes, Genesis and God*. Cambridge: Cambridge University Press. Abbreviated *GGG*
Rolston, Holmes, III. 1999b. "Nature and Culture in Environmental Ethics." Pages 151–158 in Klaus Brinkmann, ed., *Ethics: The Proceedings of the Twentieth World Congress of Philosophy*, vol. 1. Bowling Green, OH: Philosophy Documentation Center.
Rolston, Holmes, III. 1999c. "A Managed Earth and the End of Nature?" *Research in Philosophy and Technology* 18:143–164.
Rolston, Holmes, III. 2001. "Natural and Unnatural, Wild and Cultural," *Western North American Naturalist* 61:267–276.
Rolston, Holmes, III. 2002a. "Justifying Sustainable Development: A Continuing Ethical Search," *Global Dialogue* (Centre for World Dialogue, Nicosia, Cyprus) 4(no. 1):103–113.
Rolston, Holmes, III. 2002b. "Environmental Ethics in Antarctica," *Environmental Ethics* 24:115–134.
Rolston, Holmes, III. 2003. "Naturalizing and Systematizing Evil." Pages 67–86 in Willem B. Drees, ed., *Is Nature Ever Evil? Religion, Science and Value*. London: Routledge.
Rolston, Holmes, III. 2004. "Caring for Nature: From Fact to Value, from Respect to Reverence," *Zygon: Journal of Religion and Science* 39:277–302.
Rolston, Holmes, III. 2005a. "Genes, Brains, Minds: The Human Complex." Pages 10–35 in Kelly Bukeley, ed., *Soul, Psyche, Brain: New Directions in the Study of Religion and Brain-Mind Science*. New York: Palgrave Macmillan.
Rolston, Holmes, III. 2005b. "F/Actual Knowing: Putting Facts and Values in Place," *Ethics and the Environment* 10:137–174.
Van Valen, Leigh M. 1991. "How Far Does Contingency Rule?" *Evolutionary Theory* 10:47–52.
Venter, J. C. et al. 2001. "The Sequence of the Human Genome," *Science* 291(16 February):1304–1351.
Wilson, Edward O. 1992. *The Diversity of Life*. Cambridge, MA: Harvard University Press.
Wilson, Edward O. 2002. *The Future of Life*. New York: Alfred A. Knopf.

THE WORK OF HOLMES ROLSTON, III: ABBREVIATED BIBLIOGRAPHY

BOOKS

Biology, Ethics, and the Origins of Life. (Boston: Jones and Bartlett, 1995). Belmont, CA: Wadsworth Publishing Co., 1997). Edited anthology from Conference on Biology, Ethics, and the Origins of Life, held at Colorado State University, September 1991. Contributors: Thomas R. Cech, Lynn Margulis and Dorion Sagan, Niles Eldredge, Michael Ruse, Francisco J. Ayala, Langdon Gilkey, Charles Birch.

Philosophy Gone Wild (Buffalo, NY: Prometheus Books, 1986). 269 pages. A collection of essays in environmental ethics. Paperbound edition 1989.

Science and Religion—A Critical Survey (New York: Random House, 1987; McGraw-Hill, 1989; Harcourt Brace, 1997). 358 pages. (Temple University Press, Philadelphia, hardbound, 1987). (Ft. Worth, TX: Harcourt Brace College, 1997). Reissued with new introduction: Philadelphia Templeton Foundation Press 2006.

Environmental Ethics: Values in and Duties to the Natural World (Philadelphia: Temple University Press, 1988). 400 pages. Paperbound edition, 1989.

Conserving Natural Value (New York: Columbia University Press, 1994). Published in electronic format by Columbia University Press Online Books, 1997.

Genes, Genesis and God: Values and their Origins in Natural and Human History (New York: Cambridge University Press, 1999) Gifford Lectures, University of Edinburgh, 1997–1998.

ARTICLES BY TOPIC AREA

Value theory

"Is There an Ecological Ethic?" *Ethics: An International Journal of Social and Political Philosophy* 85(1975):93–109. Also published in *Philosophy Gone Wild*.

"Values in Nature," *Environmental Ethics* 3(1981):113–128. Also published in *Philosophy Gone Wild*.

"Are Values in Nature Subjective or Objective?" *Environmental Ethics* 4(1982):125–151. Also published in *Philosophy Gone Wild*.

"Values Gone Wild," *Inquiry* 26(1983):181–207. Also published in *Philosophy Gone Wild*.

"Human Standing in Nature, The: Fitness in the Moral Overseer." In Wayne Sumner, Donald Callen, and Thomas Attig, eds., *Values and Moral Standing* (Bowling Green, OH: Bowling Green State University Studies in Applied Philosophy, 1986), vol. 8, pp. 90–101.

"Disvalues in Nature," *The Monist* 75(1992):250–278.

"Biophilia, Selfish Genes, Shared Values." In Stephen R. Kellert and Edward O. Wilson, eds., *The Biophilia Hypothesis: A Theoretical and Empirical Inquiry* (Washington: Island Press, 1993), pages 381–414.

"Value in Nature and the Nature of Value." In Robin Attfield and Andrew Belsey, eds., *Philosophy and the Natural Environment* (Cambridge: Cambridge University Press, 1994), pages 13–30. Royal Institute of Philosophy, Annual Supplement Volume.

"Can and Ought We to Follow Nature?" *Environmental Ethics* 1(1979):7–30. Also published in *Philosophy Gone Wild*.

"Nature, the Genesis of Value, and Human Understanding," *Environmental Values* 6(1997):361–364.

"Naturalizing Values: Organisms and Species." In Louis P. Pojman, ed., *Environmental Ethics: Readings in Theory and Application*, 3rd ed. (Belmont CA: Wadsworth Publishing Co., 2001), pages 76–86.

Reason and emotion in ethics

"Hewn and Cleft from this Rock," *Main Currents in Modern Thought* 27(1971):79–83. Also published in *Philosophy Gone Wild as "Meditation at Precambian Contact."*

"Pasqueflower, The," *Natural History* (Magazine of the American Museum of Natural History) 88 (no. 4, April 1979): 6–16. Reprinted in *Philosophy Gone Wild.*

"Nature and Human Emotions." In Fred D. Miller, Jr., and Thomas W. Attig, eds., *Understanding Human Emotions* (Bowling Green, OH: Bowling Green State University Studies in Applied Philosophy, 1979), volume 1, pages 89–96. Reprinted in *Philosophy Gone Wild.*

"Lake Solitude: The Individual in Wildness," *Main Currents in Modern Thought* 31(1975):121–126. Reprinted in *Philosophy Gone Wild.*

Theology and the environment

"Community: Ecological and Ecumenical." In *The Iliff Review* 30(1973):3–14 (Iliff Theological Seminary, Denver).

"Wildlife and Wildlands: A Christian Perspective." In Dieter T. Hessel, ed., *After Nature's Revolt: Eco-justice and Theology*. (Minneapolis: Fortress Press, 1992), pages 122–143. First published in *Church and Society* 80 (no. 4, March/April 1990):16–40.

"Creation and Recreation: Environmental Benefits and Human Leisure." In B. L. Driver, Perry J. Brown, and George L. Peterson, eds., *Benefits of Leisure* (State College, PA: Venture Publishing, Inc., 1991), pages 393–403.

"Does Nature Need To Be Redeemed?" *Zygon: Journal of Religion and Science* 29(1994):205–229. Also in *Horizons in Biblical Theology* 14 (no. 2, 1993):143–172.

"Environmental Ethics: Some Challenges for Christians." In Harlan Beckley, ed., *The Annual: Society of Christian Ethics* (Washington: Georgetown University Press, 1993), pages 163–186.

"Creation: God and Endangered Species." In Ke Chung Kim and Robert D. Weaver, eds., *Biodiversity and Landscape* (New York: Cambridge University Press, 1994), pages 47–60.

"Bible and Ecology, The," *Interpretation: Journal of Bible and Theology* 50(1996):16–26.

"Ecological Spirituality," *American Journal of Theology and Philosophy* 18(1997):59–64.

"Evolutionary History and Divine Presence," *Theology Today* (Princeton) 55(1998):415–434.

"Kenosis and Nature." In John Polkinghorne, John, ed., *The Work of Love: Creation as Kenosis* (London: SPCK, 2001 and Grand Rapids, MI: Eerdmans, 2001), pages 43–65.

"Religion and Values" In J. Wentzel Vrede Van Huyssteen, Editor-in-Chief, *Encyclopedia of Science and Religion*, volume 2 (New York: Macmillan Reference), pages 722–724.

"Naturalizing and Systematizing Evil." In Willem B. Drees, ed., *Is Nature Ever Evil? Religion, Science and Value* (London: Routledge, 2003), pages 67–86.

"Caring for Nature: From Fact to Value, from Respect to Reverence," *Zygon: Journal of Religion and Science* 39 (no. 2, 2004):277–302.

Environmental aesthetics

"Beauty and the Beast: Aesthetic Appreciation of Wildlife." In D. J. Decker and G. Goff, *Valuing Wildlife Resources: Economic and Social Perspectives* (Boulder, CO: Westview Press, 1987), pages 187–207.

"Does Aesthetic Appreciation of Landscapes Need to be Science-Based?" *British Journal of Aesthetics* 35(1995):374–386.

"Landscape from Eighteenth Century to the Present." Volume 3. In Michael Kelly, ed., *Encyclopedia of Aesthetics*, volume 3 (New York: Oxford University Press, 1998) pages 93–99 .

"Aesthetic Experience in Forests," *Journal of Aesthetics and Art Criticism* 56(1998):157–166.

"Aesthetics in the Swamps," *Perspectives in Biology and Medicine* (University of Chicago; Johns Hopkins University) 43(2000):584–597.

"From Beauty to Duty: Aesthetics of Nature and Environmental Ethics." In Arnold Berleant, ed., *Environment and the Arts: Perspectives on Environmental Aesthetics* (Aldershot, Hampshire and Burlington, VT: UK: Ashgate Publishing, 2002), pages 127–141.

Natural resource policy

"Beyond Recreational Value: The Greater Outdoors." In Laura B. Szwak, ed., *Americans Outdoors: A Literature Review* (Washington: Government Printing Office, 1987). Paper commissioned by President's Commission on Americans Outdoors.

"Values Deep in the Woods," *American Forests* 94, nos. 5 & 6 (May/June 1988): 66–69.

"Human Values and Natural Systems," *Society and Natural Resources* 1(1988):271–283.

"Biology Without Conservation: An Environmental Misfit and Contradiction in Terms." In David Western and Mary C. Pearl, eds., *Conservation for the Twenty-first Century* (New York: Oxford University Press, 1989), pages 232–240.

"Forest Ethic and Multivalue Forest Management, A," co-authored with James Coufal, College of Environmental Science and Forestry, State University of New York, Syracuse, *Journal of Forestry* 89(no. 4, 1991):35–40.

"Using Water Naturally," Natural Resources Law Center, University of Colorado, Western Water Policy Project, Discussion Series Paper No. 9, 1991.

"Fishes in the Desert—Paradox and Responsibility." In W. L. Minckley and James E. Deacon, eds., *Battle Against Extinction: Native Fish/Management in the American West* (Tuscon: University of Arizona Press, 1991), pages 93–108.

"Wilderness Idea Reaffirmed, The," *Environmental Professional* 13(1991):370–377. Reprinted in Lori Gruen and Dale Jamieson, eds., *Reflecting on Nature: Readings in Environmental Philosophy* (New York: Oxford University Press, 1994), pp. 265–278.

"Nature, Spirit, and Land Management." In Beverly L. Driver, Daniel Dustin, Tony Baltic, Gary Eisner, and George Peterson, eds., *Nature and the Human Spirit: Toward an Expanded Land Management Ethic* (State College, PA: Venture Publishing Co., 1996). Anthology published by a U.S. Forest Service task force, pages 17–24.

"What Is Responsible Management of Private Rangeland?" In Larry D. White, ed., *Private Property Rights and Responsibilities of Rangeland Owners and Managers* (College Station, TX: Texas A&M University, 1995), pages 39–49.

"Restoration." In Willian Throop, ed., *Environmental Restoration* (Amherst, NY: Humanity Books, Promethus Press, 2000), pages 127–132.

National parks

"Biology and Philosophy in Yellowstone," *Biology and Philosophy* 5(1990):241–258.

"Yellowstone: We Must Allow It To Change," *High Country News* 23 (no. 10, June 3, 1991):12–13.

"Life and the Nature of Life—in Parks." In David Harmon and Allen D. Putney, eds., *The Full Value of Parks: From the Economic to the Intangible* (Lanham. MD: Rowman and Littlefield, 2003), pages 103–113.

Endangered species, biodiversity, and wildlife

"Duties to Endangered Species," *BioScience* 35(1985):718–726. Reprinted in *Philosophy Gone Wild*.

"Duties to Ecosystems." In J. Baird Callicott, ed. *Companion to a Sand County Almanac* (Madison, WI: University of Wisconsin Press, 1987), pages 246–274.

"On Behalf of Bioexuberance," *Garden* 11, no. 4 (July/August 1987): 2–4, 31–32. Reprinted in *The Trumpeter* (Canada) 5, no. 1 (Winter 1988):26–29.

"The Nonhuman Dimensions in Wildlife," *Human Dimensions in Wildlife*, 8, no. 2 (Spring 1989): 6–8.

"Treating Animals Naturally?" *Between the Species* 5(1989):131–137.

"Endangered Species and Biodiversity: Ethical Issues." In Warren T. Reich, ed. *Encyclopedia of Bioethics*, Revised Edition (New York: Macmillan Library Reference, Simon and Schuster, 1995), pages 671–675.

"The Moral Case for Saving Species," *Defenders: The Conservation Magazine of Defenders of Wildlife* 73 (no. 3, Summer 1998):10.

"Duties to Wild Animals," In Marc Bekoff and Carron A. Meaney, eds., *Encyclopedia of Animal Rights and Animal Welfare* (Westport, CT: Greenwood Press, 1998), pages 362–364.

"Biodiversity and Endangered Species." In Dale Jamieson, ed., *A Companion to Environmental Philosophy*, (Oxford: Blackwell Publishers, 2001), pages 402–415.

"*In Situ* and *Ex Situ* Conservation: Philosophical and Ethical Concerns." In Edward O. Guerrant, Jr., Kathy Havens, and Mike Maunder, eds. *Ex Situ Plant Conservtion: Supporting Species in the Wild.* Society for Ecological Restoration International and Center for Plant Conservation (Washington, DC: Island Press, 2004), pages 21–39.

Environment, business, law

"Just Environmental Business." In Tom Regan, ed., *Just Business: New Introductory Essays in Business Ethics* (New York: Random House, 1984). Also published in *Philosophy Gone Wild.*

"Valuing Wildlands," *Environmental Ethics* 7(1985):23–48. Also published in *Philosophy Gone Wild.*

"Engineers, Butterflies, Worldviews," *The Environmental Professional* 9(1987):295–301.

"Property Rights and Endangered Species," *University of Colorado Law Review* 61(1990):283–306.

"Life in Jeopardy on Private Property." In Kathryn A. Kohm, ed., *Balancing on the Brink of Extinction: The Endangered Species Act and Lessons for the Future* (Washington, D. C.: Island Press, 1991), pages 43–61.

"Whose Woods These Are. Are Genetic Resources Private Property or Global Commons?" *Earthwatch*, vol. 12, no. 3 (March/April 1993): 17–18.

"Enforcing Environmental Ethics: Civic Law and Natural Value." In James P. Sterba, ed., *Social and Political Philosophy: Contemporary Perspectives* (London: Routledge, 2001), pages 349–369.

International environmental ethics

"Can the East Help the West to Value Nature?" *Philosophy East and West* 37(1987):172–190.

"Respect for Life: Can Zen Buddhism Help in Forming an Environmental Ethic?" In *Zen Buddhism Today*, No. 7, September 1989, pp. 11–30. Annual Report of the Kyoto Zen Symposium, Kyoto Seminar for Religious Philosophy, Institute for Zen Studies, Hanazono College and Kyoto University.

"Science-Based vs. Traditional Cultural Values in a Global Ethic." In J. Ronald Engel and Joan Engel, eds., *Ethics of Environment and Development.* (London: Belhaven Press and Tuscon: University of Arizona Press, 1990), pages 63–72.

"Environmental Protection and an Equitable International order: Ethics after the Earth Summit." In Donald A. Brown, compiler, *Proceedings of the Interdisciplinary Conference Held at the United Nations on the Ethical Dimensions of the United Nations Program on Environmental and Development, Agenda 21* (Camp Hill, PA: Earth Ethics Research Group, 1994), pages 267–284.

"People, Population, Prosperity, and Place." In Noel J. Brown and Pierre Quibler, eds., *Ethics and Agenda 21: Moral Implications of a Global Consensus* (New York: United Nations Publications, United Nations Environment Programme, 1994), pages 35–38.

"Environmental Protection and an Equitable International Order: Ethics after the Earth Summit," *Business Ethics Quarterly* 5(1995):735–752.

"Global Environmental Ethics: A Valuable Earth." In Richard L. Knight and Sara F. Bates, eds., *A New Century for Natural Resources Management* (Washington, DC: Island Press, 1995), pages 349–366.

"Feeding People versus Saving Nature." In William Aiken and Hugh LaFollette, eds., *World Hunger and Morality*, 2nd ed. (Englewood Cliffs, NJ: Prentice-Hall, 1996), pages 248–267.

"Nature and Culture in Environmental Ethics." In Klaus Brinkmann, ed., *Ethics: The Proceedings of the Twentieth World Congress of Philosophy*, vol. 1 (Bowling Green, Ohio: Philosophy Documentation Center, 1999), pages 151–158.

"Justifying Sustainable Development: A Continuing Ethical Search," *Global Dialogue* (Centre for World Dialogue, Nicosia, Cyprus) 4(1) 2002:103–113.

Science and religion

"Methods in Scientific and Religious Inquiry," *Zygon* 16(1981):29–63.
"Joining Science and Religion." In Robert John Russell, William R. Stoeger, and George V. Coyne, eds., *John Paul II on Science and Religion: Reflections on the New View from Rome* Rome: (Vatican City State: Vatican Observatory Foundation, 1990; in U.S.: University of Notre Dame Press, pages 83–94.
"Religion in an Age of Science; Metaphysics in an Age of History," commissioned longer critical review of Ian Barbour, *Religion in an Age of Science: The Gifford Lectures*, vol. 1 (San Francisco: Harper and Row, 1990) in *Zygon:* 27(1992):65–87.
"Science and Christianity." In Donald W. Musser and Joseph L. Price, eds., *The New Handbook of Christian Theology* (Nashville: Abingdon Press, 1992), pages 430–432. Revised edit in 2003.
"Order and Disorder in Nature, Science, and Religion." In George W. Shields and Mark Shale, eds., *Science, Technology and Religious Ideas: Proceedings of the Institute for Liberal Studies*, vol. 4 (Frankfort, KT: Institute for Liberal Studies, Kentucky State University, 1994), pages 1–14.
"Science, Religion, and the Future." In W. Mark Richardson and Wesley J. Wildman, eds., *Religion and Science: History, Method, Dialogue* (London: Routledge, 1996), pages 61–82.
"Biodiversity and Spirit," *Science and Spirit* 11(4) (2000):34.
"Biological Diversity." In J. Wentzel Vrede Van Huyssteen, Editor-in-Chief, *Encyclopedia of Science and Religion*, (New York: Macmillan Reference, Thomson/Gale), page 62.
"Nature versus Nurture." In J. Wentzel Vrede Van Huyssteen, Editor-in-Chief, *Encyclopedia of Science and Religion*, volume 2. (New York: Macmillan Reference, Thomson/Gale), pages 607–609.

Environmental ethics education

"Science Education and Moral Education." *Zygon* 23(1988):347–355.
"Earth Ethics: A Challenge to Liberal Education," In J. Baird Callicott and Fernando José R. da Rocha, eds., *Earth Summit Ethics: Toward a Reconstructive Postmodern Philosophy on the Atlantic Rim* (Albany: State University of New York Press, 1996), pages 161–192.
"Environmental Ethics in the Undergraduate Philosophy Curriculum." In Jonathan Colett and Stephen J. Karakashian, eds., *The Environment: Conservation of Biodiversity, and Sustainable Development: A Multidisciplinary Guide for College Teachers* (Washington, DC: Island Press, 1996), pages 206–234.
"Environmental Science and Environmental Advocacy." In Anders Norgren, *Science, Ethics, Sustainability: The Responsibility of Science in Attaining Sustainable Development*, Centre for Research Ethics, University of Uppsala, Sweden. *Acta Universitatis Upsaliensis, Studies in Bioethics and Research Ethics 2* (Uppsala, Sweden, Centre for Research Ethics, 1997), pages 137–153.
"Science, Advocacy, Human and Environmental Health," *The Science of the Total Environment* 184(1996):51–56.

Environmental ethics overviews

"Rights and Responsibilities on the Home Planet," *Yale Journal of International Law* 18 (no. 1, 1993):251–279.
"Ethics on the Home Planet." in Anthony Weston, ed, *An Invitation to Environmental Philosophy*. New York: Oxford University Press, 1999, pages 107–139.
"Challenges in Environmental Ethics." In Michael E. Zimmerman, J. Baird Callicott, George Sessions, Karen J. Warren, and John Clark, eds., *Environmental Philosophy: From Animal Rights to Radical Ecology*. (Englewood Cliffs, NJ: Prentice-Hall, 1993), pages 135–157.
"Winning and Losing in Environmental Ethics." In Frederick Ferré and Peter G. Hartel, eds., *Ethics and Environmental Policy: Theory Meets Practice* (Athens: University of Georgia Press, 1994), pages 217–234.
"Land Ethic at the Turn of the Millennium, The," *Biodiversity and Conservation* 9(2000):1045–1058.

Miscellaneous areas in environmental philosophy

"River of Life, The,: Past, Present, and Future." In Ernest Partridge, ed., *Responsibilities to Future Generations* (Buffalo, NY: Prometheus Books, 1981), pages 123–132. Reprinted in *Philosophy Gone Wild.*

"Preservation of Natural Value in the Solar System, The." In Eugene C. Hargrove, ed., *Beyond Spaceship Earth: Environmental Ethics and the Solar System* (San Francisco: Sierra Club Books, 1986), pages 140–182.

"Ethical Responsibilities toward Wildlife," *Journal of the American Veterinary Medical Association* 200(1992):618–622.

"Nature for Real: Is Nature a Social Construct?" In Timothy D. J. Chappell, ed., *The Philosophy of the Environment* (Edinburgh: University of Edinburgh Press, 1997), pages 38–64.

"Down to Earth: Persons in Place in Natural History." In Andrew Light and Jonathan M. Smith, eds., *Philosophy and Geography III: Philosophies of Place* (Lanham, MD: Roman and Littlefield, 1998), pages 285–296.

"Technology versus Nature: What is Natural." *CPTS Ends and Means: Journal of the University of Aberdeen Centre for Philosophy, Technology & Society* 2(no. 2, Spring 1998):3–14.

"Respect for Life: Counting What Singer Finds of No Account." In Dale Jamieson, ed., *Singer and His Critics* (Oxford, UK: Blackwell Publishers, 1999), pages 247–268.

"Natural and Unnatural, Wild and Cultural," *Western North American Naturalist* 61(2001):267–276.

"Environmental Ethics in Antarctica," *Environmental Ethics* 24(2002):115–134.

"F/Actual Knowing: Putting Facts and Values in Place," *Ethics and the Environment* 10 (2)(2005): 137–174.

"Genes, Brains, Minds: The Human Complex." In Kelly Bulkeley, ed., *Soul, Mind, Brain: New Directions in the Study of Religion and Brain-Mind Science* (New York: Palgrave Macmillan, 2005).

INDEX

The International Library of Environmental, Agricultural and Food Ethics

1. H. Maat: *Science Cultivating Practice.* A History of Agricultural Science in the Netherlands and its Colonies, 1863-1986. 2002 ISBN 1-4020-0113-4
2. M.K. Deblonde: *Economics as a Political Muse.* Philosophical Reflections on the Relevance of Economics for Ecological Policy. 2002 ISBN 1-4020-0165-7
3. J. Keulartz, M. Korthals, M. Schermer, T.E. Swierstra (eds.): *Pragmatist Ethics for a Technological Culture.* 2003 ISBN 1-4020-0987-9
4. N.P. Guehlstorf: *The Political Theories of Risk Analysis.* 2004 ISBN 1-4020-2881-4
5. M. Korthals: *Before Dinner. Philosophy and Ethics of Food.* 2004 ISBN 1-4020-2992-6
6. J. Bingen and L. Busch (eds.): *Agricultural Standards: The Shape of the Global Food and Fibre System*. 2006 ISBN 1-4020-3983-6
7. C. Coff: *The Taste of Ethics: An Ethic of Food Consumption*. 2006 ISBN 1-4020-4553-0
8. C.J. Preston and Wayne Ouderkirk (eds.): *Nature, Value, Duty: Life on Earth with Holmes Rolston III*, 2007 ISBN 1-4020-4877-7